Nonlinear Transistor Model Parameter Extraction Techniques

Achieve accurate and reliable parameter extraction using this complete survey of state-of-the-art techniques and methods. A team of experts from industry and academia provides you with insights into a range of key topics, including parasitics, instrinsic extraction, statistics, extraction uncertainty, nonlinear and DC parameters, self-heating and traps, noise, and package effects.

Learn how similar approaches to parameter extraction can be applied to different technologies. A variety of real-world industrial examples and measurement results show you how the theories and methods presented can be used in practice. Whether you use transistor models for evaluation of device processing, need to understand the methods behind the models you use in circuit design, or you want to develop models for existing and new device types, this is your complete guide to parameter extraction.

Matthias Rudolph is the Ulrich-L.-Rohde Professor for RF and Microwave Techniques at Brandenburg University of Technology, Cottbus, Germany. Prior to this, he worked at the Ferdinand-Braun-Institut, Leibniz-Insitut für Höchstfrequenztechnik (FBH), Berlin, where he was responsible for the modeling of GaN HEMTs and GaAs HBTs and for heading the low-noise components group.

Christian Fager is an Associate Professor at Chalmers University of Technology, Sweden, where he leads a research group focussing on energy efficient transmitters and power amplifiers for future wireless applications. In 2002 he received the Best Student Paper Award at the IEEE International Microwave Symposium for his research on uncertainties in transistor small-signal models.

David E. Root is Agilent Research Fellow and Measurement and Modeling Sciences Architect at Agilent Technologies, Inc., where he works on nonlinear device and behavioral modeling, large-signal simulation, and nonlinear measurements for new technical capabilities and business opportunities. He is a Fellow of the IEEE and he was a co-recipient of the 2007 IEEE ARFTG Technology Award.

The Cambridge RF and Microwave Engineering Series

Series Editor
Steve C. Cripps, Distinguished Research Professor, Cardiff University

Peter Aaen, Jaime Plá, and John Wood, *Modeling and Characterization of RF and Microwave Power FETs*
Dominique Schreurs, Máirtín O'Droma, Anthony A. Goacher, and Michael Gadringer, *RF Amplifier Behavioral Modeling*
Fan Yang and Yahya Rahmat-Samii, *Electromagnetic Band Gap Structures in Antenna Engineering*
Enrico Rubiola, *Phase Noise and Frequency Stability in Oscillators*
Earl McCune, *Practical Digital Wireless Signals*
Stepan Lucyszyn, *Advanced RF MEMS*
Patrick Roblin, *Nonlinear RF Circuits and the Large-Signal Network Analyzer*
John L.B. Walker, *Handbook of RF and Microwave Solid-State Power Amplifiers*

Forthcoming
Sorin Voinigescu, *High-Frequency Integrated Circuits*
David E. Root, Jason Horn, Mihai Marcu and Jan Verspecht, *X-Parameters*
Richard Carter, *Theory and Design of Microwave Tubes*
Anh-Vu H. Pham, Morgan J. Chen, and Kunia Aihara, *LCP for Microwave Packages and Modules*
Nuno Borges Carvalho and Dominique Scheurs, *Microwave and Wireless Measurement Techniques*

Nonlinear Transistor Model Parameter Extraction Techniques

Edited by

MATTHIAS RUDOLPH
Brandenburg University of Technology, Cottbus, Germany

CHRISTIAN FAGER
Chalmers University of Technology, Sweden

DAVID E. ROOT
Agilent Technologies, Inc.

CAMBRIDGE
UNIVERSITY PRESS

CAMBRIDGE UNIVERSITY PRESS
Cambridge, New York, Melbourne, Madrid, Cape Town,
Singapore, São Paulo, Delhi, Tokyo, Mexico City

Cambridge University Press
The Edinburgh Building, Cambridge CB2 8RU, UK

Published in the United States of America by Cambridge University Press, New York

www.cambridge.org
Information on this title: www.cambridge.org/9780521762106

First published 2012

Printed in the United Kingdom at the University Press, Cambridge

A catalog record for this publication is available from the British Library

Library of Congress Cataloging in Publication data
Nonlinear transistor model parameter extraction techniques / edited by Matthias Rudolph,
Christian Fager, David E. Root.
    p.   cm. – (Cambridge RF and microwave engineering series)
Includes bibliographical references and index.
ISBN 978-0-521-76210-6 (hardback)
1. Transistors – Mathematical models.   2. Electronic circuit design.   I. Fager, Christian.
II. Root, David E.   III. Rudolph, Matthias, 1969–   IV. Title.   V. Series.
TK7871.9.N66   2011
621.3815′28 – dc23   2011027239

ISBN 978-0-521-76210-6 Hardback

For Katrin, Elisabeth and Jakob
M.R

For Maria and Sonja
C.F

For Marilyn, Daniel, and Alex
D.E.R

Modeling is at the heart of any modern design process and improvements in transistor modeling have made a significant, but often unrecognized, contribution to the wireless revolution impacting our daily lives. The authors and contributors have collaborated across academic and company boundaries to bring together the latest techniques in a comprehensive and practical review of transistor modeling. This book is destined to become the "go to" reference on the subject.

Mark Pierpoint, Agilent Technologies

Without accurate component models, even the most powerful circuit simulator cannot provide meaningful results. The old saying, "Junk in-Junk out," summarizes the process. The textbook, *Nonlinear Transistor Model Parameter Extraction Techniques*, contains a wealth of theoretical and practical information. It should be read by every designer of active RF/microwave circuits and devices.

Les Besser, Author of COMPACT and Founder of Besser Associates

Nonlinear Transistor Model Parameter Extraction Techniques is an excellent book that covers this extremely important topic very well . . . the editors have done a thorough job in putting together a complete summary of the important issues in this area. For a range of device technologies, the main themes that need to be addressed including measurement, extraction, DC and non-linear modelling, noise modelling, thermal issues and package modelling are covered in a clear but detailed manner, by experts in each area. I would highly recommend this title.

John Atherton, WIN Semiconductors

Contents

List of contributors

Kristoffer Andersson
Chalmers University of Technology

Manfred Berroth
University of Stuttgart

William J. Clausen
RFMD, Greensboro

Giovanni Crupi
University of Messina, Italy

Jens Engelmann
Ferdinand-Braun-Institut, Leibniz-Institut
für Höchstfrequenztechnik

Christian Fager
Chalmers University of Technology

Matthias Ferndahl
Chalmers University of Technology

Mats Fredriksson
Skyworks Solutions, Inc.

Joseph M. Gering
RFMD, Greensboro

Jason Horn
Agilent Technologies, Inc.

Juntao Hu
Skyworks Solutions, Inc.

Masaya Iwamoto
Agilent Technologies, Inc.

Olivier Jardel
3–5 Labs, France

Faramarz Kharabi
RFMD, Greensboro

Kai Kwok
Skyworks Solutions, Inc.

Bin Li
Skyworks Solutions, Inc.

John R.F. McMacken
RFMD, Greensboro

Maciej Myslinski
K.U.Leuven, Belgium

Sonja R. Nedeljkovic
RFMD, Greensboro

Raymond Quéré
University of Limoges, France

David E. Root
Agilent Technologies, Inc.

Matthias Rudolph
Brandenburg University of
Technology

Franz-Josef Schmückle
Ferdinand-Braun-Institut, Leibniz-Institut
für Höchstfrequenztechnik

Dominique Schreurs
K.U.Leuven, Belgium

Hongxiao Shao
Skyworks Solutions, Inc.

Raphael Sommet
CNRS-University of Limoges, France

Jean-Pierre Teyssier
University of Limoges, France

Jianjun Xu
Agilent Technologies, Inc.

Yingying Yang
Skyworks Solutions, Inc.

Peter Zampardi
Skyworks Solutions, Inc.

Preface

Designing microwave circuits today means relying on numerical circuit simulation. While not a substitute for one's own skills, knowledge, and experience, a designer must be able to count on the adequacy of circuit simulation tools to accurately simulate the circuit performance. Circuit simulators themselves are generally up to the challenge. However, there is a perpetual quest for good transistor models to use with the simulator, because models are usually the limiting factor in the accuracy of a simulated design. This is due to the continuous evolution of transistor technology, requiring the models to keep up, and also to the increasing demands placed on the models to perform with respect to wider classes of signals, operating conditions (e.g., temperature), and statistical variation. Circuit designers therefore often face the challenge of adapting the models that are provided with simulators to better describe the actual transistor that is being used in the design. This is achieved by characterizing the transistor, mainly by measurement, but also by electromagnetic and/or thermal simulation. Finally, model parameter values must be extracted from this data before the model can be used at all in a design.

As transistor modeling is a key to circuit design, many publications are available on the models for any type of transistor, ranging from model documentation in simulator products, to application notes and scientific papers in technical conferences and journals; but it seems that much less is published on how the respective model parameters can be determined.

It is the aim of this book to provide a comprehensive overview of transistor model parameter extraction. The basic premise is that parameter extraction, on one hand, is at least as important as the physics-based development of the model formulation itself. On the other hand, extraction approaches, even for quite different technologies, are often based on the same ideas and assumptions. Therefore, the book is intended to give a broad perspective, focusing on one issue and concept after the other, but not restricting any particular concept to a single type of transistor.

The book is based on a workshop presented at the IEEE Microwave Theory and Techniques Society International Microwave Symposium in 2009, organized by the editors. Each chapter of the book corresponds to an individual talk at the workshop. The range of the topics presented covers almost all challenges in parameter extraction, from DC to small-signal parameters, how to integrate small-signal parameters to obtain large-signal quantities such as charge and current, how to determine extrinsic element values, transistor package modeling and self-heating, dispersion effects, noise, statistics of a transistor process, and an overview of measurement techniques for extraction and validation.

The editors would like to thank all the authors for their contributions. We enjoyed the workshop presentations very much and felt that publishing this knowledge in an appropriate way would constitute a valuable contribution to the field. The estimate of the workload to transform a talk into a book chapter turned out to be initially a little optimistic, so we especially appreciate the authors' extra efforts.

We would also like to thank Cambridge University Press, especially Dr. Julie Lancashire and Ms. Sarah Matthews, for their support of the project. It was always a pleasure to work with the staff at Cambridge.

Cottbus, Germany; Göteborg, Sweden; and Santa Rosa, CA, USA
M.R., C.F., D.E.R.

1 Introduction

Matthias Rudolph

Brandenburg University of Technology

If one is about to design a circuit, one certainly relies on a circuit simulation tool that provides us with the capability to determine circuit performance with high accuracy without even fabricating a prototype. We expect the simulation to provide us with the numerical algorithm that is capable of accurately calculating the relevant variables, such as currents, voltages, noise, distortion products, etc. At least as important is the description of the components that will be used, since ultimately the simulation can never be more accurate than the models of the components used. Component models commonly are provided as drag-and-drop components in modern circuit simulators. At least for established technologies, accurate models are available for passive and active components. All problems solved?

Unfortunately not. The models, especially compact transistor models, are parametrized. It is a big step from the general-purpose model that is capable of describing, say, SiGe heterojunction bipolar transistors (HBTs) in general to the specific model for a specific transistor of a specific size, from a specific foundry, that one plans to use in the actual design.

But why bother? One would expect the foundry selling the transistor also to provide us with a valid model.

In reality the situation is more like the following:

1. Some vendors simply do not provide their customers with appropriate models. Either one gets just plain data sheets providing some figures of merit and printed *S*-parameters. Or quite often, even for the most advanced transistors, only very basic SPICE-type model parameters are provided. While these models are available in literally all circuit simulators, their accuracy is often quite limited, since these models only describe the very basic transistor behavior.
2. If one deals with the most advanced generation of devices, the model might not yet be determined. This happens frequently in research.
3. Often, it turns out that the model provided by the device manufacturer is generally of good accuracy, but unfortunately not for the special application one is aiming at. Therefore, a model parameter refinement is required.

This book presents an overview of the different aspects and methods of model parameter extraction. In this context, we assume and take for granted that the model chosen is, in general, accurate. This means that the underlying mathematical formulation is able to account for all effects that are observed. The method behind how a parameter for a specific model is extracted depends on which physical effect it describes rather than what specific model is used. This holds even for different types of device. Hence, the chapters in this book address the relevant issues in parameter extraction topic by topic. Specific device technologies are only addressed in order to provide examples, while the methods in general can be applied also to other transistor technologies.

1.1   Model extraction challenges

Before parameter extraction is discussed in detail, some basic issues shall be addressed: first, what it means that a model is accurate. To state that a model possibly predicts device performance very well is, however, not sufficient. It needs to do so under realistic working conditions. Simulation of realistic cases is usually numerically much more challenging than the standard test cases commonly used to judge model accuracy. Whether or not a simulation converges depends not only on the numerical solver the simulator uses, but also model description and model parameter values extracted. These aspects, which parameters to determine, and which ranges of values are advisable therefore also need to be discussed in this introduction. The final section will address how to choose a suitable type of transistor for the modeling procedure.

1.1.1   Accuracy

If one asks for the model of a specific transistor, this is what one usually gets: a set of parameters, and, hopefully, some comparison between the simulation results with measurement. Typically, output IV-curves, and S-parameters at a few bias points are shown. Is it safe to trust the model, provided that the agreement of simulated and measured data we have seen is perfect? The obvious answer: not necessarily. Whether the model parameter set is valid depends on a number of questions that will be discussed in the following in more detail: what is the application one wants to simulate? How accurate is the measurement? How about performance variations of nominally identical transistors?

1.1.1.1   Circuit application

So far, we spoke of an accurate model, but, as with everything on earth, a model cannot be just completely accurate. All we will get is a model that is accurate within certain limits. Model validity can only be guaranteed within a certain range of, e.g., bias points, frequencies, temperatures, or output power levels. These limits need to match the targeted circuit type. Some examples:

1. True small-signal operation requires only the bias point to be predicted well, together with the S-parameters in the respective frequency range. Of course, this model is restricted to one bias point. If another bias point is to be simulated (e.g., in order to minimize power consumption or to improve noise performance), accuracy cannot be assumed a priori without proving first how well the model will match.

2. Weakly nonlinear operation can be assumed in the case of a low-noise amplifier. The transistor will only be subject to low-power signals, but the small-signal condition

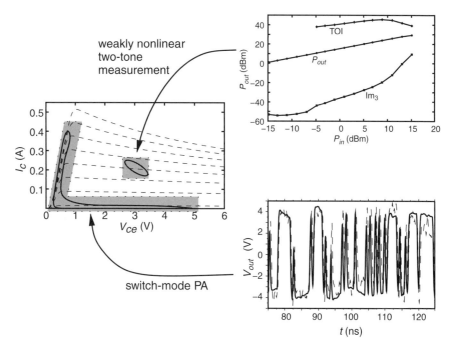

Figure 1.1 Depending on the circuit application, a model is required to be accurate in different operation areas, while others can be neglected. Shown here are two extreme cases: slightly nonlinear class-A operation, and switching operation.

might be violated which calls for a nonlinear model. The common two-tone measurement for the determination of the third-order intercept points characterizes this operation condition. Besides the bias point, S-parameters also need to be predicted well within a limited operating area around the quiescent bias point. This area is shown in gray in Figure 1.1. While high accuracy is required here, no specific accuracy requirements apply outside. Again: if the bias point is shifted, the model needs to be re-examined and possibly the parameters need to be adjusted.

3. Power compression results from output voltages reaching the minimum, or currents approaching the maximum. Starting from the weakly nonlinear operation, the voltage and current swings increase with increasing transistor output power. The range of currents and voltages, where the model needs to be accurate, therefore increases. However, things change significantly once the voltage dynamically drops below the knee voltage, and the current swings into the saturation (bipolar transistors) or linear (field-effect transistors) region. In this region, the transistor is working in a different mode than in class A, and a different subset of parameters is used to describe it in the model. Therefore, a model that is very accurate in class-A operation can completely fail to predict the compression behavior.

4. In order to extend the last statement: if any additional nonideal effect comes into play, for example breakdown or self-heating, the respective model parameters need to be determined and the model is to be verified by appropriate measurements. No compact

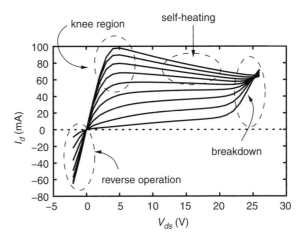

Figure 1.2 Schematic output-IV curves of a FET. Some critical areas are highlighted where the electrical behavior is dominated by physical effects that are not significant elsewhere.

model can be expected to predict all physical effects. All need to be characterized first and the respective model parameters need to be determined.

5. A special case are switch-mode power amplifiers, such as class-E power amplifiers. For this application, the transistor model needs to be highly accurate in on and off states. The common class-A operation area is in the switch-mode case only shortly touched during the switching event. A gray area in Figure 1.1 shows the area where high model accuracy is required for this operation condition.

6. Another special case is the resistive mixer. The field effect transistor (FET) is biased at a drain-source voltage of zero. The model therefore needs to describe the transistor performance in forward and in reverse mode. It needs to be accurate not only for a certain voltage swing at the gate, but also for a certain voltage swing at the drain around zero volts. At negative drain voltages, drain and source terminals exchange their functionality, with drain becoming the effective source and vice versa. This reverse operation condition is commonly ignored completely during model parameter extraction.

At this point, it is pretty obvious that a semiconductor manufacturer who provides us with a model parameter set cannot be expected to anticipate all possible modes of operation. Commonly, the parameter sets are determined for the generic case of class-A types of operation at the optimum bias point of the respective device. Therefore, without careful assessment of model accuracy, no more than an estimate of circuit performance can be expected from using the model. If the mode of operation is too far from class A, such as in the cases of resistive mixer or switch-mode amplifier, the model might even fail to provide a rough estimate.

Figure 1.2 illustrates possible sources of error. Depending on the mode of operation, different physical effects come into play such as breakdown or self-heating. These might not be present during parameter extraction, but the model cannot extrapolate into this region, unless these were already considered during device characterization and parameter extraction. Physical device behavior differs from a low heat-dissipation class-A operation significantly, and additional mathematical formulas with additional parameters are required.

The model is of course not only restricted regarding the range of currents and voltages. It can always only be valid for a specified range of frequencies. Towards higher frequencies, more and more parasitic effects become important, until the lumped device description fails and it is required to treat the transistor like a distributed element. But also, the mathematical approximations of the underlying model are commonly based on more or less sophisticated approximations, especially regarding the implementation of time-delay or transit-time effects. Therefore, compact models are commonly valid only below the transit frequency of a transistor. But also towards lower frequencies, the model might not be fully correct. Self-heating, for example, and memory effects caused by trapping of electrons in FETs are slow processes that basically impact the electrical performance only up to the lower MHz region, or even less. In classical microwave applications, it is not of interest which thermal time constants are present in a device, as long as self-heating is controlled by the dynamical bias, and as long as the device is isothermal at microwave excitation. Therefore, models commonly just apply a simple low-pass to these slow effects, in order to separate DC from microwave range. Therefore, there is a gap in the frequency range regarding model accuracy: the model is valid at and around DC, and in the microwave range below transit frequency. It is not valid in the kHz to MHz range, and around and above the transit frequency. This can have severe implications, e.g., when simulating wideband modulated signals that excite these frequencies.

Unfortunately, it would not be fair to claim that ignorance or laziness on the part of the company providing the model is the root cause of the problem. Anticipating the possible modes of operation during parameter extraction will not fully solve the issue. Why? Because accuracy of a model is not an absolute measure, but a relative one. For example, in the case of a saturated class-AB amplifier, there is a rather high voltage and current swing. High accuracy is achieved when current and voltage swing are correctly predicted, and the deduced quantities such as output power and generation of harmonics matches measured quantities. This, however, does not mean that the model is still perfect if the input power is reduced and high linearity at a certain bias point is regarded. Small inaccuracies here and there might not play an important role in the large-signal case, or can even cancel out completely. They do, however, become important when the amplitude is reduced.

Although it is not the purpose of this book to discuss the accuracy of the model mathematics, and it was stated before that we will take for granted that the models are well suited for our applications, it is simply impossible to formulate a compact model that is highly accurate on all scales at a time. Necessarily, models rely on formulas that are idealized to a certain extent. This limits the complexity of the model, and is required anyway, since a whole class of transistors needs to be described. Thus, the formulas approximate the real behavior. Determining the model parameters therefore means to optimize the range of operation without compromising accuracy.

1.1.1.2   Measurement uncertainty

Parameter extraction and model verification is performed on the basis of measured data. Measured data, however, is determined only with a certain accuracy, or, equivalently, has always a certain amount of uncertainty. There are basically two ways in which uncertainty impacts measured data.

1. Random errors. These cause noisy curves that scatter around the correct one. Sometimes the impact of this type of uncertainty is pretty obvious, especially when the measured data resembles a cloud of points rather than points on a line. But depending on the way the data is visualized, also random errors might not be directly seen.
2. Systematic errors. These are much more difficult to identify from judging the measured data. In contrast to the random error, a systematic error rather shifts the whole curve that still looks much more precise than it is in reality. This kind of uncertainty arises, e.g., if the attenuation of cables, bias-tees, or wafer probes is not fully known.

Precondition to working with measurement data is therefore knowledge of the degree of accuracy that the measurement equipment can provide. Especially, systematic errors should be kept in mind. Moreover, the full cycle of parameter extraction and model verification relies on data that is often determined through different measurement setups that might have different degrees of uncertainty. The model performance, finally, should not be expected to match all measured curves exactly, even if this might be possible to achieve. As an example, consider a load-pull measurement that gives the output power as a function of input power at a certain load condition. If the output power is determined with an accuracy of 1 dB, and the simulation data is about 0.5 dB off – it is not very useful to try to improve the model until measurement and simulation curve match exactly. Indeed, the simulation curve is already within the range of uncertainty. No real improvement in accuracy is gained from changing parameter values in order to reduce the difference of measured and simulated curves [1].

When deriving an algorithm for extraction, it is also worthwhile to consider the fact that we are not dealing with exact data. Therefore, simply applying exact mathematics – neglecting the measurement uncertainties – certainly would not yield the expected result. For example, consider extracting a resistance R from a I–V measurement. One would not just apply one voltage, measure one current, and expect to calculate R with highest possible accuracy. In fact, a number of measurements would be performed, resulting in different values of R. If the variation is reasonably low, one would conclude that the measurement was correct, and define the average of the measured values to be the true value of R. Instead of relying on the exact formula $R = U/I$, it is in many cases advantageous to take into account that the uncertainty results in $U - R \cdot I = r$, where r represents the residual error. Rather than ignoring the uncertainty, least-squares estimation can be employed to minimize r based on a number of measurements. Least-squares estimation has a number of advantages. Basically, it attempts to minimize the quadratic error r^2 relying on a number of measurements that is higher than the number of unknowns.

$$A x + d = r, \tag{1.1}$$

with the column vector x containing the unknowns, the matrix A which contains a row of known values for each measurement, and a known vector b. The vector r contains the errors and is to be minimized. The number of measurements must exceed the number of unknowns in order to take advantage of the averaging effect. The quadratic error of the equation must be minimized, which reads

$$x^{\mathrm{T}} A^{\mathrm{T}} A x + 2(A^{\mathrm{T}} d)^{\mathrm{T}} x + d^{\mathrm{T}} d = r^{\mathrm{T}} r \stackrel{!}{=} \mathrm{Min}, \qquad (1.2)$$

with ^{T} denoting transposed matrices. The minimum is found by differentiating and finding the roots of the derivative. This yields a linear equation to be solved:

$$(A^{\mathrm{T}} A) x + (A^{\mathrm{T}} d) = 0. \qquad (1.3)$$

Noisy measurement data will not only lead to noisy extracted parameters. It can even cause a complete failure of the extraction attempt. The question is, how the uncertainty of the initial data is transformed through the parameter extraction algorithm. It should be kept in mind that the error contained in the data usually has different properties than the exact data would have. For example, the error of two measurements can be correlated or uncorrelated, while the exact data of the two measurements has a certain fixed mathematical relation to each other. Second, the error can be of an absolute value, or be relative to the magnitude of the measured data.

As a result, the following mathematical operations are especially critical:

1. Division or multiplication with small numbers of high uncertainty. It happens quite often that one term in a mathematical equation cancels out, since a large number is multiplied with a very small one. However, if the small value is significantly affected by uncertainty, the term will not necessarily cancel out in reality. Extremely dangerous also are operations like dividing two values each of which is extremely small, e.g., close to dividing zero by zero.
2. Subtraction of theoretically almost equal numbers. For example, it will not work to determine a parameter from the difference of two very large numbers, if the parameter value is so small that it is in the range of the relative error of the measurement data.

The propagation of measurement uncertainty through the extraction therefore calls for extraction routines that are tolerant of errors. The formulas to be employed need to be analyzed in this respect, and often an approximation is superior to an exact formulation.

A good example concerns the extraction of the intrinsic base resistance of an HBT, R_{b2}. In HBTs, base resistance and base-collector capacitance are usually divided into two parts. After de-embedding of the extrinsic parameters, R_{b2} still needs to be determined, which is not too simple, since it is bridged by the extrinsic part of the base-collector capacitance. However, after some calculation, R_{b2} can be given as a function of Y-parameters [2]:

$$R_{b2} = \frac{\mathrm{Re}\,\{ac^*\}}{\mathrm{Re}\,\{|Y|\,c^*\}}, \qquad (1.4)$$

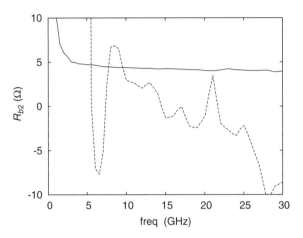

Figure 1.3 Impact of measurement uncertainty on parameter extraction at the example of the intrinsic base resistance of an HBT. Solid line: approximated formula, dashed line: exact formula.

where the $*$ denotes the complex conjugate, and with

$$a = Y_{12} + Y_{22}$$

$$c = Y_{11} + Y_{12} + Y_{21} + Y_{22}$$

$$|Y| = Y_{11} Y_{22} - Y_{12} Y_{21}.$$

This formula has been applied to a standard GaAs-HBT, with an emitter area of $3 \times 30\,\mu m^2$ and an f_t of about 35 GHz. The value of R_{b2} is approximately 4 Ohm. However, applying this exact formula fails completely, as shown by the dashed line in Figure 1.3. The extracted values scatter between ± 200 Ohm at a few GHz, which is not shown in the figure. But even at higher frequencies, the extracted values are in the range of -10 to 10 Ohm.

Instead, the following simple approximation yields good results:

$$R_{b2} \approx \mathrm{Re}\left\{\frac{1}{Y_{11}}\right\} \qquad \text{for high frequencies.} \tag{1.5}$$

The result obtained with this approximation is shown as a solid line in Figure 1.3. Since the approximation holds for higher frequencies, the curve rather converges towards the value of R_{b2}, but the variation is very low over a frequency range of more than 15 GHz.

This example was based on a well-calibrated measurement of S-parameters, providing smooth curves also after transformation into Y-parameters. It is thus quite impressive that the curve obtained from a few mathematical operations through equation (1.4) scatters by approximately two orders of magnitude around the target value.

1.1.1.3   Process variations

Consider the case that, due to a miracle, one would be able to determine all model parameters of a transistor without any uncertainty due to measurement, model simplifications, and the like. Even now, the model will still be affected by a non-negligible amount of uncertainty.

Unfortunately, we are facing the fact that no two samples of a transistor are absolutely identical. No one would expect that, of course, but this issue is often overlooked, especially if one has no access to the foundry data gathered in order to monitor process stability. It is therefore highly recommended to care about this issue, at least if the model is not meant to be used just for the same transistor that has been modeled before. Commonly, we expect the model to describe all the transistors from one dedicated technology. This is especially true with respect to monolithic integration, where the model is used to design chips and transistors that will be processed after modeling and circuit design has been done.

The impact of process variations is different, depending on technology. It is expected to be low for a mainstream silicon production line and high for advanced transistor technologies that are not yet mainstream. In any case it is worthwhile to be aware of the issue.

Process variations are of different origin, for example:

- random fluctuations of dimensions, doping concentrations, and the like. These uncertainties in process technology add up and result in a stochastic variation of the parameters around a mean value. As in the case of noise, it is a good guess to assume that the parameters probability density function (pdf) is bell-shaped;
- systematic differences, depending on the location on the wafer. Wafers are large, transistor geometries are really small. Therefore, not all places on the wafer are treated equally. It is to expect that parameters show a certain slope as a function of wafer radius;
- run-to-run differences. A new batch of wafers is processed slightly differently from the previous batches. Environmental conditions in the cleanroom slightly change, new chemicals might be used, and the like. The result is a shift in certain parameters for the new transistors compared to the ones that were previously processed.

1.1.2 Numerical convergence

Model accuracy is unfortunately not the only issue to take care of during parameter extraction. The choice of parameter values may also impact simulation speed. Poorly extracted model parameter values have the power to degrade the otherwise good numerical properties of a well-formulated model. In some cases, completely irrelevant parameters can even cause the simulation to fail in a way that the simulation software is unable to find a solution. However, the mathematical formulation of the model is not often necessarily to blame. Today, models normally are formulated in such a way that they provide good accuracy and good numerical properties. Still, the model parameters have the power to degrade the numerical properties.

Two examples will be highlighted in the following. These represent the most prominent issues: breakdown and self-heating.

1.1.2.1   Breakdown

Circuit simulation relies on an iterative solver to calculate the nonlinear differential equations describing the electrical behavior of the circuit. This is generally the case for all types of nonlinear simulation engines, no matter whether they approach the mathematical problem in time domain like SPICE or in frequency domain like harmonic balance. The mathematics behind this are quite involved, but in general, an iterative algorithm starts with guessing a solution. It then determines the error and then uses the equations and extrapolates in order to find a better guess. In our case, these equations mean the transistor model and those of the other circuit element descriptions. Convergence towards a solution depends on the mathematical formulas defining the component models. It is fast, if the extrapolation is close to the exact solution, and the second guess is better than the first.

Therefore, models are today based only on formulas that are smooth, continuous, and continuous in all derivatives, and limited in order to prevent numerical overflow. For every set of node voltages, the model formulas must always return numerically valid currents and charges. Still, model parameters possibly degrade the convergence properties of the circuit simulator.

The first issue concerns the fact that the transistor properties sometimes change very fast. Drain-source breakdown in FET devices is the most prominent example for such an effect. At a certain voltage, the drain current dramatically increases. It is not possible to guess the breakdown voltage from the shape of the output-IV curves prior to the onset of the breakdown phenomenon. Second, the current slope is very steep. Third, in physics, a DC solution beyond breakdown is nonexistent, since the device would rather melt. Thus, even if the drain current is defined by a smooth continuous formula, the current changes dramatically within a narrow range of voltages. For the numerical solver, it can effectively look like a discontinuity.

It is now time to consider an important property of iterative solvers. It is impossible to predict how the iteration will be performed. Since it is a purely mathematical operation, it is not restricted to calculation within a physically meaningful range of parameters. Even if we know in advance that the DC operating point will be around a drain voltage of 3 V and a drain current of 100 mA, with a dynamic voltage swing between 0 V and 9 V, it can well happen that the iterative solver iterates through a voltage range of ± 100 V or more before it converges. Therefore, convergence problems can even be experienced if breakdown is far from the final solution. The sketch in Figure 1.4 gives an example. If the numerical solver iteratively searches for the solution within the shaded area, convergence will be much better if the model neglects the breakdown that occurs around 25 V. When doing so, it must of course not impact model accuracy. However, this is definitively the case in this example.

In short, it can improve convergence properties, when physically meaningful effects are ignored during parameter extraction, provided that these effects are outside the voltage, current, frequency, or temperature ranges that will be simulated. This also includes that the impact of default parameters of a certain model always need to be understood even if they describe an effect that seems irrelevant at the moment. On the other hand, the tradeoff between overall accuracy of the model (describing all effects correctly) and

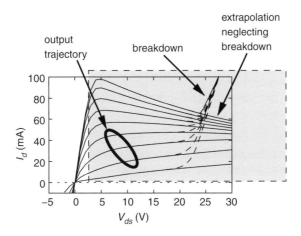

Figure 1.4 Breakdown can impact numerical stability of the simulation even if it is far off the final solution. During iterative determination of the output trajectory, voltage/current combinations are calculated within the shaded area and beyond. Neglecting breakdown has no impact on model accuracy here, but improves convergence properties.

numerical robustness (neglect what is not relevant) requires a-priori knowledge of the target application of the model. These issues require good communication between the developer and the users of the model.

1.1.2.2 Self-heating

One of the most important features of a transistor model is its ability to predict the dynamic changes of its temperature during operation. Due to power dissipation, the device heats up, which in turn changes the electrical parameters. In GaAs HBTs, for example, the current gain decreases with increasing temperature. While from the standpoint of physics, no parameter is completely insensitive to temperature, at least the most critical ones need to be defined as a function of temperature.

Self-heating is highly challenging from the numerical point of view. In general, all branch currents are directly calculated, when the node voltages are known. Most of these equations are nonlinear, of course, but self-heating is even worse: it introduces a feedback to this nonlinear system.

This feedback is caused by the fact that the branch currents are determined by node voltages and temperature – but temperature is determined by dissipated power, which is the product of voltage and current. It is not only a feedback loop; it is a nonlinear feedback loop.

Of course the whole system can become unstable, temperature and currents rise without bounds, and the simulation will fail. A nonlinear feedback system, even if this catastrophe does not strike, may have more than just one solution [3, 4]. This is not necessarily the result of a bad model implementation. Thermal runaway is an effect commonly observed in bipolar transistors. As circuit designers, we expect that the model also accounts for this effect. But again, as the simulation algorithm iteratively searches for a solution even in regions that are beyond the safe operation area, it can possibly

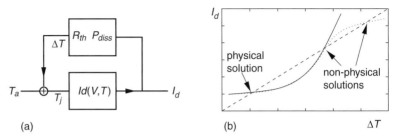

Figure 1.5 Thermal-electrical feedback loop (a) possibly causes multiple mathematical solutions (b) that may deter numerical convergence.

degrade numerical performance. Multiple solutions are a similar case. In general, these are not harmful in the real world, as physical systems always converge towards the state of lowest energy. Thus, a transistor will certainly run at 10 mA and room temperature rather than at 2 kA and 10 MK, but an iterative solver might get confused if these solutions are too close.[1]

A simple example highlighting the issue is a diode that shows self-heating [5]. The feedback loop is shown in Figure 1.5. The diode current is given by the well-known equation

$$I_D(V, T) = I_s \, e^{(V_g/V_{th0} - V_g/V_{th})} \left(e^{V/(nV_{th})} - 1 \right), \tag{1.6}$$

with voltage V, temperature T, activation energy V_g, ideality factor n, and the diffusion voltage $V_{th} = kT/q$, and $V_{th0} = 0.25$ mV.

The temperature is given by $T = T_a + \Delta T$, with ambient temperature T_a and excess temperature due to self-heating ΔT. Self-heating is given by

$$\Delta T = R_{th} \cdot I \cdot V = R_{th} P_{diss}, \tag{1.7}$$

with the thermal resistance R_{th}, current I, voltage V, and dissipated power P_{diss}. This can be rewritten as a function of current:

$$I = \frac{\Delta T}{R_{th} \cdot V}. \tag{1.8}$$

For a given voltage V, both equations (1.6) and (1.8) must be fulfilled, which is indicated in Figure 1.5b, where the dashed line refers to equation (1.8), and the solid line refers to equation (1.6). If plotting the equations as a function of ΔT, valid solutions are found at the intersections of the two curves.

Obviously, this very basic case of a single diode already provides two solutions. In terms of Figure 1.5, the reason is that equation (1.8) has a constant slope, while the diode function, equation (1.6) increases exponentially with temperature. The parameter that has the highest impact here is the activation energy parameter V_g, since it controls the slope of the current function. The situation might become even more complicated

---

[1] But there are publications on low-frequency electro-thermal oscillations, from the early days of bipolar transistors, indicating that there are multiple unstable states close together.

if the diode current is limited in order to prevent numerical overflow in the exponential function for nonphysically high temperatures. This case is shown by the dotted line in the figure.

Each and every nonlinear model faces this issue that accounting for self-heating introduces multiple solutions, FETs as well as bipolars. As this is about the convergence properties of an iterative nonlinear solver, only very general tips can be given, pointing in the direction of improved convergence properties.

1. Keep the number of feedback loops as low as possible. It is true that everything in the world is affected by temperature. But usually, there are a few key parameters in a transistor that are really significant, while the impact of keeping the others constant with respect to temperature does not affect accuracy at all.
2. Examine all the parameters, even if the model performance looks good at a first glance. A parameter that is more or less insignificant for normal operation conditions is potentially harmful in some far-off region where the solver searches for a solution.

1.1.3   Choice of the modeling transistor

It is often possible to choose among a number of different devices. When choosing which transistor to consider for modeling, the following points should be taken into consideration.

1. Verify that the transistor considered is a known good device of typical performance. If unexpected behavior is observed, it needs to be verified that it is a typical property of these transistors, not the result of imperfect technology. This requires characterization of a number of transistors and is easier performed in an industrial environment where a relevant number of samples is available.
2. Select a device where the performance is determined by the active transistor, and not by the extrinsic parasitics. Best suited are reasonably small devices measured on-wafer, or on-chip by means of wafer probes. Thereby the reference planes of the measurement can be placed as close to the device as possible.
3. The layout of this initial modeling transistor should be similar to the layout of the transistors for which the model will be used later on. The idea of beginning with a small device is that an initial set of parameters is generated rather easily, providing a good starting point for the derivation of the parameters describing larger devices. Therefore, the small device must not be not too different from the other devices, otherwise it would not be possible to scale parameters. Small and large transistors need to share the same basic layout, e.g., the emitter finger dimension and base contact configuration in HBTs, or gate length and width in the case of high electron nobility transistors (HEMTs).
4. The device must behave like a lumped device in initial characterization. For large devices, this precondition is easily violated, when phase differences arise due to the gate or drain feed structure. Large power devices can also be thermally uneven due to hot-spot formation, even if it might not be severe enough to degrade device

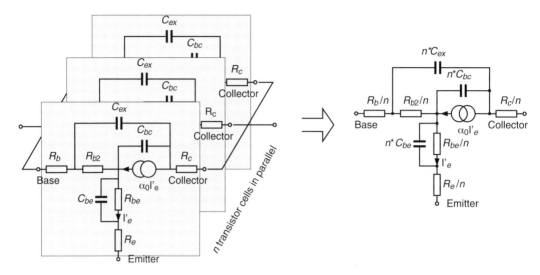

Figure 1.6 Impact of transistor size on the impedance level at the example of the HBT equivalent circuit. When the transistor size is increased by a factor n, capacitances increase and resistance values decrease accordingly.

performance absolutely. These nonideal distributed effects will be described on a higher level model later on, first through a distributed model relying on a number of nonlinear compact transistor models that are properly interconnected, and then possibly through condensing this model again into a compact description; but in initial device characterization, these effects cannot be identified unambiguously and accounted for in the model.

5. Microwave measurement is commonly performed in a 50 Ohm environment. For extraction purposes, it is best to choose the device size in such a way that its input and output impedances are close to this value.

The absence of distributed effects, the low impact of the parasitic extrinsic network, and the right impedance level are hence the reasons why most nonlinear models are verified for small devices measured on-wafer. The impact of transistor size is highlighted at the example of an HBT in Figure 1.6. Larger transistors can be understood as parallel connections of smaller transistors. Usually, microwave transistors are realized exactly this way, relying on a small basic cell. However, the resistances decrease, and capacitances increase proportional to transistor size. The impact of this effect on S-parameters is shown in Figure 1.7 that shows measurements of a one, two, and ten-finger HBT with an area of $3 \times 30\,\mu\text{m}^2$ per emitter. The total base resistance ($R_b + R_{b2}$ in Figure 1.6) decreases from 5.5 Ohm to 0.6 Ohm, while the total base-collector capacitance ($C_{bc} + C_{ex}$ in Figure 1.6) increases from 40 fF to 0.4 pF. Since these values need to be determined from the 50 Ohm measurement, the advantage of the small device is obvious. It is simply impossible to determine a base resistance of about 0.5 Ohm with reasonable accuracy. Any resistance from 0 to 1 Ohm would yield perfect model accuracy in terms of

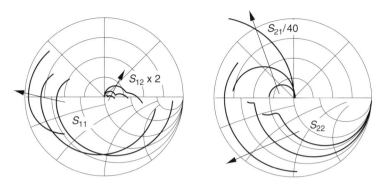

Figure 1.7 Impact of transistor size on measured S-parameters of GaAs HBTs for
$f = 100\,\text{MHz}, \ldots, 50\,\text{GHz}$. Parameter is the number of emitter fingers, 1, 2, 10 emitters of
$3 \times 30\,\mu\text{m}^2$.

S-parameters in a 50 Ohm environment. Therefore, determining the base resistance from
a smaller device and scaling it up for a larger device provides much more reliable, and
physically meaningful, parameter values.

1.2   Model extraction workflow

After all these preliminary basic thoughts on model accuracy and extraction uncertainties, and how the parameters possibly impact the simulation speed, let us come back to the similarly very basic problem that modern models come with high numbers of parameters. Often, over a hundred parameters need to be known in order to fully describe a transistor model. How should all these values be determined?

Of course it is at best tedious, and commonly near impossible, to determine all parameters from a single measurement. On the other hand, it is a property of a well-formulated model that its parameters have a distinct meaning such as, for example, *BF* which stands for current gain in the Gummel–Poon bipolar transistor model. *BF* is a good example in this respect. It describes the ideal current gain, without any of the Early, Kirk, self-heating, or base recombination effects that make the current gain of a real transistor a function of bias. Nevertheless, the ideal *BF* is usually easily determined from a Gummel plot measurement, that is a setup where the nonideal effects are minimized. The Gummel plot measurement will be addressed in Chapter 3. In short, it fixes the base-collector voltage at 0 V, and keeps currents low: not a suitable bias point for circuit design, but reasonable for parameter extraction.

Starting from this ideal parameter value, measurements are performed in order to characterize all the secondary effects of the transistor. Ideally, these effects are described by parameters that are not interdependent if their physical origin is independent. Thus, ideally, one can work one's way through the model, characterizing one effect after the other, and determining one parameter after the other.

It is the purpose of this book to guide the reader through all these steps.

- Chapter 2 will start with the DC part of the model. The DC performance is, however, closely linked to self-heating. Extraction of DC parameters is required as it is commonly used to characterize the nonlinear current sources (and resistances) that will be needed for a large-signal model. Together with the thermal parameters, these are an important part of the entire model.
- Chapter 3 addresses an issue that can hardly be overestimated: how to characterize the extrinsic part of the model. Extrinsic elements, by definition, describe everything around the active device that degrades the transistor performance. It is, basically, the pad capacitances, feed inductances, and feed resistances including the resistance of the metal–semiconductor interface. Without sound knowledge of these parameters, any further analysis and parameter extraction of transistor will fail.
- Chapter 4 assumes that the extrinsic elements were successfully de-embedded. Now it is time to determine the small-signal equivalent circuit parameters. In contrast to what one would expect of first glance, there is still research required and on-going in this field. As the devices keep changing, the algorithms need to be adapted. This chapter will focus on methods that allow the impact of measurement uncertainty on extracted parameter uncertainty to be estimated. This approach is very useful, since it provides, in the end, not only a device parameter, but also the range in which it might vary.
- Chapter 5 addresses the task to go from a number of small-signal equivalent-circuits to a large-signal nonlinear model. This step is mathematically intricate as it requires the determination of multidimensional functions (charges and currents) for the elements just from the knowledge of its derivatives (capacitances and small-signal resistances). In the case of the currents, the DC measurements can be taken into account, but in the case of the capacitance, no similar data can be measured. Modern approaches based on large-signal data that bypass the need for linear equivalent circuits are presented in addition.
- Chapter 6 assumes that the transistor chip is fully characterized. Power transistors, however, consist of a parallel connection of many transistors within a package. This package significantly impacts the performance of a power transistor. It will therefore be addressed how the electromagnetic and thermal performance of a package can be characterized and modeled for circuit design.
- Chapter 7 discusses transistor dispersive effects. Dispersion, also known as memory effect, is present in many devices, but significantly determines the performance of the emerging GaN-based HEMT devices. A transistor showing dispersion "remembers" previous voltage levels for a certain time. Current-voltage-relations therefore are different in static (DC) and for any kind of dynamic radio frequency (RF) operation. The chapter considers measurement, modeling, and parameter extraction techniques allowing these dispersive effects to be accurately accounted for.
- Chapter 8 discusses the measurements needed for accurate modeling. Measurement, of course, is the basis of all modeling. It gives us the basis for the parameter extraction and provides us with the data to verify the model. The chapter also reviews techniques

to derive abstract mathematical models directly from large-signal measurement techniques.

- Chapter 9 brings our attention to the fact that there is always a certain amount of process spread in the manufacturing of transistors and integrated circuits. In order to guarantee high yield in production, the designer needs to anticipate these variations. A single model for a typical device will not be enough. This chapter addresses the question of how to determine the statistics of the model parameters and how to implement it in the usual transistor models.

- Chapter 10 concludes the book by taking a look at transistor noise modeling. It shows how the noise sources within a model can systematically be determined relying on the correlation matrix approach.

The reader will discover that the concepts and styles of writing somehow differ between the chapters. The reason is that some of the topics are well established, while others are either new issues or not discussed as often. For example, determining extrinsic and intrinsic parameters is well established. Even if the reader is not familiar with the question of how to get the parameters, it can be assumed that the question why they are needed is understood. The same holds for the physical background of the equivalent circuits and for the S-parameter measurement involved.

Dispersion effects, and to a certain extent, noise is different in that respect. First, it requires measurements to be done that were not common in microwave labs just a few years ago. Also, the underlying physics is involved, and the question of how to model the effects is also still a matter of research. Therefore, the topic is carefully introduced, including a broad discussion of the underlying physics.

The chapters differ also in the ratio of theory and principles to practical experimental data. For example, Chapter 9 concerning the process variations lives from the example. Huge amounts of measurements are investigated, which gives the reader not only the theory, but also a sense of what might be relevant in practice.

In conclusion, we hope that this book provides the reader with an appropriate discussion of all relevant issues one is facing in model parameter extraction.

References

[1] C. C. McAndrew, "Practical modeling for circuit simulation," *IEEE J. Solid-State Circuits*, vol. 33, no. 3, pp. 439–448, Mar. 1998.

[2] F. Lenk and M. Rudolph, "New extraction algorithm for GaAs-HBTs with low intrinsic base resistance," *IEEE MTT-S Int. Microw. Symp. Dig.*, 2002, pp. 725–728.

[3] S. A. Maas, "Ill conditioning in self-heating FET models," *IEEE Microw. Wireless Compon. Lett.*, vol. 12, no. 3, pp. 88–89, Mar. 2002.

[4] A. E. Parker and S. A. Maas, "Comments on 'Ill conditioning in self-heating FET models'," *IEEE Microw. Wireless Compon. Lett.*, vol. 12, no. 9, pp. 351–352, Sept. 2002.

[5] M. Rudolph, "Uniqueness problems in compact HBT models caused by thermal effects," *IEEE Trans. Microw. Theory Tech.*, vol. 52, no. 4, pp. 1399–1403, May 2004.

2 DC and thermal modeling: III–V FETs and HBTs

Masaya Iwamoto, Jianjun Xu, and David E. Root

Agilent Technologies, Santa Rosa, California, USA

2.1 Introduction

A brief overview and selected topics on DC and thermal compact modeling of III–V compound semiconductor transistors are discussed. Specifically, characteristics of field-effect transistors (GaAs MESFET and pseudomorphic HEMT (pHEMT)) and heterojunction bipolar transistors (GaAs and InP HBTs) are presented. Basic DC characteristics of III–V FETs and HBTs are reviewed, and parameter extraction techniques and methods pertinent to compact modeling are shown. Properly extracted parameters, in turn, can be used for process control and optimization by monitoring device characteristics within and across many wafers. Thermal modeling pertinent to compact modeling is also presented with two approaches demonstrated: physics-based and measurement-based. Finally, device reliability evaluation is presented as an example where both DC parameters and thermal modeling are extensively utilized.

The DC model serves as a foundation for any compact model. In the most fundamental sense, DC simulation results based on an accurate model provide detailed bias point information for circuits such as current density, output voltage, and power dissipation to enable robust and well-optimized circuit designs. Furthermore, the DC model serves as the foundation for RF characteristics such as S-parameters. For example in an HBT, S_{11} and S_{21} at low frequencies are related to the small-signal model parameters r_π and g_m, respectively. For an FET, S_{22} and S_{21} at low frequencies are related to the small-signal model parameters g_{ds} and g_m, respectively. Additionally, g_m plays a role in RF figures of merit such as f_T and f_{max} (e.g., $f_T = g_m/(2\pi C_{in})$). With respect to nonlinear device characteristics, the detailed bias dependence of g_m (and g_{ds}) influences the harmonic and intermodulation distortion at low and mid frequencies relative to the maximum frequency limits of the device. Finally, DC characteristics influence large-signal behavior such as P_{1dB} and P_{sat}, since the knee characteristics (at low voltage and high current) and the turn-on (at high voltage and low current) determine the gain compression behavior at large output swing conditions.

A thermal model is typically added to the DC model with thermal scaling (ambient temperature dependence) to account for self-heating (power dissipation dependence). Thermal scaling characteristics of a compact model are critical for circuit applications that have a wide range of operating temperature requirements (e.g., $-40\,°C$ to $+100\,°C$). Self-heating is the temperature increase in the intrinsic part of the device owing to the power dissipation, and is significant in applications where the current density (e.g.,

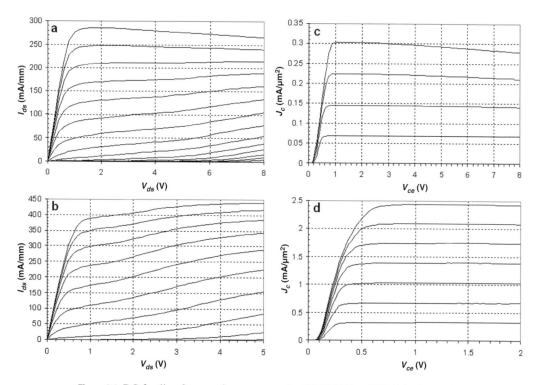

Figure 2.1 DC family of curves for representative III–V FET and HBT devices. (a) 0.35 μm GaAs MESFET: V_{gs} swept from −2.4 to +0.2 V in 0.2 V steps. (b) 0.25 μm AlGaAs/InGaAs pHEMT: V_{gs} swept from −1.6 to +0.2 V in 0.2 V steps. (c) GaAs Power HBT: I_b swept from 55 to 220 μA in 55 μA steps (d) InP high-speed HBT: I_b swept from 20 to 160 μA in 20 μA steps

digital/mixed circuits) and voltage (e.g., power amplifiers) are high. Accurate prediction of self-heating is important for robust circuit design, where the lifetime of a device has a strong dependence on temperature. This topic of device reliability evaluation will be discussed in more detail in the latter part of this chapter.

2.2   Basic DC characteristics

Output characteristics of a 0.35 μm GaAs MESFET and 0.25 μm AlGaAs/InGaAs pHEMT are shown in Figures 2.1a and 2.1b, respectively. The equivalent plots for GaAs power HBT and InP high-speed digital HBT are shown in Figures 2.1c and 2.1d, respectively. The general shape of the two types of device (FETs versus HBT) are very similar, but are typically modeled differently for compact modeling.

The drain current equation for FET compact models typically is based on the Curtice– Van Tuyl [1, 2] formulation, which has the form,

$$I_{ds} \approx f(V_{gs}) \times g(V_{ds}) \times \tanh(\alpha V_{ds}), \tag{2.1}$$

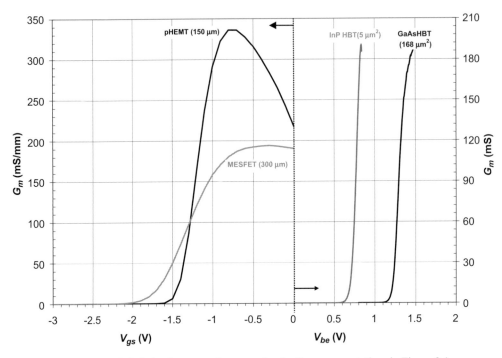

Figure 2.2 DC-derived transconductances for the four representatives in Figure 2.1.

where a hyperbolic tangent function describes resistive behavior in the "triode" region (at low voltage in the vicinity of $V_{ds} = 0$) and the plateau behavior of the drain current in the "saturation" region at higher drain voltages.

Collector current equation for III–V HBT compact models typically leverages the Gummel–Poon (homojunction) BJT model [3] formulation, which has the form,

$$I_{ce} \approx ISf \left(\exp \left(V_{be}/(Nf \times V_t) \right) - 1 \right) - ISr \left(\exp \left(V_{bc}/(Nr \times V_t) \right) - 1 \right) \qquad (2.2)$$

that can be interpreted as the difference of two "diode" equations controlled by base-emitter and base-collector voltages. The two equations interacting with each other results in a resistive behavior in the low voltage region ("saturation" region) and a relatively flat current behavior at higher collector voltages ("forward-active" region).

Although the equations for the two types of device are modeled differently, in a broad physical sense, they operate in a very similar manner. The low-voltage region is a resistor-like region (i.e., conducting channel for a FET, and two forward-biased "on" diodes for an HBT), while the higher voltage region describes a velocity-limited current behavior (i.e., electrons traveling at saturated velocity in the channel under the gate for the FET, and electrons traversing the collector depletion region at saturated velocity for the HBT).

DC transconductances as a function of V_{gs} (computed from taking the first derivative of I_{ds} versus V_{gs} data) are shown for the FET and HBT in Figure 2.2. Again, the general turn-on characteristics between the two devices are similar, although the output current dependence of the input voltage (V_{gs} for a FET, and V_{be} for an HBT) are modeled differently.

For the FET, this dependence is typically modeled with a polynomial or an empirical formulation which is embodied in the term $f(V_{gs})$ in equation (2.1). Furthermore, $g(V_{ds})$ empirically accounts for the V_{ds} bias dependence apart from the tanh contribution. The empirical formulations are necessitated by the fact that the input–output current characteristics of a FET are very dependent on the epitaxial layer design and processing of the devices (e.g., quantum-well channel (HEMT) versus doped channel (MESFET), single versus dual channel designs for an HEMT, Schottky layer recess details, pulsed doped versus uniformly doped, etc.) which result in varying transistor characteristics.

For the HBT, the input–output current relation is described by the much simpler "diode" equation, which is basically the first term of equation (2.2). A more simple, but physical, description is possible for the HBT, since the general shape of the characteristics is consistent (or stays near-ideal) across epitaxial design variations and material technologies. Furthermore, major characteristics of the HBT are strongly influenced by the material properties rather than the process or device design; for example, the forward turn-on voltage is largely influenced by the bandgap of the narrowband gap base material, and maximum current density is determined by the collector material's saturation velocity for a fixed collector thickness and doping profile (which are usually controlled tightly with epitaxial growth).

2.3 FET DC parameters and modeling

Some of the general features of an AlGaAs/InGaAs pHEMT device are shown in Figure 2.3. Figure 2.3a shows the drain current and gate current plotted against drain current at various values of gate voltage. Figure 2.3b, in turn, shows drain and gate currents plotted against gate voltage at various values of drain voltage. Some of the important DC parameters extracted or inferred from the data are transcribed on these plots.

The parameters from Figure 2.3a include Idss (I_{ds} with gate shorted at a specified V_{ds} value), Vknee (drain voltage at which the saturated current decreased by, for example, 10% in the triode region), resistance near $V_{ds} = 0$ in the "on" condition Ron, and BVds (drain-source breakdown voltage at a fixed gate current target). It is apparent that there are many values of BV_{ds} (on and off state) since BV_{ds} is a function of the current in the channel. Figure 2.4 shows more detailed characteristics of drain-gate breakdown voltage (i.e., $V_{ds} - V_{gs}$), where the measurement is taken with a fixed gate current condition (e.g., $I_g = -0.1\,\text{mA/mm}$ and $I_g = -1\,\text{mA/mm}$) and the forced drain current is swept. Both the BV_{ds} locus on Figure 2.3a and BV_{dg} versus V_{gs} plot of Figure 2.4 show that off-state and on-state breakdown values can be significantly different. In particular, the on-state breakdown characteristics should be well characterized and modeled for robust RF power amplifier designs. Additionally, significant gate leakage current during operation (both DC and RF) is likely to affect the long-term reliability of the device.

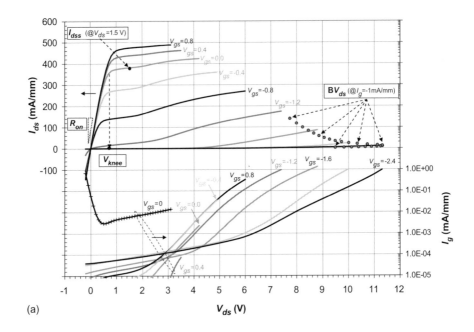

(a)

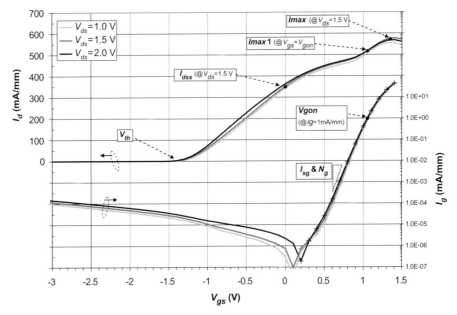

(b)

Figure 2.3 Representative DC parameters for a pHEMT device: (a) I_d (left axis) and I_g (right axis) versus V_{ds}. Positive I_g values denoted with "+"; (b) I_d (left axis) and I_g (right axis) versus V_{gs}. Positive I_g values denoted with "+".

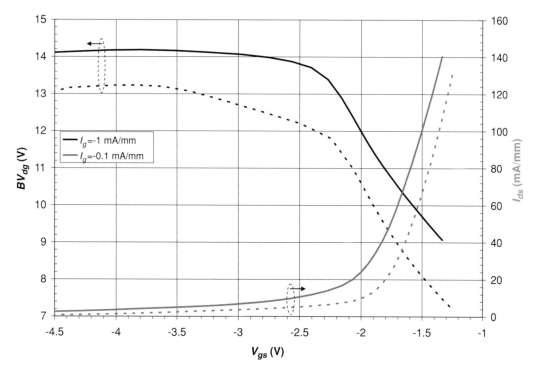

Figure 2.4 Drain-source breakdown voltage (V_{ds}) and I_d versus V_g at the two I_g conditions: -0.1 and -1.0 mA/mm. Measurement is taken by fixing I_g and sweeping I_d. V_{gs} and V_{ds} are monitored.

The FET parameters from Figure 2.3b include Vth (the threshold or turn-on voltage), ISg (saturation current coefficient for gate diode current), Ng (ideality factor for gate diode current), Vgon (gate current turn on voltage), and Imax (the maximum drain current). Imax can be alternatively defined at a fixed gate voltage such as at V_{gon} (or when the gate diode is forward-biased), shown as Imax1 in Figure 2.3b. A GaAs MESFET example in Figure 2.5 shows more details on how V_{th} can be extracted from the drain current versus gate voltage plot. Typically, a drain current target in specified (1 mA/mm in this case), and the gate voltage at which the drain current condition is met is reported back. As seen in the plots, V_{th} can be strongly drain voltage dependent, and an accurate model should account for this behavior.

The parameters described so far describe the general shape (or rough boundaries) of the bias space, and the first derivatives of the DC data (i.e., g_m and g_{ds}) are typically used to further fine-tune the model parameters. Such data is shown in Figure 2.6, for DC g_m and DC g_{ds}. Due to the fact that most III–V FET models use an empirical definition of the drain current (e.g., polynomial functions), most of the time extracting the DC model is spent with these two plots, which is critical to achieve the desired tradeoff in the accuracy in the gain (S_{21} at low frequencies) and output match (S_{22} at low frequencies) within a determined range of bias voltages.

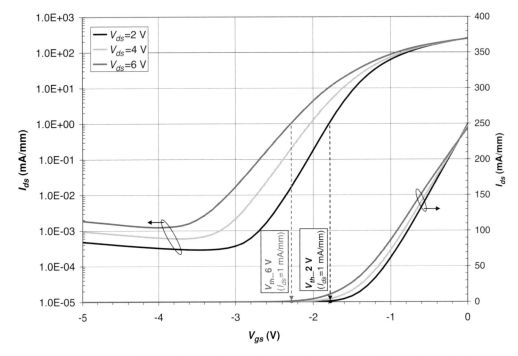

Figure 2.5 Log I_d (left axis) and linear I_d (right axis) versus V_{gs} and the definition of V_{th} at 2 and 6 V V_{ds}.

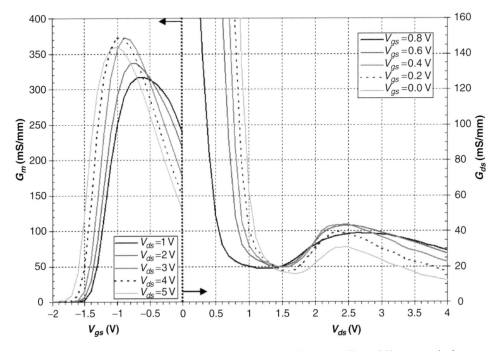

Figure 2.6 DC-derived g_m (left axis) and g_{ds} (right axis) versus V_{gs} and V_{ds}, respectively. The various lines represent different bias conditions described in the legend.

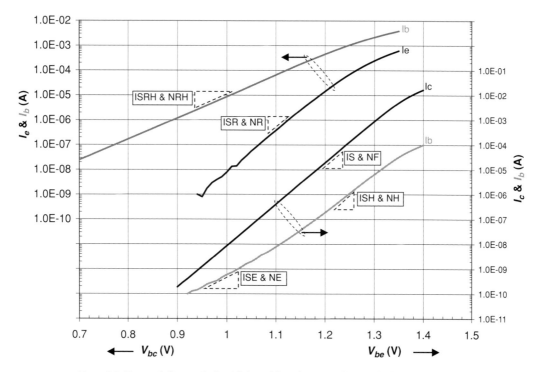

Figure 2.7 Forward Gummel plot (right axis) and reverse Gummel plot (left axis). The DC parameter values next to the triangles represent the saturation current (I_{Sx}) and ideality factor (N_x) of the "diode" equation, that are related to the y-intercept and the slope, respectively, on this log-linear plot.

Some examples of commercially available III–V models used in the industry include the Angelov [4] and EEFET (EEHEMT) [5] models. Reference [6] gives a broader overview of large-signal III–V FET models developed before the 1990s.

2.4 HBT DC parameters and modeling

Forward and reverse Gummel plots for a GaAs HBT device are shown in Figure 2.7. The saturation currents (IS_x) and ideality factors (N_x) for each of the current components are annotated on the plots. Similar to the gate diode current parameters of the FET, the saturation current is the vertical axis intercept and the ideality factor is related to the slope on a log-linear I–V plot.

Three current components are typically considered. The main collector–emitter current, I_{ce}, has already been shown in (2.1). The base–emitter diode current (which is a function of the base–emitter voltage) is the sum of the high current ("ideal") and low current ("nonideal", or "$n = 2$") terms.

$$I_{be} = I_{be\_high} + I_{be\_low} = ISH\left(\exp\left(V_{be}/(NH \times V_t)\right) - 1\right)$$
$$+ ISE\left(\exp\left(V_{be}/(NE \times V_t)\right) - 1\right). \qquad (2.3)$$

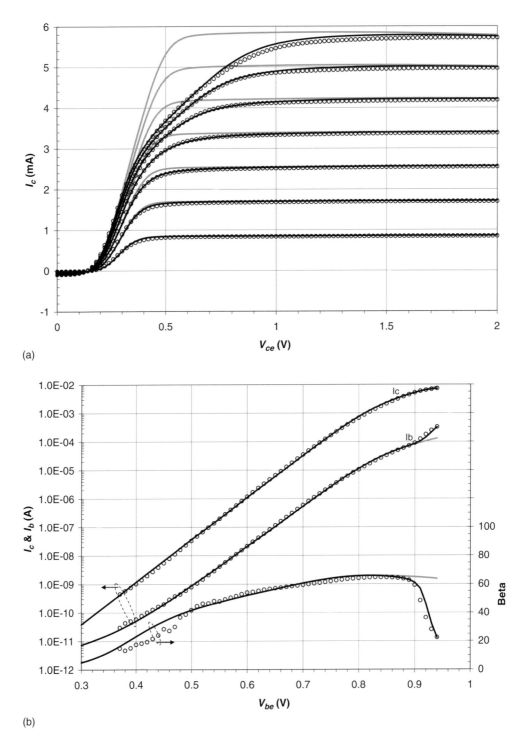

(a)

(b)

Figure 2.8 Data for $1 \times 3\,\mu\text{m}^2$ InP HBT exhibiting significant high current compression effects at low voltage ("soft-knee" effect). Measurements (symbols), simulation model with soft-knee effect (dark solid lines), and simulation model without soft-knee effect (light solid lines). (a) $I_c - V_{ce}$ family of curves with forced I_b; (b) forward Gummel plot, and Beta versus V_{be}.

Similarly, the base–collector diode current (which is a function of the base–collector voltage) is the sum of the high and low currents (note: low-current term is not shown in Figure 2.7):

$$I_{bc} = I_{bc\_high} + I_{bc\_low} = ISCH \left(\exp\left(V_{bc}/(NCH \times V_t)\right) - 1\right)$$
$$+ ISC \left(\exp\left(V_{bc}/(NC \times V_t)\right) - 1\right). \tag{2.4}$$

The current gain, Beta, is an important parameter used to describe the performance of the device, but in recent bipolar models such as VBIC [7] and HICUM [8], Beta is not a model parameter, since it is strongly bias dependent. However, Beta can be inferred from the quotient, I_{ce}/I_{be}, using equations (2.2) and (2.3). Similarly, the forward turn-on voltage, Vbef, is also an important device parameter, but it is not a model parameter and is inferred from equation (2.2) at a target value of the collector current.

The DC currents described so far are shared with Si-based BJT models (in which many are based on the charge-control model of Gummel–Poon [3]), and further enhancements are typically implemented in HBT models which take into account heterojunction and high-current effects. For example, the UCSD model [9] takes into account the conduction band barrier (blocking) effects in both the junctions at the base-emitter (single heterojunction) and base-collector (double heterojunction). The Agilent HBT [10] model empirically models the high current effect that results in a "soft-knee" in the low collector voltage region [11]. Figures 2.8a and 2.8b show a Gummel plot and I_c-V_{ce} family of curves showing the consequences of modeling the "soft-knee" effect.

It is still common to see Si-based models such as VBIC (includes self-heating) and Gummel–Poon (no self-heating) models used for III–V HBT designs. However, there are III–V specific compact models available commercially such as the aforementioned UCSD and Agilent HBT models, and also the FBH HBT model [12].

2.5   Process control monitoring

Taking statistical data of extracted transistor device parameters (both DC and S-parameters) is standard procedure in the wafer fabrication process, widely referred to as process control monitor (PCM). Since DC measurements are fast and efficient, it is typical to measure more than 20 devices (sites) per wafer, which is enough for statistical analysis.

Figure 2.9 shows individual values of three pHEMT DC parameters (*gm*, *Idss*, and *Vth*). Plotting the data by reticle locations is helpful in showing the spatial dependence of device performance, which sometimes exposes the variability issues of a process within a wafer. From this data, statistical values representing the entire wafer such as the mean and standard deviation of the parameters can be calculated. This is a more convenient way to view variations of a wafer, especially when inter-wafer variations are monitored from lot to lot. Figure 2.10 is a box and whisker plot of FET parameter V_{th} plotted across 53 wafers and multiple lots, which shows that this model parameter (and device characteristic) can vary significantly over multiple wafers.

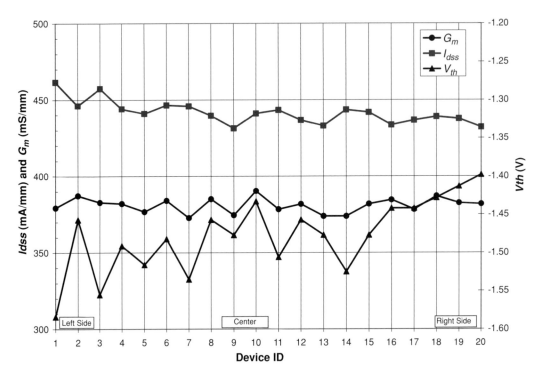

Figure 2.9 DC-derived transconductance (G_m), threshold voltage (V_{th}), and drain current at $V_{gs} = 0$ (I_{dss}) data from a AlGaAs/InGaAs pHEMT wafer for 20 sites.

PCM data can be used outside of the intended *process control* purposes. For example, circuit performance variations can be predicted via statistical device modeling (Chapter 9). In the area of device reliability, it is now very common for PCM data to be used to predict SiN capacitor lifetime [13] through the breakdown voltage and voltage ramp rate. More recently, it has been shown that current gain, Beta, is a predictor of infant (short-term) failure rates in GaAs HBTs [15] when thousands of devices can be measured for a single wafer, since some forms of infant failure are dependent on substrate dislocation density. It has been demonstrated that outlier devices can be used to estimate the infant failure rate of high integration level HBT ICs (e.g., >500 HBTs).

2.6   Thermal modeling overview

Examples of the ambient temperature dependence of an AlGaAs/InGaAs pHEMT device (I_{ds} versus V_{ds}) and an InP HBT device (I_c versus V_{be}) are shown in Figure 2.11. In general, FET compact models take into account the temperature-dependent electron mobility in the channel (in which current decreases with temperature at a fixed V_{gs}), while the HBT compact models take into account the temperature dependence of the diffusion current (in which current increases with temperature at a fixed V_{be}). III–V transistors reside on low thermal conductivity substrates (e.g., GaAs $=0.46$ W/K·cm^{-1},

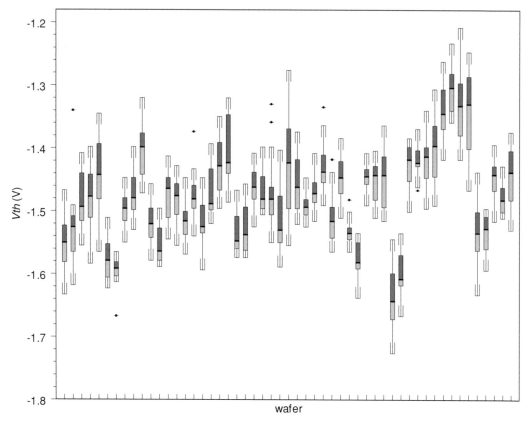

Figure 2.10 Box and whisker plot of parameter V_{th} over 53 AlGaAs/InGaAs pHEMT wafers. The device in Figure 2.9 is the fourth device from the left.

InP $= 0.68$ W/K·cm^{-1}, Si $= 1.5$ W/K·cm^{-1}) that makes ICs susceptible to self-heating effects when power dissipation is high. This is relevant in applications such as GaAs HBT-based power amplifiers, microwave GaAs FET amplifiers, and InP HBT-based digital circuits.

A variety of thermal modeling techniques, dependent on the intended application, are used for III–V transistors. For detailed analysis at the single device level, physical modeling [16] and analytical modeling [17] techniques are used to find the hotspot of a device. This is particularly relevant in the area of device epitaxial layer design and layout optimization where precise temperature dependence on position (vertical and/or horizontal) is of particular interest. For IC design applications, a thermal subcircuit model can augment a properly formulated DC compact model to calculate a single (or average) junction temperature representing the device based on the electrical interaction with the thermal material properties. Figure 2.12 shows two model topologies that will be discussed here (HBT physics-based and FET measurement-based), where they both share a similar thermal equivalent subcircuit model. The thermal subcircuit consists of the thermal resistance (related to the thermal conductivity) and thermal capacitance

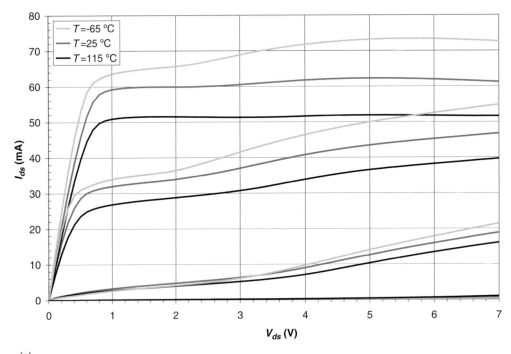

(a)

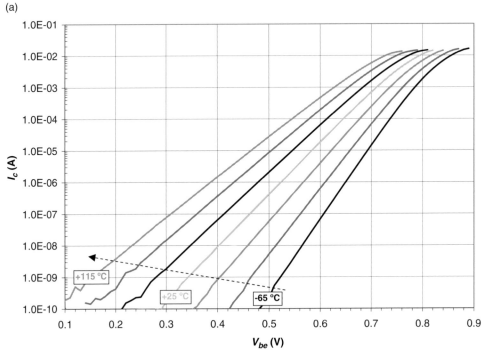

(b)

Figure 2.11 Measured ambient temperature dependence of FET and HBT devices. (a) AlGaAs/
InGaAs pHEMT I_{ds} versus V_{ds} family of curves at −65, 25, and 115 °C; (b) InP HBT Gummel
plot at −65, −35, −5, 25, 55, 85, and 115 °C.

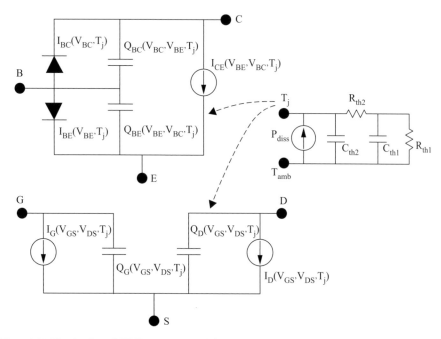

Figure 2.12 Physics-based HBT compact model and measurement-based FET compact model topologies. Also shown is the thermal subcircuit.

(that together with the thermal resistance describes the rate at which temperature changes propagate). A two-pole approximation is shown in Figure 2.12, where the intrinsic device (fast time constant) and substrate (slower time constant) can be represented with the pair of R_{th} and C_{th} values [18]. This circuit can in turn be simplified to a more popular single-pole subcircuit by setting one of the R_{th} and C_{th} pairs to 0.

2.7 Physics-based thermal scaling model for HBTs

The standard bipolar transistor main collector current can be derived by assuming that the mechanism is related to the minority carrier diffusion in the base:

$$I_c = A_E q D_n \frac{dn}{dx}, \tag{2.5}$$

where A_E is the emitter area, D_n is the electron diffusivity of the p-doped base material, and n is the minority carrier concentration. For the forward collector current term, it can be approximated as:

$$I_c \approx \left(\frac{A_E q D_n n_i^2}{p_B W_B} \right) \exp \left(\frac{q V_{be}}{\eta k T} \right), \tag{2.6}$$

where n_i is the intrinsic carrier concentration, p_B is the base doping, and W_B is the base thickness. The forward saturation current (ISf) model parameter in (2.2) is basically related to the left term of the product in (2.6). The collector current has temperature dependence in n_i^2, D_n, and the exponential term. D_n is weakly temperature dependent, while n_i^2 has very strong temperature dependence. It has the form and temperature dependence:

$$n_i^2 = N_C N_V \exp\left(-\frac{qE_g}{kT}\right) \propto T^3 \exp\left(-\frac{qE_g(T)}{kT}\right), \tag{2.7}$$

where E_g, the bandgap, itself has a temperature dependence.

In most compact models, the temperature dependence of D_n and n_i^2 are succinctly incorporated as a temperature-dependent saturation current (which becomes the pre-factor in the "diode" equation):

$$IS(T) = IS_{nom}\left(\frac{T}{T_{nom}}\right)^{XTIS} \exp\left(\frac{qE_g}{kT}\left(1 - \frac{T}{T_{nom}}\right)\right), \tag{2.8}$$

where IS_{nom} and E_g are specified at ambient temperature T_{nom}, and $XTIS$ is the temperature coefficient parameter. Furthermore, each "diode" current equation can be assigned a distinct temperature coefficient parameter, which enables flexibility in fitting the thermal dependence of the currents.

Figure 2.13a shows a progression of the DC and thermal compact modeling process for the collector current plotted on a log-linear scale. First, saturation current and ideality factors are extracted at $T_{nom} = 25\,^\circ\text{C}$ (i.e., room temperature). This result is shown in light gray where the current is the straight line, since the resistances have not been extracted. After the resistances are extracted (typically using RF techniques), the simulated results (in dark gray) are lower than the measurements. This is expected, since the self-heating effect is not modeled yet (i.e., more self-heating increases the current at a fixed V_{be}). The thermal scaling parameters consist of selecting the bandgap value and fitting the temperature coefficients ($XTIx$) for each current component. It is seen in Figure 2.13a that fitting a temperature coefficient (assuming that the bandgap has been selected) is sufficient to describe the temperature behavior of the collector current for low to mid currents at $T = -35\,^\circ\text{C}$, $25\,^\circ\text{C}$, and $85\,^\circ\text{C}$. Finally, the thermal resistance (R_{th}) is optimized to fit the collector current in the high-current (power dissipation) region. The effect of the R_{th} parameter is more evident in the I_c–V_{ce} plot of Figure 2.13b where self-heating reduces the collector current when a constant base current is forced. A useful consequence of extracting a physics-based compact model is that a key device parameter, such as R_{th}, can be easily and unambiguously extracted. R_{th} can also be extracted using DC electrical measurement techniques [13, 19], but they involve making customized measurements and numerical optimizations.

2.8   Measurement-based thermal model for FETs

The physics-based model in the previous section is an example of a bottom-up approach to device modeling. Every distinct electrical mechanism incorporated in the model has its own explicit thermal dependence (generally nonlinear) prescribed by the physics. A complementary approach is a measurement-based, or top-down, model that will

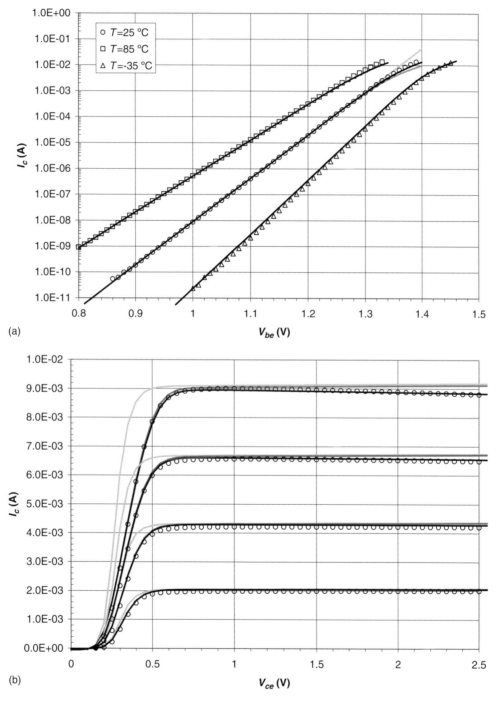

(a)

(b)

Figure 2.13 GaAs HBT data with measurement (symbols), and three different extraction scenarios: (1) Light gray solid line: DC current model with no resistance and thermal parameter extracted, (2) darker gray solid line: DC model with resistance extracted, but no thermal parameters extracted, (3) black solid line: DC model with resistance and thermal parameters extracted. (a) Gummel plot with only I_c at three different temperatures. Simulation scenarios 1 to 3 only shown at $T = 25\,^{\circ}\text{C}$. (b) I_c–V_{ce} family of curves at $T = 25\,^{\circ}\text{C}$.

be discussed here in the context of a FET thermal compact model. In this approach, the electrical characteristics are extensively measured under different thermal conditions – typically at different ambient backside or case temperatures, T_{amb}. No a-priori closed-form expressions for the electrical equations or their thermal dependence need be assumed. It is only assumed that all the observable electrical and thermal behavior of the device is contained in such data. The main issue concerns how to unravel this data.

The topology for a measurement-based dynamical self-heating model of a FET is shown in Figure 2.12. The currents are represented by the total gate (I_g) and drain (I_d) currents, defined on the independent variables V_{ds}, V_{gs}, and device or "junction temperature", T_j. Since I_d is the dominant current for device performance (and self-heating), the discussion and modeling will focus primarily on this current term. The thermal and electrical equivalent circuits are coupled in terms of temperature rise in the former driven by dissipated power in the latter, where the currents in the latter depend on the temperature of the former. The simulator solves the coupled subcircuits (and electrical networks to which the device is attached) self-consistently, to properly compute dynamic self-heating effects in response to the particular signal encountered by the device model during a simulation.

Before the thermal FET model details are discussed, extraction for the value of the thermal resistance R_{th} needs to be done, since this parameter plays a critical role in measurement-based thermal model extraction. There are various techniques to extract the FET thermal resistance including DC measurement analysis [20], and calculations based on physical/analytical thermal modeling [21]. The technique discussed here is probably the most conceptually straightforward [22], but requires more sophisticated equipment capable of pulsed-IV measurement. When fast pulsed-IV data is taken at a quiescent bias point where trapping and thermal effects are minimal (e.g., $V_{gs} = 0$; $V_{ds} = 0$), the drain current data represent an isothermal result (equal to the ambient temperature) at the respective V_{gs} and V_{ds} points. Such measurements can be taken at various ambient temperatures, which result in a collection of isothermal-IV data that correspond to the respective ambient temperature. DC data is also taken at the same bias points, and overlaid on the pulsed IV data. An example of this is shown in Figure 2.14, where V_{gs} is fixed at the same value of 0.6 V, and the DC data is taken at an ambient temperature of 25 °C. The intersections of the DC and pulsed-IV data are significant, since the ambient temperatures of the pulsed-IV data at the intersections are approximately the junction temperature of the DC data. The intersections can then be plotted on a temperature versus DC power dissipation, and the slope of this plot (dT/dP) is an approximation of the thermal resistance (and forcing the condition that temperature is equal to the ambient temperature at zero power dissipation).

After the thermal resistance (or at least an initial guess) is found, electrical data should be taken at bias points of various T_{amb}. The difficulty is that under these static (DC operating point) measurement conditions, the junction temperature, T_j, is completely determined by the electrical data at each operating point through the simple equation:

$$T_j = T_{amb} + R_{th} \sum_{i=1}^{2} \tilde{I}_i \left(V_{gs}, V_{ds}, T_{amb} \right) \cdot V_i, \tag{2.9}$$

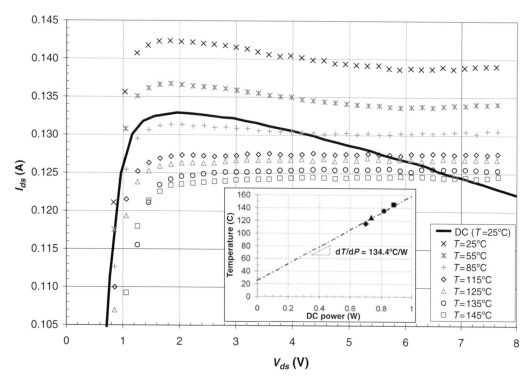

Figure 2.14 Pulsed IV data taken at various ambient temperatures with DC I–V data at $T = 25\,°C$. The pulse width is 0.5 μs with quiescent bias point of $V_{ds} = 0$ and $V_{gs} = 0.5\,V$. The inset shows the extraction of the thermal resistance inferred from the intersection of the DC and pulsed-IV data.

($i = 1$ and $i = 2$ refer to the gate and drain, respectively) where T_j is *not* an independent variable as in the case of a physically based model. The "∼" over the current in (2.9) emphasizes that this is the measured value depending on the voltages and also on the parameter of the experiment, T_{amb}, the backside temperature. The objective of the measurement-based modeling exercise is to convert this "nonisothermal" data into isothermal model functions defined on the proper set independent variables, V_{ds}, V_{gs}, and T_j so the self-consistent electrothermal model can be implemented with the same equivalent electrical and thermal circuits as before. Procedurally, this is accomplished by first computing T_j from each set of measured currents using (2.9), and then fitting the model functions (constitutive relations) on the new set of variables V_{ds}, V_{gs}, T_j. At this point, T_j can be considered an independent variable and associated with the "isothermal" current source in the equivalent circuit. Mathematically, this is equivalent to defining a new model function, $I(V_{gs}, V_{ds}, T_j)$, by equation (2.10)

$$I\left(V_{gs}, V_{ds}, T_j\right) = \tilde{I}\left(V_{gs}, V_{ds}, T_{amb}\right), \qquad (2.10)$$

together with the condition in (2.9).

For practical implementation, artificial neural networks (ANNs) [23] can be used to fit the functions defined by (2.9) and (2.10) on the discrete set of values of I corresponding to the values of terminal voltage and junction temperature. If the set of backside values, T_{amb}, are uniformly distributed, the corresponding values of T_j will not have a uniform distribution. This is not a problem for fitting with neural networks, since they can be trained just as easily on scattered data as on gridded data. Using other basis functions or table-based approaches to the fitting may require an additional step to put the new T_j independent variables on a grid. The nonlinear change of variables means the shape versus bias for fixed T_j of the recovered model function, $I(V_{gs}, V_{ds}, T_j)$, is different from that of the nonisothermal data of $\tilde{I}(V_{gs}, V_{ds}, T_{amb})$ for fixed T_{amb}.

It is important to make sure that the set of measurement data be taken over a sufficiently wide range of backside temperatures, T_{amb}, so that any resulting T_j required by the model in a dynamical simulation is within the range for which (2.10) was trained; that is, we do not want the model extrapolating (to either higher or lower junction temperatures) during a simulation. An appropriate backside temperature can be estimated, given a known thermal resistance value, R_{th}, by solving equation (2.9) (considered for this purpose as an implicit nonlinear equation) for T_{amb} given a fixed T_j. In practice, it is generally sufficient to measure IV characteristics and S-parameters (for the charges) versus bias voltages at three backside temperatures, from well below the nominal 25 °C (e.g., 0 °C or even below, if needed) up to the hottest expected condition of use, often above 85 °C.

Example results of the artificial neural network self-heating model are shown in Figure 2.15 for a 0.125 μm gate length AlGaAs/InGaAs pHEMT device. The model is able to predict simultaneously the self-heating in the high-power dissipation region of I_{ds} versus V_{ds} in Figure 2.15a, and the ambient temperature dependence of the subthreshold current in Figure 2.15b. The great benefit of this measurement-based approach is that it naturally incorporates the detailed thermal dependence on the device model, regardless of the mechanisms involved. Such models automatically include the effects of mobility reduction with temperature that causes a decrease of the drain current, the dominant effect of self-heating. At the same time, such models exhibit the decrease of pinch-off voltage (V_{th} becomes more negative) with increasing temperature that causes the increase of drain current in the subthreshold and just beyond the threshold voltage regions of operation. Physics-based or specific empirical/phenomenological models, with their assumed thermal dependence, have been known to neglect some of these different effects that can cause significant errors in simulated model performance.

2.9   Transistor reliability evaluation

Maintaining reliable performance over a long period of time (e.g., >30 years) is an important characteristic for any manufacturable III–V device integrated circuit (IC) technology. To evaluate the reliability of a device, it is essential to stress a representative population of devices at an elevated temperature to accelerate the wear out of the device, while monitoring the performance of the device [24]. In general, accelerated temperature

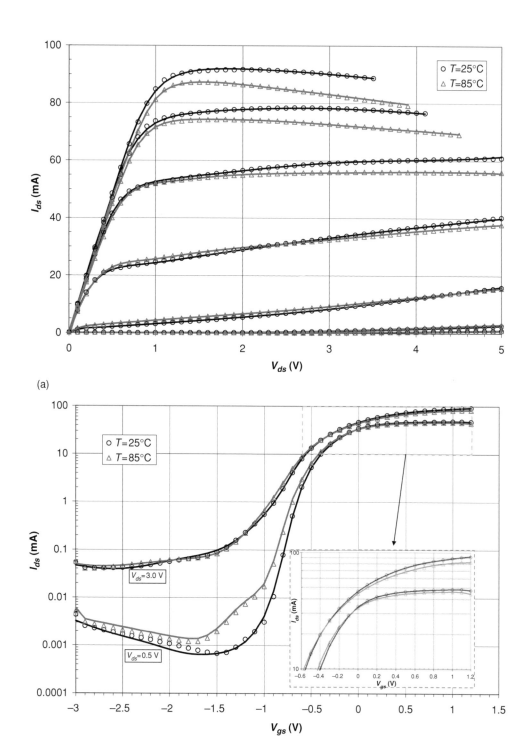

Figure 2.15 DC validation of the measurement-based thermal model at $T = 25\,°C$ and $T = 85\,°C$: (a) I_{ds} versus V_{ds} family of curves; (b) I_{ds} versus V_{gs} with the detail between 10 and 100 mA shown in the inset.

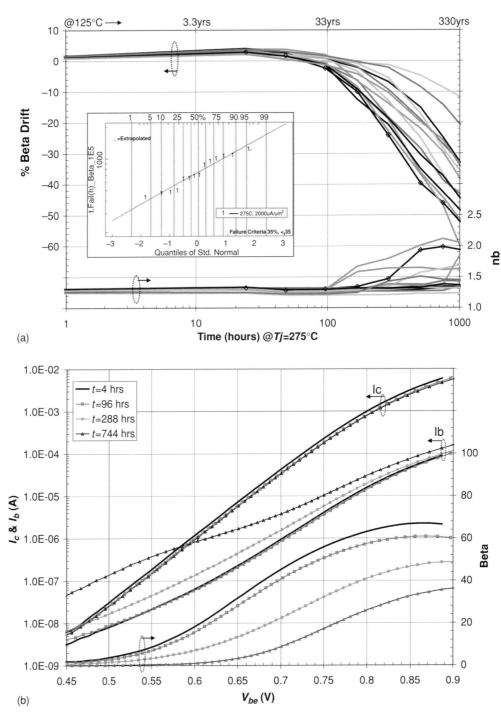

(a)

(b)

Figure 2.16 Example stress test result of 14 InP HBT devices stressed at $T_j = 275\,^\circ$C. The parametric data shown was taken at room temperature. (a) % Beta drift and base current ideality factor (n_b) versus time. The top x-axis shows equivalent time at $T_j = 125\,^\circ$C assuming a device activation energy of 1.0 eV (which is a representative of this technology). (b) Forward Gummel plot and Beta versus V_{be} snapshot at various stress times for device denoted by symbols in (a).

stress is possible, since time to failure between the use-condition ($time_{use}$) and stress-condition ($time_{stress}$) are "scaled" by temperature via an activation energy (based on Black's law for electromigration [25]):

$$time_{use} = time_{stress} \cdot \exp\left(\frac{E_a\left(1/T_{use} - 1/T_{stress}\right)}{k}\right) \cdot \left(\frac{I_{stress}}{I_{use}}\right)^{nj}, \qquad (2.11)$$

where E_a is the activation energy of the failure mechanism and n_j is the current acceleration factor. T_{use} and T_{stress} are the temperatures at the use and stress conditions, respectively.

Figure 2.16a shows a plot of Beta and base current ideality versus time of 14 InP HBTs devices stressed at $T_j = 275\,°C$. For reliability evaluation, it is mandatory to measure a large number of devices since, as Figure 2.16a shows, there is a distribution or variability in the drift of the parameters. The parametric measurements are taken at room temperature at each interval in order to monitor various DC parameters. Although it is typical for the main failure mechanism to be the current gain Beta for the HBT, there are other DC parameters such as leakage currents and resistances that may drift over time. Figure 2.16b shows a detailed forward Gummel plot of one of the devices in Figure 2.16a (the data with the symbols) which gives insights into the failure mechanism of this particular device. It is apparent that the low-current base current term (represented by ISE and NE in equation (2.3)) is the HBT DC model parameter that is drifting with stress. From a physical device point of view, the failure of this device may, for example, be due to the effective emitter "ledge" degrading over time.

A single temperature and bias point stress experiment, as in Figure 2.16a, is sufficient to extract a mean time to failure (MTTF) and a distribution parameter sigma (σ) assuming a log-normal cumulative distribution function for the failures, or cumulative failure function (CFF):

$$CFF(t) = \int_0^t \frac{1}{t\sigma\sqrt{2\pi}} \exp\left[-\frac{1}{2\sigma^2}\left(\ln\frac{t}{MTTF}\right)^2\right] dt. \qquad (2.12)$$

The parameter σ enables estimates of time-to-failures at the percentiles other than 50 (where the 50th percentile is the MTTF).

In order to extract E_a and n_j of equation (2.11), it is necessary to run a set of stress experiments at various temperatures (typically at three temperatures) and current densities. Figure 2.17, taken from reference [26], shows the results of an InGaP/GaAs HBT reliability evaluation in which the failure distribution was fit to calculate MTTF at the various stress conditions assuming a *fixed* σ for all four test conditions, and then the MTTFs at each stress condition in turn were used to extract E_a (via an Arrhenius plot) and n_j.

With representative E_a and n_j known for a process, it is possible to extrapolate the MTTF at $time_{use}$ of a single device at any temperature and bias conditions (i.e., at T_{use} and I_{use}) when an MTTF at $time_{stress}$ is known by using (2.11). Furthermore, to calculate the MTTF of an IC, where multiple devices are biased at various junction temperatures and bias current densities, the log-normal distribution parameter, σ, is needed. Since σ

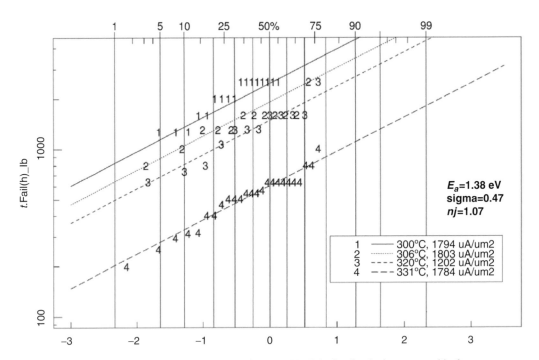

Figure 2.17 Failure distribution plot of a GaAs HBT device for devices stressed in four temperature and current conditions (from reference [26]). This data is used to extract the activation energy (E_a), log-normal distribution (σ), and current acceleration parameter (n_j) as denoted in the figure.

describes the distribution of the failure times of an individual device, cumulative failures of the IC can be calculated once the junction temperature and current densities for individual devices are known. The temperature and current densities of each device can be found, for example, by running DC and thermal simulations using models described in the previous section. Figure 2.18 shows an example of a lifetime calculation (% failures versus time) of a 75 transistor GaAs HBT divider circuit with different ambient temperature, E_a, and σ scenarios. The ambient temperature variation is a practical issue where package design and robustness influence the lifetime of an IC. The dependence of the IC lifetimes on E_a and σ show that any errors or uncertainty in the extraction of these reliability model parameters will result in significantly different IC lifetime estimates. This application example emphasizes that accurate *DC and thermal modeling* (to estimate bias conditions and junction temperatures) and *parameter extraction* (to define the failure criteria) are important in device reliability evaluation, where various simulation and extraction techniques can be used together to estimate the lifetime of an IC.

Acknowledgments

The authors would like to thank the following individuals: Dr. Bob Yeats of Agilent Technologies (now retired) for concepts and ideas on III–V device reliability evaluation

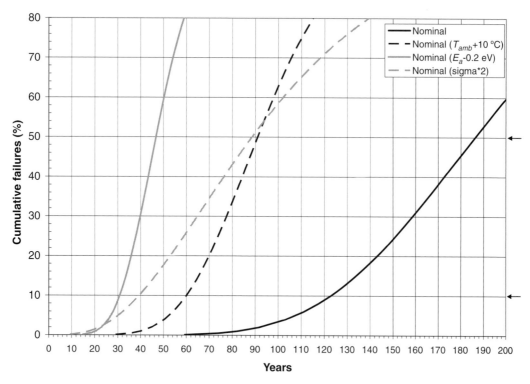

Figure 2.18 Cumulative failures versus time for a 75 HBT divider IC at $T_{amb} = 95\,°C$, using the following "nominal" reliability test parameters: MTTF $= 100\,h$, $\sigma = 0.5$, $E_a = 1.2\,eV$, $n_j = 2.0$, $T_j = 316\,°C$, $J_e = 1.8\,mA/\mu m^2$. The plot shows four cases: (1) nominal, (2) T_{amb} higher by $10\,°C$ than the nominal ($T_{amb} = 105\,°C$), (3) activation energy lower by $0.2\,eV$ than the nominal ($E_a = 1.0\,eV$), (4) log-normal distribution parameter twice that of the nominal ($\sigma =1.0$).

techniques presented here, and Professor Peter Asbeck of the University of California at San Diego for information on HBT model device equations. We would also like to thank the Agilent's wafer fabrication facility personnel in Santa Rosa, California, USA for making the devices discussed in this chapter.

References

[1] W. R. Curtice, "A MESFET model for use in the design of GaAs integrated circuits," *IEEE Trans. Microw. Theory Tech.,* vol. 28, pp. 448–456, 1980.

[2] R. Van Tuyl and C. Liechti, "Gallium arsenide digital integrated circuits," Techn. Rep. AFL-TR-74-40, Air Force Avionics Lab., Mar. 1974.

[3] H. K. Gummel and R. C. Poon, "An integral charge control model of bipolar transistors," *Bell Syst. Tech. J.,* vol. 49, pp. 827–852, May–June 1970.

[4] I. Angelov, H. Zirath, and N. Rosman, "A new empirical nonlinear model for HEMT and MESFET devices," *IEEE Trans. Microw. Theory Tech.,* vol. 40, 1992, pp. 2258–2266.

[5] Agilent Technologies, "Circuit components: nonlinear devices," *Advanced Design System 2008 Manual*, pp. 261–308, 2008.

[6] J. M Golio, *Microwave MESFETs & HEMTs*, Norwood, MA: Artech House, 1991, ch. 2.

[7] C. C. McAndrew, J. A. Seitchik, D. F. Bowers, M. Dunn, M. Foisy, I. Getreu, M. McSwain, S. Moinian, J. Parker, D. J. Roulston, M. Schroter, P. van Wijnen, and L. F. Wagner, "VBIC95, The Vertical Bipolar Inter-Company Model," *IEEE J. Solid/state Circuits*, vol. 31, pp. 1476–1483, Oct. 1996.

[8] M. Schröter and T-Y. Lee, "Physics-based minority charge and transit time modeling for bipolar transistors," *IEEE Trans. Electron Devices*, vol. 46, pp. 288–300, Feb. 1999.

[9] M. Rudolph, *Introduction to Modeling HBTs*, Norwood, MA: Artech House, 2006, pp. 232–250.

[10] M. Rudolph, *Introduction to Modeling HBTs*, Norwood, MA: Artech House, 2006, pp. 250–263.

[11] S. Tiwari, "A new effect at high currents in heterostructure bipolar transistors," *IEEE Electron Device Lett.*, vol. 9, pp. 142–144, Mar. 1999.

[12] M. Rudolph, *Introduction to Modeling HBTs*, Norwood, MA: Artech House, 2006, pp. 263–276.

[13] B. Yeats, "Inclusion of topside metal heat spreading in the determination of HBT temperatures by electrical and geometrical methods," *IEEE GaAs IC Symp.*, pp. 59–62, 1999.

[14] B. Yeats, "Assessing the reliability of silicon nitride capacitors in a GaAs IC process," *IEEE Trans. Electron Devices*, vol. 45, pp. 939–946, Apr. 1998.

[15] K. Alt, T. Shirley, C. Hutchinson, M. Iwamoto, B. Yeats, B. Gierhart, M. Bonse, R. Shimon, F. Kellert, and D. D'Avanzo, "Use of a transistor array to predict infant transistor mortality rate in InGaP/GaAs HBT technology," *Reliability of Compound Semiconductor Workshop*, Oct. 2008.

[16] J. C. Li, T. Hussain, D. A. Hitko, P. A. Asbeck, and M. Sokolich, "Characterization and modeling of thermal effects in sub-micron InP HBTs," *IEEE Compound Semiconductor IC Symp.*, pp. 65–68, Nov. 2005.

[17] D. H. Smith, A. Fraser, and J. O'Neil, "Measurement and prediction of operating temperatures for GaAs ICs," *Proc. SEMITHERM*, pp. 1–20, Dec. 1986.

[18] W. R. Curtice, V. M. Hietala, E. Gebara, and J. Laskar, "The thermal gain effect in GaAs-based HBTs," *IEEE Int. Microw. Symp. Dig.*, pp. 639–641, June 2003.

[19] D. E. Dawson, A. K. Gupta, and M. L. Salib, "CW measurement of HBT thermal resistance," *IEEE Trans. Electron Devices*, vol. 39, pp. 2235–2239, 1992.

[20] B. Yeats, *1992 GaAs Rel Workshop*, paper IV–I.

[21] D. E. Dawson, "Thermal modeling, measurements and design considerations of GaAs microwave devices," *IEEE GaAs IC Symp.*, pp. 285–290, 1994.

[22] K. A. Jenkins and K. Rim, "Measurement of the effect of self-heating in strained-silicon MOSFETs," *IEEE Electron Device Lett.*, vol. 23, pp. 360–362, June 2002.

[23] Q. J. Zhang and K. C. Gupta, *Neural Networks for RF and Microwave Design*, Norwood, MA: Artech House, 2000.

[24] J. V. DiLorenzo and D. D. Khandelwal, *GaAs FET Principles and Technology*, Dedham, MA: Artech House, 1982, ch. 6.

[25] J. R. Black, "Electromigration – a brief survey and some recent results," *IEEE Trans. Electron Devices,* vol. 16, pp. 338–347, 1969.

[26] B. Yeats, P. Chandler, M. Culver, D. D'Avanzo, G. Essilfie, C. Hutchinson, D. Kuhn, T. Low, T. Shirley, S. Thomas, and W. Whiteley, "Reliability of InGaP-Emitter HBTs," *GaAs Manufacuring Technology Conf.*, 2000, pp. 131–135.

3 Extrinsic parameter and parasitic elements in III–V HBT and HEMT modeling

Sonja R. Nedeljkovic, William J. Clausen, Faramaiz Kharabi, John R.F. McMacken, and Joseph M. Gering

RFMD, Greensboro, North Carolina, USA

3.1 Introduction

A model formulation that is founded on device physics and optimized for good convergence in different simulator platforms is a first step in providing good, advanced models that will satisfy designers' needs. The degree of extraction complexity will depend directly on the choice of model, but every modeling procedure, regardless of model complexity, relies on accurate device measurements, reliable de-embedding techniques, and methods for parameter extraction.

The scope of this chapter is extrinsic parameter extraction. Separating the device parameters from the probe pads' parasitic elements is the first step in the modeling procedure. Thus, the second section will analyze the measurement calibration and parasitic de-embedding techniques that are applicable in III–V HBT and HEMT modeling with an overview of RF test structures for device S-parameter measurements.

After the reference planes are moved from the probe pads to the device, the extraction of extrinsic parameters is the next step. The third section will give an overview of extrinsic parameter extraction methods used in HBTs while the fourth section will cover methods used in GaAs-based pHEMTs and GaN-based HEMTs.

Extrinsic parameter methods described in Sections 3.3 and 3.4 are mostly applicable for a unit cell device. Large transistor arrays are built up from smaller unit cells that are replicated to achieve a required emitter area (HBT) or gate periphery (pHEMT/HEMT). Accurate modeling of large arrays depends not only on an accurate unit cell model, but also on the choice of the scaling approach and the inclusion of parasitics that are generated by interconnect metals, vias, and bondpads, which modify the impedances of the large array. The results of a practical and accurate model scaling approach will be compared with large signal measured data for a multicell array in Section 3.5.

3.2 Test structures with calibration and de-embedding

Accurate device measurements are as essential to the proper extraction of extrinsic parameters and parasitic elements of a device model as they are to the extraction of the active portion of the model. For III–V HBT and HEMT modeling, the main challenge with these device measurements is in making broad-frequency, S-parameter measurements of just the device under test (DUT) with any pads or access connections removed. While this is also important with DC measurements, most semiconductor parameter analyzers and power supplies have separate force and sense lines, which make accurate, Kelvin-style measurements more straightforward. In those cases where Kelvin measurements are not practical, the judicious use of Ohm's Law can often de-embed DC measurements sufficiently for device modeling. Therefore, this section will deal exclusively with S-parameter measurements for device modeling.

There are generally two types of device-level test structure for S-parameter measurements, and these are demonstrated in Figure 3.1. Both structures accommodate ground-signal-ground (GSG), on-wafer probing. Figure 3.1a shows a more compact structure that is coplanar waveguide-(CPW-)based. It does not require a connection to the backside of the wafer, so it tends to be more flexible as to when in the fabrication process the device is tested. Also, being smaller, it enables more devices to be included in the available wafer "real estate". Figure 3.1b shows a larger structure that is microstrip-(MS-)based. This version does require through-wafer vias to make ground connections. The advantage of this structure is that it provides a uniform transmission line running up to the device. This allows a more accurate, on-wafer calibration to be used to bring the error-corrected measurement planes closer to the DUT.

S-parameter measurements for both the CPW- and MS-styled test structures involve a two-tiered approach. The first step uses conventional error correction techniques to extend a network analyzer's calibrated reference plane to the solid lines shown in Figure 3.1. The second step uses a simplified circuit to represent the remaining connections from the calibrated reference planes to the DUT shown by the boxes in Figure 3.1. So, to first order, the tradeoff between these two structures is between the size of the structure and the accuracy of the DUT S-parameters, since the less accurate second tier has less to de-embed in the MS case. It should be noted that the use of a CPW style for the compact structure and an MS style for the uniform transmission line structure are conventional choices; however, other configurations can be used if convenient to the application. Also, since the goal of any test structure is to have as little disturbance to the transmission line feeding the DUT, in both structures, care is taken to preserve a 50 Ohm CPW or MS for as much of the structure as possible.

There are many approaches to performing probe-tip or on-wafer calibrations for the first tier. Reference [1] has discussions on many of these. Several of these methods are implemented directly in modern vector network analyzers (VNAs) or in commercial or free software packages. The two common techniques used in laboratories are the short-open-load-thru (SOLT) technique and the line-reflect-reflect-match (LRRM) technique.

SOLT is a direct mapping technique, where the impedances of the standards are assumed to be exactly known. The error correction translates the raw VNA measurements to the calibrated reference plane in such a way that the standards will translate to their known values. Any inaccuracy in defining these standards results in inaccuracies in the error-corrected results. The SOLT technique is often used to give a calibration to the tips of the RF probes using a commercial, ceramic calibration substrate and as such is more commonly used with the CPW-style structure from above. By using a precision

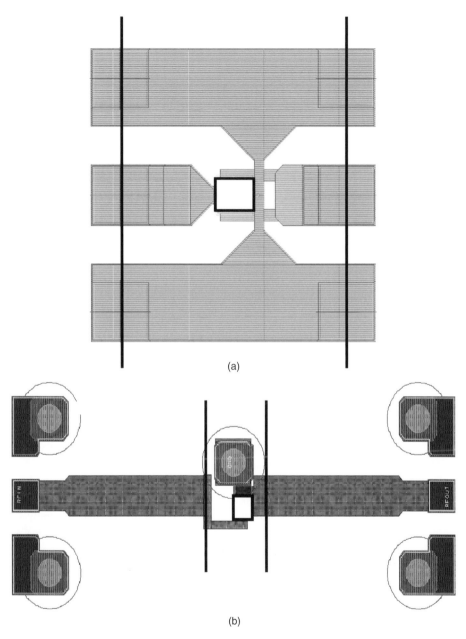

(a)

(b)

Figure 3.1 Two device-level, GSG, S-parameter test structures employing (a) a coplanar waveguide and (b) a microstrip approach. The lines denote the typical network analyzer calibration reference planes, and the box highlights the device under test.

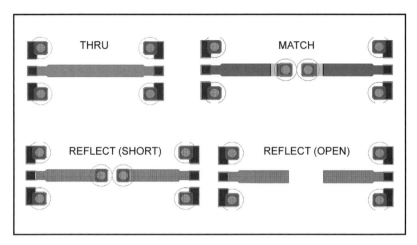

Figure 3.2 LRRM calibration set for the microstrip test structures.

calibration substrate, the accuracy of the standard definitions is much improved, since metal patterns are reproducibly made and load resistors can be laser trimmed to the exact desired value. Unfortunately, this improved precision of the standards comes at a price. The type of substrate material will influence the electric fields in the vicinity of the probe tips and thus the electrical discontinuity being corrected by the calibration. Having different substrate materials for the SOLT standards and the DUT will introduce errors at higher frequencies [2]. Also, if the standards on the calibration substrate are too close to the probe tips, differences in the pattern of the standards can influence the electric fields at the probe tips and cause additional errors [3]. Lastly, since the SOLT standards are assumed to be perfectly known, variations in probe placement or contact resistance will add further errors [4]. Fortunately, the errors mentioned above are mainly a problem at higher frequencies, so the ceramic substrate, SOLT calibration enjoys good success up to the 15–20 GHz range. Also, the SOLT calibration is incorporated into essentially every VNA, making it convenient to implement, especially since standard definitions are typically provided with a calibration substrate matched to a given VNA.

The LRRM technique is an internally consistent error correction technique, where the standards do not have to be precisely known (except for the real part of the match impedance, which is approximated by its DC resistance). Redundancies in the standard measurements allow both the error terms and the remaining standard values (including the imaginary part of the match impedance) to be calculated in the calibration process. LRRM is less prevalent than the SOLT and is not directly implemented in VNAs but is accomplished with external software. Nevertheless, it has been thoroughly investigated and compared to other techniques [5–8]. While the LRRM can also be used to provide a good probe-tip calibration with a commercial calibration substrate, it is more often used to provide a true on-wafer calibration with a test structure like the MS style discussed above. Figure 3.2 shows a set of LRRM standards for the MS structures. For this set, the line is a "zero length" thru, the reflects are shorts and opens, and the matches are nominally 50 Ohm thin-film resistors. It is worth noting that this set of standards can be augmented with different length lines, which would enable it to be used as a

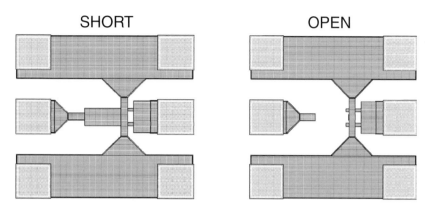

Figure 3.3 Short and open de-embedding structures for the coplanar waveguide test structures.

thru-reflect-line (TRL) calibration set. While the TRL is the most accurate, it is frequency-band limited and, with multiple lines, typically takes a lot of space. The LRRM is a good compromise that provides an internally consistent calibration with a reduced wafer footprint. In fact, since only one match standard it actually used [8], only one of the two match structures need be placed on the wafer. Lastly, by using on-wafer standards that are far removed from the RF probe contact points, the MS structure with an LRRM calibration eliminates several of the errors discussed above with the SOLT approach, which allows the MS/LRRM approach to operate to higher frequencies.

Once an accurate VNA calibration has been achieved, the second tier de-embedding can be used to remove any residual pad or interconnect effects. While this de-embedding can be applied after any calibration, in a practical sense, if the DUT is configured properly with the MS structures, no further de-embedding is needed. This is obviously a convenient approach and should be used with the MS structures when available. As such, much of the focus on de-embedding is with the CPW-style structures discussed previously.

The most common de-embedding is accomplished with two "dummy" test structures: an open and a short [9]. Figure 3.3 shows these de-embedding structures for the CPW test structure. The most common open–short de-embedding is the "Y-parameter / Z-parameter subtraction" approach. It assumes the embedding network for the pads and interconnects shown in Figure 3.4. The technique is called a Y-parameter / Z-parameter subtraction, because it is conveniently implemented by converting the measured S-parameters of the DUT, short, and open to Y-parameters and Z-parameters and subtracting the parameter matrices at each measurement frequency. This approach makes two fundamental assumptions. First, the circuit topology of Figure 3.4 represents the dominant, pad, and interconnect effects. While any passive circuit branch can be represented with a three-element pi or tee network, the circuit for Figure 3.4 approximates the branches with two elements. Second, the short and open are assumed to be ideal. If the DUT is electrically small, this is typically a reasonable assumption, but care must be taken if one intends to measure physically large transistor arrays. The algorithm for this de-embedding approach is given in Table 3.1.

Table 3.1 Algorithm for the open–short de-embedding.

Convert measures S-parameters to Y-parameters.	$S_{MEAS-DUT} \rightarrow Y_{MEAS-DUT}$ $S_{MEAS-OPEN} \rightarrow Y_{MEAS-OPEN}$ $S_{MEAS-SHORT} \rightarrow Y_{MEAS-SHORT}$
Correct the DUT and short with the open.	$Y_{2-DUT} = Y_{MEAS-DUT} - Y_{MEAS-OPEN}$ $Y_{2-SHORT} = Y_{MEAS-SHORT} - Y_{MEAS-OPEN}$
Convert the intermediate Y-parameters to Z-parameters.	$Y_{2-DUT} \rightarrow Z_{2-DUT}$ $Y_{2-SHORT} \rightarrow Z_{2-SHORT}$
Correct the intermediate DUT with the intermediate short.	$Z_{DE-EMBEDDED-DUT} = Z_{2-DUT} - Z_{2-SHORT}$
Convert the de-embedded DUT back to S-parameters.	$Z_{DE-EMBEDDED-DUT} \rightarrow S_{DE-EMBEDDED-DUT}$

Table 3.2 Equations for the open–short de-embedding circuit elements.

$$Y_C = -Y_{MEAS-OPEN(1,2)} = -Y_{MEAS-OPEN(2,1)}$$
$$Y_A = Y_{MEAS-OPEN(1,1)} - Y_C$$
$$Y_B = Y_{MEAS-OPEN(2,2)} - Y_C$$
$$Z_C = Z_{2-SHORT(1,2)} = Z_{2-SHORT(2,1)}$$
$$Z_A = Z_{2-SHORT(1,1)} - Z_C$$
$$Z_B = Z_{2-SHORT(2,2)} - Z_C$$

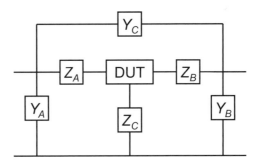

Figure 3.4 Embedding network used with the "Y-parameter / Z-parameter subtraction" de-embedding approach.

Occasionally, it is useful to create the equivalent circuit representation given in Figure 3.4. Table 3.2 gives the equations to calculate the elements in Figure 3.4 from parameter matrices given in algorithm of Table 3.1. Also, since the dominant portion of the circuit elements is the reactive component, it is useful at times to fit the admittances and impedances to capacitances and inductances: $Y = j\omega C$ and $Z = j\omega L$.

As mentioned before, the circuit topology of Figure 3.4 is one choice. One simple alternative is to swap the order of the admittances and impedances to create a short–open de-embedding versus the open–short described here. The short–open order can be preferred at times when the major capacitive parasitic to be de-embedded is overlap capacitance close to the device. At times, more elaborate circuit topologies may be

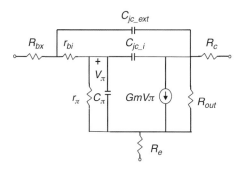

Figure 3.5a Small signal T-model equivalent circuit.

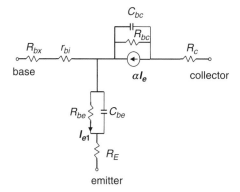

Figure 3.5b Small signal π-model equivalent circuit.

needed for a given application, which may require alternate or additional standards [10–14].

This section has discussed RF test structures for device S-parameter measurements along with calibration and de-embedding techniques. With the proper combination of VNA calibration techniques and device de-embedding techniques, accurate device measurements can be obtained and used as the starting point for device modeling.

3.3   Methods for extrinsic parameter extraction used in HBTs

3.3.1   Equivalent circuit topology

There are a number of small-signal equivalent circuits that are used for HBT modeling and they are classified into two categories: T-topology and π-topology (Figures 3.5a and 3.5b) [15–17]. These two topologies can be different in extrinsic element placement, parasitic element location, and presentation of base impendence. The π-topology can easily be related to small-signal (Y-parameter) measurements, while the T-topology more closely matches the physical construction of the HBT.

Large signal equivalent circuit, on the other hand, utilizes a π-model topology and it is based on charge instead of capacitance and current instead of resistance. The small-signal

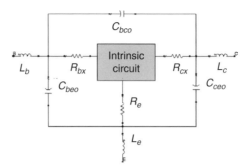

Figure 3.6 Equivalent circuit after pad parasitic de-embedding.

resistances in the large signal equivalent circuit are modeled by diode current-voltage characteristics for the corresponding junctions. Based on references [18–21], it is evident that one can obtain a small-signal model from the linearized large signal circuit and the requirement that the total derivative of the charge sources yields the small-signal capacitances and time delays is satisfied.

The subject of this section is the extrinsic parameter extraction and that procedure will require the use of both small-signal circuit topologies. The operating regime of the HBT will determine which equivalent circuit will be used in parameter extraction.

The de-embedding procedure described in Section 3.2 will bring the reference planes from the probe pads to the device. After pad capacitances and lead inductances are de-embedded, the equivalent circuit of an HBT can be presented as shown in Figure 3.6. The circuit is divided into two parts: an outer part with extrinsic elements that are bias independent and an inner part with intrinsic elements that are bias dependent. Accurate extraction of the extrinsic elements values is important, since they have significant influence on the extraction of intrinsic elements.

Extrinsic elements are terminal resistances/impedances and overlap capacitances. According to associated terminals, terminal resistances/impedances are classified as emitter resistance, base impedance, and collector resistance.

3.3.2   Physical description of contact resistances and overlap capacitances

Terminal resistances/impedances are composed of the three components: the metal line, the metal/semiconductor contact, and the contact between semiconductor material and the intrinsic device. The third component accounts for the semiconductor material in the emitter cap (between the emitter and metal contact) as well as for the base and subcollector material between the contacts and the inner transistor [15]. The physical significance of these elements in the small-signal equivalent circuit has been shown [17].

Current flow patterns are not the same for all three contacts. The emitter has vertical current flow contact and the contact capacitance is negligible. The collector and base have horizontal current flow which introduces contact capacitance [22,23], which can be neglected for collector contact because larger base contact resistance causes a significant portion of the current to flow through capacitive elements at higher frequencies.

Emitter and collector resistances R_E and R_C are modeled as Ohmic resistances and both contain a component due to contact and several epitaxial layers:

$$R_E = R_{E(epi)} + R_{EE} + R_{Em}, \tag{3.1}$$

where $R_{E(epi)}$ is emitter epitaxial resistance, R_{EE} is emitter contact resistance, and R_{Em} is resistance of emitter metal [15].

The extrinsic collector resistance is divided in three parts: $R_{C(epi)}$, $R_{SC(epi)}$, R_{CC}, which represent resistance due to the n-collector, the n^+ access region, and the collector contact. The intrinsic collector resistance is represented by R_{ci}, which characterizes the distribution effect of the base–collector junction at the collector side.

$$R_{CX} = R_{C(epi)} + \frac{R_{CC}}{2} + \frac{R_{SC(epi)}}{2}. \tag{3.2}$$

As mentioned above, base impedance consists of base contact impedance and intrinsic base impedance. Extrinsic base resistance consists of a contact resistance R_{BB}, metal resistance associated with base interconnect metal R_{Bm}, and an access resistance $R_{Bx(epi)}$ formed by the extrinsic epitaxial layer:

$$Z_B = \frac{R_{Bx(epi)}}{2} + \frac{Z_{BB}}{2} + \frac{R_{Bm}}{2}. \tag{3.3}$$

This formulation of base impedance assumes that HBT has two base contacts. If the base contact is nonalloyed (metal is placed directly on base semiconductor layer), then the interfacial layer is formed between the base contact metal and base semiconductor layer. This interfacial layer contributes to the base contact capacitance C_B, which affects the real part of base contact impedance that will decrease with increasing frequency. In the case of alloyed contact, this capacitance can be neglected and base contact impedance will be treated as resistance:

$$R_B = \frac{R_{Bx(epi)}}{2} + \frac{R_{BB}}{2} + \frac{R_{Bm}}{2}. \tag{3.4}$$

In Figure 3.6, the extrinsic base resistance is represented as R_{bx} with all three resistances from equation (3.4) lumped together.

Overlap capacitance is most often determined by the small overlapping area of the interconnect layer and the contact metal. Depending on the way in which interconnect metals are routed, besides emitter–collector overlap capacitance; there could be other overlapping capacitances for a given HBT technology as shown in Figure 3.6.

In order to extract overlap capacitances, HBT needs to be reverse biased. Under these conditions, HBT can be represented in simple π-topology (Figure 3.7) and contains only capacitances. Measured S-parameters are converted to Y-parameters which are used to calculate capacitances:

$$C_{be} = \mathrm{Im}(Y11 + Y21)/2\pi f$$

$$C_{bc} = -\mathrm{Im}(Y12)/2\pi f \tag{3.5}$$

$$C_{ce} = \mathrm{Im}(Y22 + Y12)/2\pi f.$$

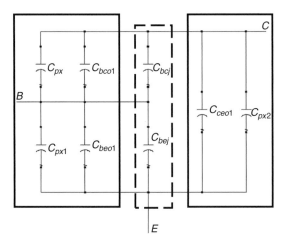

Figure 3.7 Equivalent circuit for the reverse biased device; junction capacitances in dashed line box and bias-independent capacitances in solid line box.

The extraction is performed at lower frequency where the effect of resistances and inductances is negligible ($\sim$1–3 GHz).

From equation (3.5), it is evident that the total capacitance consists of both junction and the bias-independent capacitances, and it is necessary to separate these two components in the extrinsic parameter model extraction. Bias-independent capacitances represent the sum of pad capacitances and overlap capacitance. Pad capacitances are extracted during pad parasitic de-embedding described in Section 3.2, thus should be subtracted from the bias-independent portion of the total capacitance. From references [15, 18], the junction capacitance due to the depletion region in an abrupt p–n junction can be expressed as a function of the applied reverse voltage:

$$C_j = \frac{A}{\sqrt{\frac{2}{q\varepsilon_s N}\Phi_b}},\tag{3.6}$$

where A is the area of junction, N is the doping level (cm^{-3}), ε_s is the dielectric constant, and Φ_b is the built-in voltage.

Thus, total capacitance for one junction (base-emitter or base-collector) can be expressed as follows:

$$C_{tot} = C_{const} + \frac{C_j}{\sqrt{1 - \frac{V_b}{\Phi_b}}},\tag{3.7}$$

where V_b is the applied reverse voltage, C_j is the junction capacitance, and C_{const} is the bias-independent portion of total capacitance (C_{tot}) for a given junction.

In order to separate junction capacitances from the bias-independent components, the linear representation of equation (3.7) can be used where on the X-axis the inverse of square root will be plotted and on the Y-axis the total capacitance.

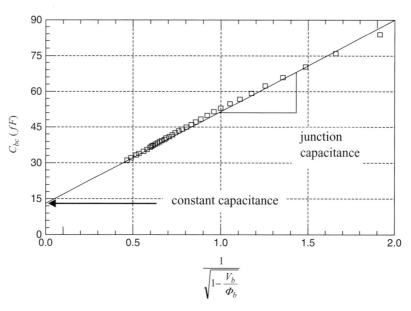

Figure 3.8 Linear representation of equation (3.8) showing linear fit used to extract bias-independent (Y-intercept) and junction capacitance (slope).

The plot on Figure 3.8 shows linear representations on measured data. In models that are used for HBT/BJT modeling [24–29], this equation is represented as:

$$C_{tot\_model} = C_{const} + Cj \cdot \left(1 - \frac{V_b}{V_j}\right)^{-Mj}, \tag{3.8}$$

where C_j is the junction capacitance, C_{const} is the sum of overlap and pad capacitances, V_j is the built-in potential, and M_j is the grading factor.

Junction capacitance is extracted by iteratively varying parameters M_j and V_j until the resulting curve fits to a straight line extracted from measured data (Figure 3.8).

Thus, the extrapolated intercepts at the ordinate of these lines give the value of bias-independent capacitance, while the slope of the linear curve will represent the bias-dependent, junction capacitance assuming that the values used for built-in potentials (base-emitter and base-collector) are close to the ones predicted by theory [23, 30]. Pad capacitances are subtracted from the extrapolated intercepts and the overlap capacitances are calculated. These capacitances will be de-embedded during the intrinsic parameter extraction.

3.3.3 Extrinsic resistance and inductance extraction

The most common methods for parasitic resistance extraction are "flyback" [15, 31, 32] (based on DC measurement) and "open collector" [30, 32, 33] (based on S-parameter measurement).

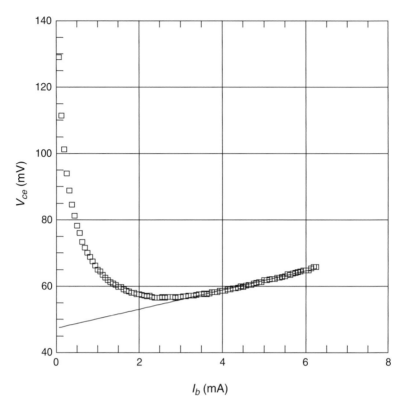

Figure 3.9 Emitter resistance extraction from the DC flyback method: from the slope of V_{ce} versus I_b when the collector is left floating ($I_c = 0$).

Using the flyback method, emitter resistance is extracted by determining the dependence of the collector–emitter voltage V_{ce} on the base current I_b when the collector is left floating ($I_c = 0$).

Under these conditions, the current will flow from base to emitter and the collector–emitter voltage, V_{ce} will be equal to the turn-on voltage [32]:

$$V_{ce\_Ic=0} = R_e \cdot I_b + V_{th} \cdot \ln\left[\left(\frac{I_b}{I_s}\right)^{n_f} \cdot \left(\frac{I_{SC}}{I_b}\right)^{n_c}\right], \tag{3.9}$$

where $V_{th} = (kT)/q$ is the thermal voltage, I_s and I_{sc} are saturation currents, and n_f and n_c are the ideality factors of collector–emitter and base–collector current, respectively.

From equation (3.9), it is expected to see that V_{ce} would decrease at lower I_b, but then it will increase at higher base currents as the term $R_e I_b$ will become dominant. Thus, at higher base currents, the dependence of V_{ce} on base current is linear and the slope of the linear fit represents emitter resistance as shown in Figure 3.9.

To extract collector resistance from the DC measurement, it is necessary to maintain a zero emitter current and this can be done with a synchronized sweep of the base and collector currents forcing $I_b = -I_c$. Under this condition, V_{ce} will be shifted in respect

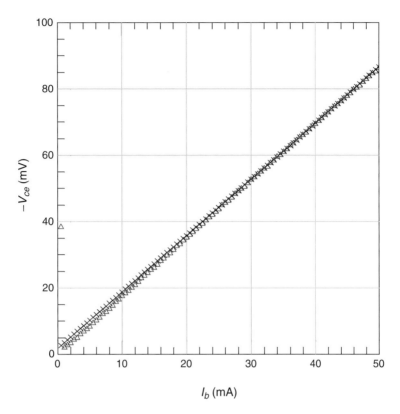

Figure 3.10 Collector resistance extraction linear line fit of V_{ce} versus I_b.

to turn-on voltage by $R_c I_c$:

$$V_{ce} = R_e \cdot I_e + R_c \cdot I_c + V_{th} \cdot \ln\left[\left(\frac{I_c + I_b}{I_s}\right)^{n_f} \cdot \left(\frac{I_{SC}}{I_b}\right)^{n_c}\right]. \tag{3.10}$$

Collector resistance is extracted from the slope of the linear line fit (Figure 3.10) of the collector voltage dependency on the base current.

The method based on S-parameter measurement is more practical, since with the single measurement, it is possible to extract all terminal resistances as well as parasitic inductances.

In "open collector" measurements, base–collector and base–emitter junctions are forward-biased, the same as described in the DC flyback method, the difference being that the collector is open for the DC but not for the RF signal. Thus, pushing the device into saturation by forcing a high DC base current, the intrinsic transistor is reduced to a pair of forward-biased diodes showing a low impedance with high junction capacitances and low junction dynamic resistances (Figure 3.11a) that will short the capacitances, thus canceling the intrinsic Z-matrix (Figure 3.11b). That is the reason why the imaginary parts of the Z-parameters of the equivalent circuit are dominated by the parasitic inductances of the device.

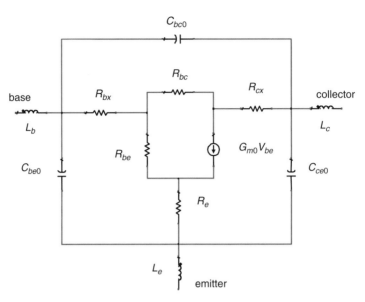

Figure 3.11a T-equivalent circuit for the open collector regime including low junction dynamic resistances.

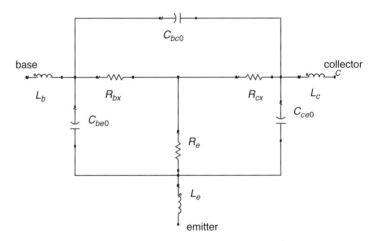

Figure 3.11b T-equivalent circuit for the open collector regime.

In order to calculate extrinsic resistances and lead inductances, it is necessary to remove the overlap capacitances (the pad capacitances are already removed in the de-embedding procedure described in Section 3.2):

$$
\begin{aligned}
Y_{11i} &= Y_{11} - j\omega(C_{be0} + C_{bc0}) \\
Y_{12i} &= Y_{12} + j\omega(C_{bc0}) \\
Y_{21i} &= Y_{21} + j\omega(C_{bc0}) \\
Y_{22i} &= Y_{22} - j\omega(C_{bc0} + C_{ce0}).
\end{aligned}
\tag{3.11}
$$

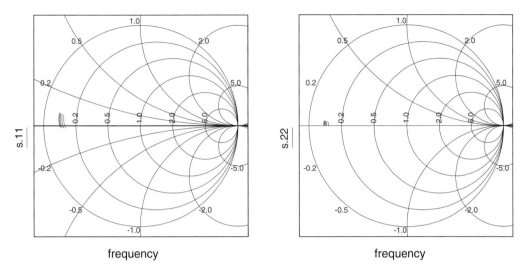

frequency   frequency

Figure 3.12 S-parameters in the open collector regime.

These Y-parameters are converted to Z-parameters that are in turn used for extrinsic parameters extraction.

In an open-collector regime, the HBT equivalent circuit (Figure 3.11a) may be modeled by a simple T-circuit with its Z-matrix written as follows [23]:

$$Z_{11} = R_b + R_e + \frac{R_{be}}{1+G_{m0} \cdot R_e} + j\omega \cdot (L_b + L_e)$$

$$Z_{12} = R_e + \frac{R_{be}}{1+G_{m0} \cdot R_{be}} + j\omega \cdot L_e$$

$$Z_{21} = R_e + (1 - G_{m0} \cdot R_{bc}) \cdot \frac{R_{be}}{1+G_{m0} \cdot R_{be}} + j\omega \cdot L_e \qquad (3.12)$$

$$Z_{22} = R_c + R_e + \frac{R_{be}}{1+G_{m0} \cdot R_{be}} \cdot \left(1 + \frac{R_{be}}{R_{bc}}\right) + j\omega \cdot (L_c + L_e),$$

where G_{m0} is the DC transconductance, R_{be} and R_{bc} are the dynamic resistances of the base-emitter and base-collector junctions, respectively, and their expressions are given as follows:

$$R_{be} = \frac{n_{be}kT}{qI_{be}}$$

$$R_{bc} = \frac{n_{bc}kT}{qI_{bc}}. \qquad (3.13)$$

Under the open collector bias condition, measured S-parameters will have only real and inductive components proving that both junctions and corresponding differential capacitances are shunted (Figure 3.12).

The extrinsic resistances are determined at low frequency from the real parts of the calculated Z-parameters:

$$real(Z_{11} - Z_{12}) = R_b$$

$$real(Z_{12}) = R_e + \frac{R_{be}}{1+G_{m0} \cdot R_{be}} \qquad (3.14)$$

$$real(Z_{22} - Z_{21}) = R_c + \frac{1}{1+G_{m0} \cdot R_{be}} \cdot (R_{bc} + G_{m0} \cdot R_{bc} \cdot R_{be}).$$

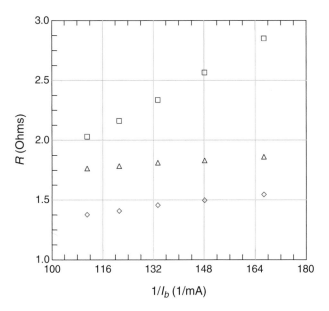

Figure 3.13a Extraction of emitter ($\diamondsuit$), extrinsic base ($\diamondsuit$) and extrinsic collector ($\triangle$) resistance from the extrapolated intercepts ($I_b \approx \infty$) from equation (3.14).

As mentioned above, at high current densities the total base resistance R_b tends asymptotically to the access base resistance, the R_{be} and R_{bc} became very small, and the real part of $Z_{12}, Z_{21}, Z_{22} - Z_{21}$ increases linearly as a function of $1/I_b$; thus, the equivalent circuit can be represented as shown in Figure 3.11b. The extrapolated intercepts at the ordinate ($I_b - \infty$) of these lines give the values of R_e, R_b, and R_c (Figure 3.13a).

If the parasitic inductance of the device itself is included in the model, it would be necessary first to remove lead inductance (extracted from the short de-embedding structure) and proceed with a calculation of terminal resistances from the real part and device inductance from the imaginary part of de-embedded Z-parameters (Figure 3.13b):

$$\text{Im}(Z_{11} - Z_{12}) = L_b$$
$$\text{Im}(Z_{12}) = L_e \qquad (3.15)$$
$$\text{Im}(Z_{22} - Z_{21}) = L_c.$$

The temperature dependence for resistance is found by fitting the resistance versus temperature based on the model formula [24–29]:

$$R - T = R \cdot \left(\frac{T_{amb}}{T_{nom}} \right)^X, \qquad (3.16)$$

where T_{amb} is the ambient temperature and T_{nom} is the nominal temperature in the model (usually it is 25 °C) and X is the temperature coefficient that is extracted as a slope from Figure 3.14 for each resistance, respectively.

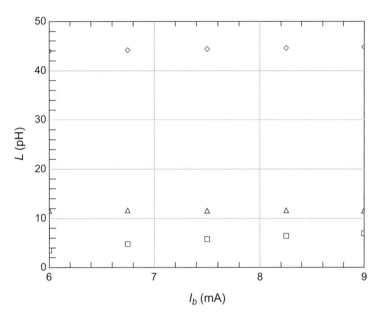

Figure 3.13b Inductance calculated from equation (3.15) at high frequency (above 5 GHz): L_e ($\diamond$), L_b ($\triangle$), L_c ($\diamond$).

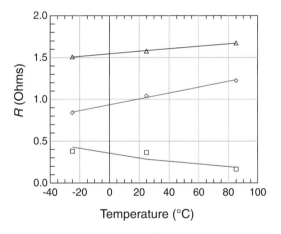

Figure 3.14 Extraction of resistance temperature coefficient, resistances over temperature: emitter ($\diamond$), extrinsic base ($\diamond$), and extrinsic collector ($\triangle$).

To conclude, an extrinsic parameter extraction based on an open collector measurement provides:

- values for all external resistances;
- resistance extraction that, in most cases, will give unique solutions, which is not the case always with the DC flyback method;

- extraction of inductance from the open collector measurement that can be used to determine device inductance (after the lead inductance is de-embedded);
- the temperature dependence of terminal resistances.

The disadvantages of this method are:

1. For some devices, forcing a high base current can be damaging, as the base is very thin; thus, it will not be possible to shunt out the effect of dynamic resistances and junction capacitances.
2. The current flow in the open collector measurement is different from the device active regime. It will not produce error in the R_{rx} extraction, but the R_e value should be cross-checked in some other extractions [34].

Values for extrinsic resistances could also be obtained from PCM data and these values can be used as starting values.

3.4   Methods for extrinsic parameter extraction used in HEMTs

Direct extraction techniques for determining equivalent circuit parameters from S-parameter data have become well published in recent years for FETs and HEMTs [35–37]. Typical small signal model extraction relies on sets of S-parameter data representing different areas of operation of the FET or HEMT to isolate certain properties of the device. The hot FET data, which implies that the drain voltage is nonzero, is used to extract the intrinsic components. In order to have consistent parameter sets for the active device, the extrinsic bias-independent components need to be de-embedded first. This requires zero drain voltage measurements (cold FET) with the gate in forward conduction and the channel pinched off. The subthreshold data is used to determine the parasitic capacitance or pad capacitance. The forward gate bias data is used to extract the extrinsic parameters that include the gate, source, and drain resistance and inductance.

In Figure 3.15, the representation of the basic small signal circuit is shown. Typical extraction of intrinsic and extrinsic parameters is carried out by converting S- to Y- and Z-parameters. Initially, the pinched FET S-parameters are converted to Y-parameters for extraction of shunt pad capacitances. Afterwards, forward gate cold FET converted Z-parameters are used to extract R_g, R_d, R_s, L_g, L_d, and L_s.

3.4.1   Cold FET technique

In the cold-FET technique, S-parameters are measured at $V_{ds} = 0$ V and $V_g \gg 0$ [37]. Setting V_{ds} to zero simplifies the equivalent circuit as in Figure 3.15. Forward-biasing the Schottky junction will overcome the capacitive effects of the gate owing to sufficient current density. Usually, $I_g \approx 150$ mA/mm (a few mA for most GaAs HEMTs), which corresponds to $V_g \leq 1$ V.

Intrinsic parameters

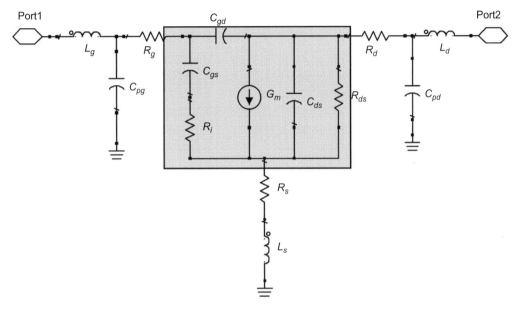

Figure 3.15 Small-signal equivalent circuit model topology.

The resulting simplified equivalent circuit shown in Figure 3.16 results in the following equations under a high gate current,

$$Z_{11} = R_s + R_g + \frac{R_c}{3} + \frac{nkT}{qIg} = j_w(L_g + L_s)$$

$$Z_{12} = R_s + \frac{R_c}{2} + j_w L_s \qquad\qquad (3.17)$$

$$Z_{22} = R_s + R_d + R_c + j_w(L_s + L_d),$$

where $1/2$ and $1/3$ are results of the distributed effects of gate current crowding around the gate edge as determined by Vogel [39] and R_c is the channel resistance. Inductances are obtained from the imaginary parts of the Z-parameters. For example, L_g is extracted from the slope of $Im\{Z_{11}\}$ with frequency. The same procedure can be repeated to determine L_d and L_s from the slope of $Im\{Z_{22}\}$ and $Im\{Z_{12}\}$ with frequency, respectively. Averaging Z-parameters over several forward-biased gate voltages, which are purely resistive and inductive, should be used to extract consistent values. In Figure 3.17, S_{11} and S_{22} are both inductive on the upper-left quadrant of the Smith chart and show the resistance as the real part on the Smith chart.

The resistances are more difficult to determine, since there are four unknown resistances and three equations. Also, the values of the parasitic resistances can be small enough that the dependence on Ig in equation (3.17) can be problematic. Simplifying the expression for Z_{12} becomes the recommended path to minimize the unknowns

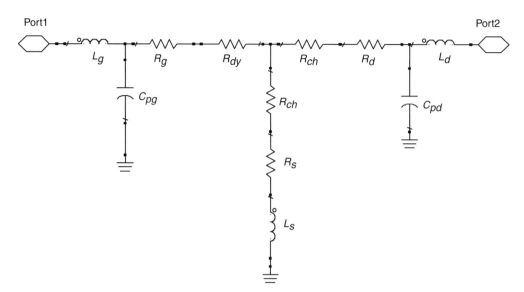

Figure 3.16 Equivalent circuit model of a cold FET measurement with $V_{ds} = 0$ and $V_{gs} \gg 0$ [38].

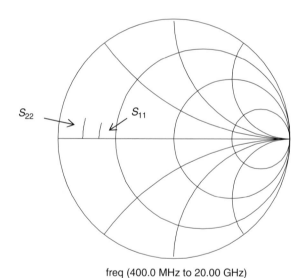

freq (400.0 MHz to 20.00 GHz)

Figure 3.17 Example: cold FET measurement of S_{11} and S_{22}.

for resistance extraction. There are several additional methods to consider for increasing confidence in consistently determining these resistances. One example is to calculate R_c from the channel technological parameters instead of optimizing the value through repeated fitting. Another procedure outlined by Yang and Long [40] was developed for source resistance extraction. This procedure monitors the forward gate bias voltage at two forced drain currents while keeping the gate current constant. This

Port1   Port2

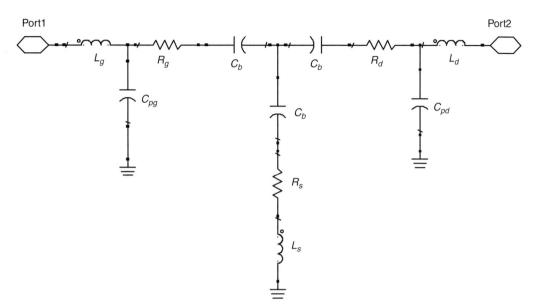

Figure 3.18 Equivalent circuit for a GaAs HEMT at pinchoff with $V_{ds} = 0$, $V_{gs} < V_{pinchoff}$.

measurement can be done without knowing the drain-to-source voltage on the device with the assumption that the $V_{ds} \gg nkt/q$ and that the channel resistance does not vary between the two drain currents by limiting V_{ds} to the linear region of operation. The differentiation between the drain currents and gate voltage is used to solve the source resistance from equation (3.18). F_2/F_1 is the correction factor ratio for the gate current owing to the drain-to-source voltage which is approximated to 0.9 for calculation.

$$R_s = \frac{V_{g2} - V_{g1} + \frac{nkt}{q} \cdot \ln\left(\frac{F_2}{F_1}\right)}{I_2 - I_1} \tag{3.18}$$

Results can be checked with the slope between the real parts of several different forward-biased gate voltages in defining Z_{11}. Also, one can use the initial DC fitting and unbiased S-parameters to check consistency. Possible problems with this method are excessive voltages (and resulting gate currents) that one has to apply to the gate in order to eliminate its capacitive effect. This can result in metal migration and reliability issues. Also, ambiguities exist as to the exact values of alpha factors (1/2 and 1/3). Lastly, for a depletion-mode HEMT, $V_g \leq 0$ V, so applying $Vg \gg 0$ V is not the normal mode of operation except under a high-power operation scenario. Overall, this method is useful for inductance extraction and initial resistance determination.

An additional S-parameter measurement is done at the pinched-FET condition, i.e., $V_{ds} = 0$ V and $V_g \ll V_{pinchoff}$. This will effectively eliminate the conductance of the channel and the intrinsic capacitances of the gate. A simplified equivalent circuit results as indicated by Figure 3.18 where C_{pg} and C_{pd} are the pad capacitances and C_b is

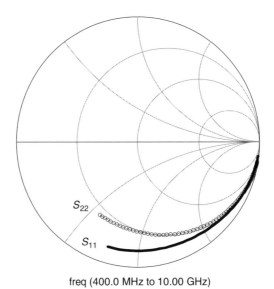

freq (400.0 MHz to 10.00 GHz)

Figure 3.19 Example: pinched FET measurement of S_{11} and S_{22}.

the fringing capacitance or depletion layer under the gate extending into the channel. The equal partitioning of C_b has also been considered for a symmetrical assumption by White and Healy [41]. For frequencies below a few GHz, the inductances and resistances have little influence on the imaginary part of the equivalent Y-matrix as shown below:

$$\text{Im}(Y_{11}) = j_w \left(C_{pg} + \frac{2}{3} C_b \right)$$

$$\text{Im}(Y_{12}) = \frac{j_w C_b}{3} \tag{3.19}$$

$$\text{Im}(Y_{22}) = j_w \left(C_{pd} + C_{ds} + \frac{2}{3} C_b \right).$$

C_{pd} and C_{pg} can be found from $\text{Im}(Y)$ averaged over frequency and bias. Notice that C_{ds} cannot be separated and will be lumped in with C_{pd}. One can assume a fixed portion of C_{pd} to be attributed to C_{ds} [38]. Figure 3.19 displays the measurement of a pinched HEMT.

3.4.2   Unbiased technique

Another technique for extrinsic parameter extraction is to measure S-parameter data at $V_{gs} = V_{ds} = 0$ V [42]. This method allows for a more realistic mode of gate operation for a HEMT. Forward-biasing the gate is not required; therefore, eliminating any damages to the gate caused by large currents is avoided. Also, all parasitic resistances can be directly obtained from the unbiased and pinched-off measurements (described earlier). Under this condition, the device equivalent circuit is shown in Figure 3.20. The following

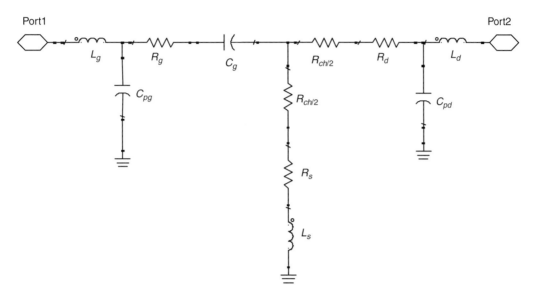

Figure 3.20 Equivalent circuit for a GaAs HEMT at V_{gs} and $V_{ds} = 0$.

equations results:

$$Z_{11} = R_s + \frac{1}{2}R_c + R_g + j_w(L_g + L_s) - \frac{1}{\omega C_g}$$

$$Z_{12} = R_s + \frac{1}{2}R_c + j_w L_s \tag{3.20}$$

$$Z_{22} = R_s + R_d + R_c + j_w(L_s + L_d),$$

where R_c is the channel resistance and C_g is the gate capacitance which can no longer be ignored. First, we can extract L_s and L_d from the imaginary parts of (Zs) at frequencies $> 5\,\text{GHz}$. Next, L_g and C_g can be extracted from Im(Z_{11}) at two frequency points. The real parts of the Z-parameters yield the following equations:

$$\text{Re}(Z_{11}) = R_s + \frac{1}{2}R_c + R_g \tag{3.21a}$$

$$\text{Re}(Z_{12}) = R_s + \frac{1}{2}R_c \tag{3.21b}$$

$$\text{Re}(Z_{22}) = R_s + R_d + R_c. \tag{3.21c}$$

For pinched-FET, the equivalent circuit also includes Ls and Cs which were ignored in the cold-FET techniques. This results in more accurate pad capacitance extractions as well as more equations to determine the resistances. Considering Ls and Rs in Figure 3.20,

$$\text{Re}(Z_{11}) = R_s + R_g \tag{3.22a}$$

$$\text{Re}(Z_{12}) = R_s. \tag{3.22b}$$

R_s can be directly determined from (3.22b) although it is very sensitive to the accuracy of the measurement calibration system and is accurate for small periphery devices using TRL calibration [2]. Alternatively, equations (3.5a) and (3.65a) can be used to find R_c and then (3.21b), (3.21a), and (3.21c) to find R_s, R_g, and R_d, respectively. Frequency characteristics of resistances can be checked and care must be taken to extract R_s at $f > 5\,\text{GHz}–6\,\text{GHz}$ where reactances of caps (X_c) will not obscure the R_s.

When the user has verified all the extrinsic values, intrinsic extraction can proceed. If the values for the parasitics are nonideal, the hot FET extraction will give poor results. These extraction results can be inadequate for matching and produce extracted values that are not consistent with the device topology. As noted earlier, the problematic extraction of the parasitic resistances can result in negative values or no solution at all.

After removing the extrinsic effects from the Y-parameters, the resulting equations are given for the intrinsic circuit. According to Dambrine *et al.* [37], low-noise devices characteristically have $D = 1$ at lower frequencies since R_i and C_{gs} have small values.

$$Y_{11} = \frac{R_i C_{gs}^2 \omega^2}{D} + j\omega \left(\frac{C_{gs}}{D} + C_{gd} \right)$$

$$Y_{12} = -j\omega C_{gd}$$

$$Y_{21} = \frac{g_m \exp\left(-j\omega\tau\right)}{1 + jR_i C_{gs}\omega} - j\omega C_{gd}$$

$$Y_{22} = \frac{1}{R_{ds}} + j\omega \left(C_{ds} + C_{gd} \right)$$

$$D = 1 + \omega^2 C_{gs}^2 R_i^2. \qquad (3.23)$$

3.4.3   GaN HEMTs exceptions

Extraction of extrinsic parameters of GaN-based HEMTs mostly follow the same procedures as those for GaAs HEMTs, with a few exceptions. GaN HEMTs have much higher conduction bands than the corresponding GaAs HEMTs. This can present problems in trying to extract parameters based on forward biasing the gate–source junction, as seen later. Additionally, depending on the layout of GaN high-power HEMTs, distributed effects might have to be added to a classical equivalent circuit. Workers [43–45] have suggested new methods in dealing with aforementioned problems that include new extraction methods for gate parameters and/or additional extraction steps in dealing with distributed effects of a multifingered GaN power cell. Also, due to excessive current density under high-drive operation, R_s might increase greatly as suggested by reference [46]. In general, starting with classical cold-FET measurements carefully tailored to GaN devices [47], one will gain at least a good starting-point in obtaining initial values for subsequent optimization of the entire small-signal extraction.

Unit Device

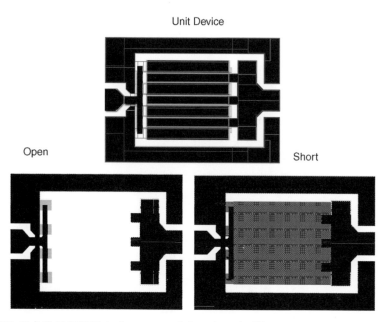

Figure 3.21 Unit GaN HEMT device, open and short de-embedding structures.

As described in Section 3.2, calibrations to the probe tips, de-embedding from dummy short and open structures from a given layout is assumed. Also, one has to be careful in choosing the proper structures in order to account adequately for the pads and any additional gate and/or source parasitics that may be specific to a layout as indicated by Figure 3.21 for a power GaN device. Next, modifications to classic methods and alternate techniques suggested for GaN HEMTs extrinsic extractions are discussed.

As pointed out earlier, work has been done [45] to modify the forward-gate portion of the cold-FET technique because of the problems in trying to eliminate the capacitance of the gate diode thru strongly forward-biasing the gate source junction. Instead, a low DC bias is applied to the gate with the drain left floating. Then, a low positive value region of I_g is chosen so that $V_{bi} > V_g > 0$ and S-parameters are measured (see Figure 3.22).

The difference between this method and the classic one lies in operating in a low I_g regime, which minimizes any damages to the gate and enables the calculation of Schottky diode capacitance (C_g) and dynamic resistance (R_0) from the resulting small-signal equivalent circuit shown in Figure 3.23. It is also similar in formulation to the unbiased method where the gate is kept at $V_{gs} = 0$ V and I_g is insignificant, so the Schottky dynamic resistance is ignored and only capacitance (C_g) is included as in Figure 3.20. Implementing this method, requires more mathematical treatment of $\text{Im}(Z_{11})$ in order to extract meaningful values for Schottky resistance and capacitance as well as R_g. Equations for the equivalent circuit at the input are more involved, as indicated in equation (3.24), because of the inclusion of R_0 and C_g of the Schottky gate

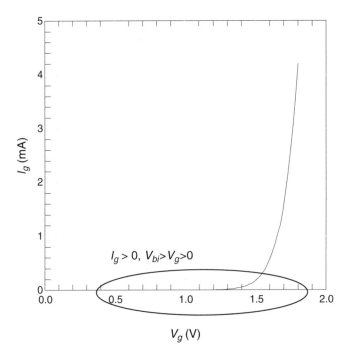

Figure 3.22 Gate IV curve of a GaN HEMT at cold-FET.

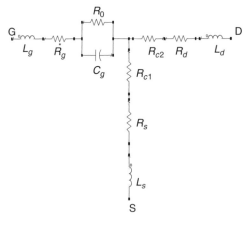

$$R_g{}^* = -R_c/6 \text{ [45]}$$

$$R_{c1} = R_{c2} = 1/2 \, {}^*R_c \text{ (channel } R) \text{ [39]}$$

Figure 3.23 Small-signal equivalent circuit of a GaN HEMT under low gate current and floating drain.

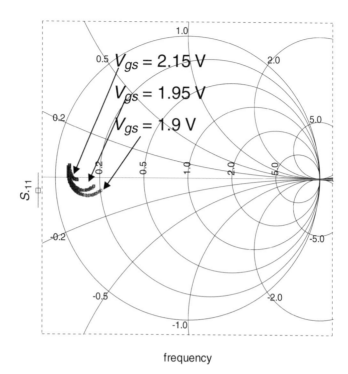

frequency

Figure 3.24 Input behavior of GaN HEMT at large gate to source currents.

and different partitioning of the channel resistance R_c (see Figure 3.23):

$$Z_{11} = R_s + 0.5\,R_c + R_g - R_c/6 + \frac{R_0}{1 + (w R_0 C_g)^2}$$

$$+ jw \left[L_g + L_s - \frac{C_g R_0^2}{1 + (w R_0 C_g)^2} \right]$$

$$Z_{12} = Z_{21} = R_s + 0.5 \cdot R_c + jwL_s$$

$$Z_{22} = R_s + R_c + R_d + j\omega(L_d + L_s). \tag{3.24}$$

This method is suggested if the classical cold-FET is not applicable. In the authors' experience, elimination of the input capacitance thru forward-biasing the gate diode of a GaN HEMT is not a "common" problem, although the significant gate current flow that results is. For example, in the case of a ~2 mm device, a gate voltage of > 2 V had to be applied in order to overcome the capacitance, as shown in Figure 3.24. In spite of the large gate current flow, inductances and resistances were extracted but were treated as initial guesses and were further refined in the next step using the unbiased technique. Typical results for a GaN device are shown in Figure 3.25.

Another issue raised by references [43] and [44] is related to the more complex layout of multifinger power GaN HEMTs. In some cases, the source air bridges used are in close proximity to the drain and gate fingers and result in significant additional extrinsic capacitances. These distributed capacitances form along the gate and drain fingers to

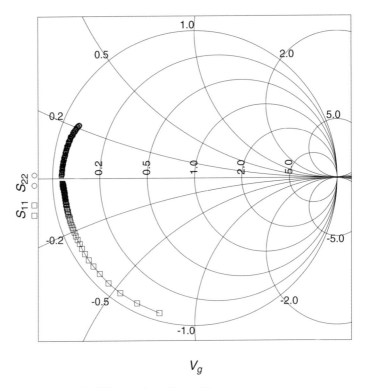

V_g

Figure 3.25 Behavior of GaN HEMT at $V_g = V_d = 0$ V.

the source. An alternate equivalent circuit suggested by reference [43] accounts for the interelectrode capacitances, C_{pgi}, C_{pdi}, and C_{gdi}, as shown in Figure 3.26.

Generally, distributed problems are layout-specific and will not become significant unless several gate fingers are used. For example, in the power cell of Figure 3.21, no airbridge is used on the drain side and the source-bar on the gate side does not significantly contribute to any distributed capacitance. Also, good choices of dummy, short and open, will eliminate any additional parasitics to the chosen reference planes of the device.

Lastly, a pinched-FET measurement applied to GaN devices should be done at deep pinch-off, which, for a typical GaN HEMT, will be ≤ -6 V. The behavior of a GaN HEMT at deep pinch-off is shown in Figure 3.27.

Typical extrinsic parameter values from a power GaN HEMT are given in Table 3.1. Notice that the values were initially extracted from the techniques discussed, then optimized in extracting the overall small-signal equivalent circuit values over bias and frequencies of interest.

Overall, extracted parameter values have to make sense for a given size device and within a self-consistent framework. Caution must be taken in not putting too much faith in one method or technique. Considering the uniqueness and application variety of GaN HEMTs and their accompanying high-power-related structure, one has to examine carefully the characteristics of the device at hand and subsequently apply one or a

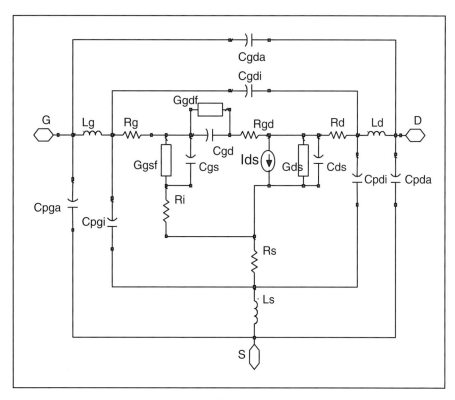

Figure 3.26 22-element active model for GaN HEMT [43].

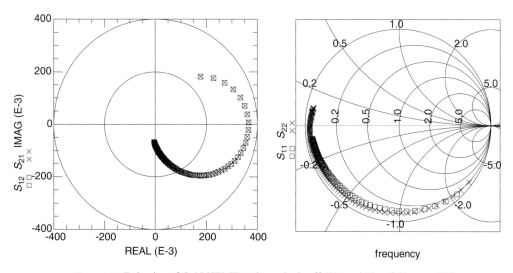

Figure 3.27 Behavior of GaN HEMT at deep pinch-off ($V_{ds} = 0$ V and $V_{gs} < -6$ V).

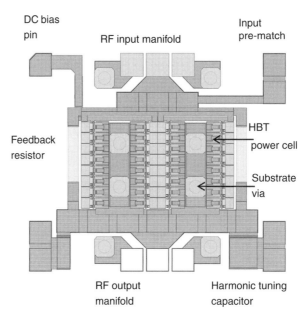

DC bias
pin

RF input manifold

Input
pre-match

HBT
power cell

Feedback
resistor

Substrate
via

RF output
manifold

Harmonic tuning
capacitor

Figure 3.28 Array of HBT devices including vias and on-chip passives.

combination of several extrinsic extraction method(s) that result in consistent small- and large-signal models.

3.5   Scaling for multicell arrays

Large transistor power amplifier output arrays are usually built up from smaller unit cells that are replicated to provide the required emitter area (HBT technology) or gate periphery (FET technology). An example of such an array is shown in Figure 3.28.

This layout represents a 32-cell HBT output stage used in evaluating technologies for cellular handset applications. The HBT power cells are grouped around four thru-substrate vias which provide access to the backside RF ground. Each device has an input blocking capacitor and bias resistor. A two-level metal interconnect is used to form the input and output manifolds that connect the cells. In addition to interconnect, there are bondpads used for packaging, and on-chip passives for feedback stabilization and harmonic tuning.

The interconnect used in this design can have a significant effect on the impedances seen by the active devices, and must be included in the model for the array. There are several approaches that may be used. The most obvious technique would be a lumped element approach, replacing interconnects with appropriate capacitors and inductors. Although this can be effective at low frequencies, it does not capture the distributed effects that may be found in microwave work. It is awkward and inaccurate to estimate the values by hand and most common commercial high-frequency CAD software is not known for the sophistication of its interconnect extraction software.

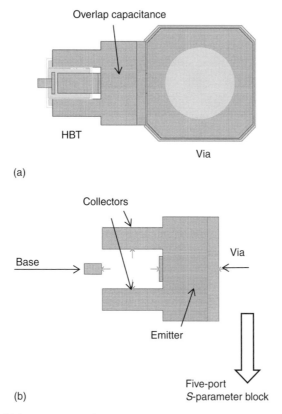

Figure 3.29 HBT with interconnect and substrate via.

Another approach would be to use microstrip transmission line models. In conjunction with a physical substrate definition, these instances can model the distributed effects in isolated lines, coupled lines, bends, and tapers. This is effective for designs which require a formal transmission line implementation, but not as useful for the more generalized layouts used in the cellular or wireless LAN bands. In these applications, distributed effects do not dominate on the relatively small die, but there is still complex coupling between lines and components that must be taken into account.

The most general solution is to use some form of planar or quasi-three-dimensional (3D) electromagnetic simulation to provide an *S*-parameter representation of the interconnect (and, if desired, any on-die passives). The interfaces between sections of interconnect and a device represented by a compact model (whether active or passive) is replaced by an EM port and the software is used to generate an *S*-parameter block that can be placed in the schematic. This technique has been applied both to HBT and FET arrays [48–51]. A simple example is shown in Figure 3.29.

Figure 3.29a is a partial layout of an HBT output cell array. The HBT and via will be replaced with compact models. These elements are deleted and the connections between them and the remaining layout are replaced by EM ports (Figure 3.29b). Note that the placement of these ports must correspond to the reference planes of the compact model.

An EM simulation is then performed over the required frequency range. If the manifold is expected to pass a DC bias current, zero Hertz frequency must be included in the frequency sweep for the simulation. This can pose a problem with some solution methods.

The results of the simulation are exported to a file containing the S-parameters for each frequency. This is usually a large multiport block with an S_{xx} entry for each port "x" and all cross terms. This file can then be imported into the schematic and the ports connected to the corresponding reference planes of the compact models.

This technique is applicable over a wide range of frequencies and can handle both distributed effects and coupling well. The traditional downside has been the considerable computer time and memory required. Fortunately, Moore's Law has brought us both faster CPUs and more memory. Software vendors have also not been idle in algorithm development and as a result, at least at the die level, complex interconnect and even complete passive elements can be analyzed on a desktop computer in times of well under an hour. The work involved in translating a physical layout into simulator input within the CAD design flow has also been streamlined.

Here, we will demonstrate a technique [51] to scale up a single power transistor model to include die-level parasitics using an EM simulator. Our approach, suggested by the regular nature of the power array core, is to develop a multicell model of several devices, including interconnect. This multicell is then replicated at the schematic level to form the array and then combined with EM simulations of the input and output manifolds to form a complete model.

Before beginning, several caveats must be mentioned. First, some effort must be made to calibrate the simulator to a particular process. This usually involves the fabrication and measurement of a variety of test structures such as transmission lines, capacitors, and inductors. Provision must also be made for including the effects of normal process variation such as metal conductivities, insulator thickness, material properties, and the final wafer thickness. Cross-sectioning of a sample of the test structures is also a good idea so that the modeling engineer knows exactly what is in hand.

Second, it should be noted that the EM field surrounding a device is not the same as what is calculated when it is removed. EM simulators work best when their ports form well-defined, continuous transmission lines with known modes of propagation. This is not the case when interior layout is cut and a port is inserted in its place. Approximations must be made and these can vary between simulators. There is no simple way around this situation; the best approach is to try different placements for a port and see what effect this has on the result.

Finally, a decision must be made on whether to include any vias in the EM simulation, or to use a lumped element model for them. Planar simulators do not always handle vias well and often have several methods for simulating them. Proper characterization work can help with this issue.

To begin the multicell development, consider the simple layout shown in Figure 3.30. This circuit contains eight HBTs grouped around a common via. Each HBT is associated with an input blocking capacitor and a DC bias resistor. We will study two approaches to this problem. In both, we will use the EM simulator for the blocking caps and parasitic interconnect, and compact models for the transistors, resistors, and via. The resulting manifold layout is shown in Figure 3.31.

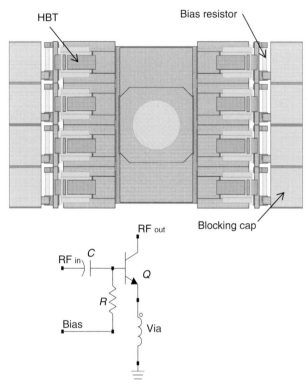

Figure 3.30 4×2 device array with blocking caps, bias resistors, and a via.

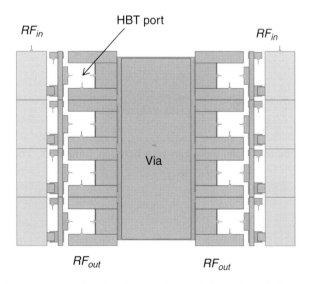

Figure 3.31 The interconnect manifold of Figure 3.30, ready for EM simulation.

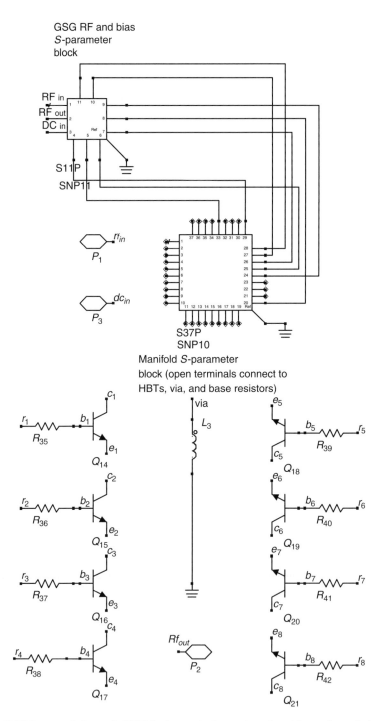

Figure 3.32 EM approach 1 – each HBT device on the layout is replaced by a schematic instance.

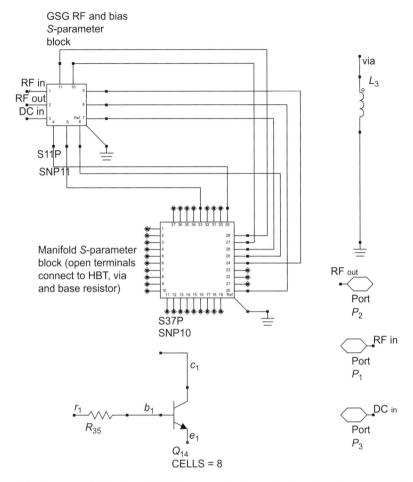

Figure 3.33 EM approach 2 – the eight HBTs are replaced by a single schematic instance with multiplicity set to 8.

In the first approach, we will simply replace each HBT in the layout with a schematic instance connected to the appropriate port on the manifold, for a total of eight instances. In the second approach, a single transistor instance with the multiplicity set to eight is used. The appropriate EM ports (eight collectors, eight emitters, etc.) will be shorted together and connected to the corresponding transistor terminal. Schematics for these are shown in Figures 3.32 and 3.33. In order to evaluate these two approaches, this test structure was fabricated with GSG microstrip launchers for the RF feeds and characterized on-wafer with power sweeps using a Maury MT4463A large signal network analyzer. The RF feeds were simulated with the EM tool and their S-parameter blocks appear on each schematic.

The decision on what approach to use depends on several factors. In general, the more individual transistor instances we have, the longer the schematic simulation time (and the greater the chance of encountering difficulties in convergence). Several issues may force

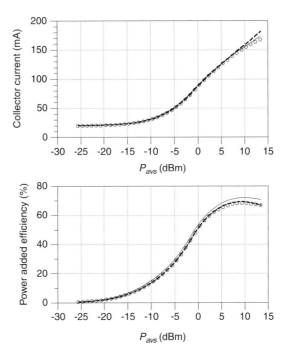

Figure 3.34a Comparison of power sweep results for the 8-cell array, collector current, and PAE (measured data circles, lumped model solid line, Approach 1 dotted line, Approach 2 dashed line).

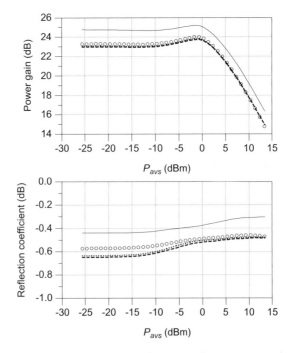

Figure 3.34b Comparison of power sweep results for the 8-cell array, power gain, and reflection coefficient (measured data circles, lumped model solid line, Approach 1 dotted line, Approach 2 dashed line).

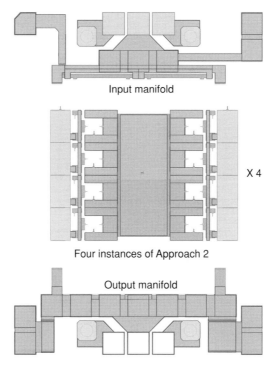

Input manifold

Four instances of Approach 2

Output manifold

X 4

Figure 3.35 Scaling approach to the 32-cell HBT array.

this choice. Significant distributed effects between device terminals across the array require separate instances. Thought must also be given to the thermal effects. If we use only one transistor instance, the thermal impedance must be scaled to give the correct effective temperature for the block. Also, if we wish to simulate the temperature profile across the array, we will need separate instances combined with a thermal coupling network.

As a comparison, we will replace the EM simulation of the second approach with a lumped element approximation of the parasitic interconnect of the layout.

The array was biased at a collector current of 30 mA, 3.5 V, and the input power was swept until the array was about 8 dB into compression. The fundamental frequency was 1.95 GHz. The results of these simulations and the measured data are presented in Figure 3.34.

In this figure, we plot collector current, power gain, power-added efficiency, and reflection coefficient against power available from the source. As expected, the lumped element method clearly has difficulties compared to the EM approaches. The power gain is high by about 2 dB and the input match is off, especially in saturation. These all introduce error into the PAE simulation. By contrast, using the EM simulator to construct S-parameter blocks of the interconnect provides a better fit. Note that there is not a great deal of difference between the two EM approaches. This is not surprising: the

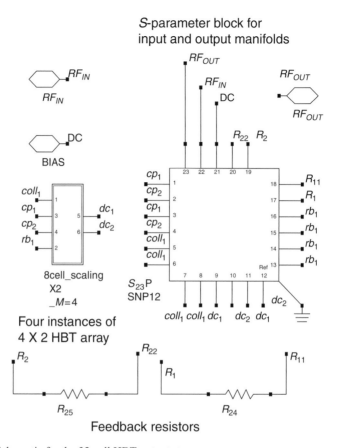

Figure 3.36 Schematic for the 32-cell HBT output stage.

array is too small to have significant distributed effects; nor would we expect a strong thermal gradient owing to self-heating at these current levels.

To show how this technique can be scaled to larger arrays, we will use Approach 2 to build a model of the output array from Figure 3.30. The schematic of Figure 3.35, without the GSG RF input and output feeds is used as a new unit multicell. This is placed in the schematic with a multiplicity of four, along with S-parameter files from the EM simulation of the input and output manifolds.

The schematic therefore contains a total of four individual transistor instances, rather than 32 (see Figures 3.35 and 3.36).

This can result in significant savings in CPU time and avoids many convergence problems encountered when large arrays of transistors are pushed deep into saturation. Note that no attempt was made to provide a thermal coupling network between the cells. For this application, we can rely on the internal thermal scaling of the self-heating model.

This 32-cell array was characterized, using the same test stand as the 8-cell structure, at 1.95 GHz with a bias of 80 mA, 3.5 V. The results are shown in Figures 3.37 and 3.38.

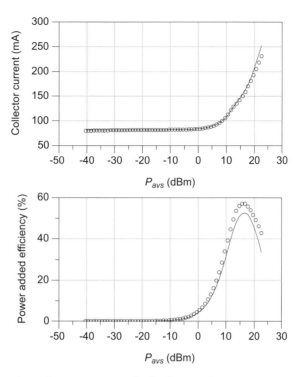

Figure 3.37a Comparison of power sweep results for the 32-cell array, collector current, and PAE (measured data circles, model solid line).

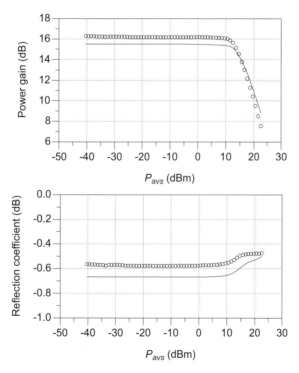

Figure 3.37b Comparison of power sweep results for the 32-cell array, power gain, and reflection coefficient (measured data circles, model solid line).

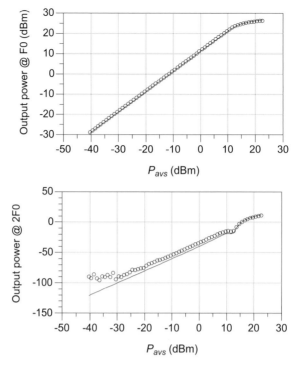

Figure 3.38a Comparison of LSNA results for the 32-cell array, output power at the fundamental and second harmonic (measured data circles, model solid line).

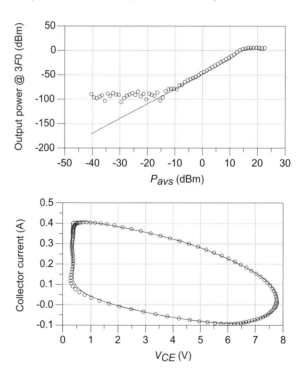

Figure 3.38b Comparison of LSNA results for the 32-cell array, output power at the third harmonic and the dynamic load line (measured data circles, model solid line).

The simulated peak PAE is within 5% of the measured value, output power within 0.7 dBm, and the third harmonic with 2.5 dBm.

The usefulness of the technique depends on several factors. Clearly, the ability to properly identify a suitable, repeatable multicell is a key. The method works best for regular arrays of devices. The designer must also have a very good idea of how far the self-heating thermal model can be scaled for a given application. This more than any other factor, will limit the size of the multicell and consequently the savings in CPU resources.

References

[1] M. Golio and J. Golio, Eds., *RF and Microwave Handbook, 2nd ed.: RF and Microwave Circuits, Measurements, and Modeling*, Abingdon, Oxon: CRC Press, 2008.

[2] R. B. Marks, "On-wafer millimeter-wave characterization," in *GAAS 98 Amsterdam Conf. Dig.*, 1998, pp. 21–26.

[3] J. C. Rautio, "A possible source of error in on-wafer calibration," in *34th ARFTG Conf. Dig.*, Dec. 1989, pp. 118–126.

[4] A. J. Lord, "Comparing the accuracy and repeatability of on-wafer calibration techniques to 110GHz," in *29th European Microw. Conf. Dig.*, Oct. 1999, pp. 28–31.

[5] R. B. Marks, "Wafer-level ANA calibration at NIST," in *34th ARFTG Conf. Dig.*, Dec. 1989, pp. 11–25.

[6] D. F. Williams, R. B. Marks, and D. Andrew, "Comparison of on-wafer calibrations," in *38th ARFTG Conf. Dig.*, Dec. 1991, pp. 68–81.

[7] A. Davidson, E. Strid, and K. Jones, "Achieving greater on-wafer S-parameter accuracy with the LRM calibration technique," in *34th ARFTG Conf. Dig.*, Dec. 1989, pp. 61–66.

[8] A. Davidson, K. Jones, and E. Strid, "LRM and LRRM calibrations with automatic determination of load inductance," in *36th ARFTG Conf. Dig.*, Dec. 1990, pp. 57–63.

[9] F. Sischka, *IC-CAP Characterization and Modeling Handbook*. Santa Clara, CA. Agilent Technologies Inc., 2002.

[10] S. Lee and A. Gopinath, "New circuit model for RF probe pads and interconnections for the extraction of HBT equivalent circuits," *IEEE Elec. Dev. Lett.*, vol. 12, pp. 521–523, Oct. 1991.

[11] E. P. Vandamme, D. M. M.-P. Schreurs, and C. van Dinther, "Improved three-step de-embedding method to accurately account for the influence of pad parasitics in silicon on-wafer RF test-structures," *IEEE Trans. Elec. Dev.*, vol. 48, pp. 737–742, Apr. 2001.

[12] L. F. Tiemeijer, R. J. Havens, A. B. M. Jansman, and Y. Bouttement, "Comparison of the 'Pad-Open-Short' and 'Open-Short-Load' deembedding techniques for accurate on-wafer RF characterization of high-quality passives," *IEEE Trans. Microw. Theory Tech.*, vol. 53, pp. 723–729, Feb. 2005.

[13] J. Tao, P. Findley, and G. A. Rezvani, "Novel realistic short structure sonstruction for parasitic resistance de-embedding and on-wafer inductor characterization," in *Proc. IEEE 2005 Int. Conf. on Microelectron. Test Structures*, Apr. 2005, pp. 187–190.

[14] R. Torres-Torres, R. Murphy-Arteaga, and J. A. Reynoso-Hernandez, "Analytical model and parameter extraction to account for the pad parasitics in RF-CMOS," *IEEE Trans. Electron. Devicer*, vol. 52, pp. 1335–1342, July 2005.

[15] W. Liu, *Handbook of III–V Heterojunction Bipolar Transistors*, New York: Wiley-Interscience, 1981.

[16] M. Dvorak and C. Bolognesi, "On the accuracy of direct extraction of the heterojunction-bipolar-transistor equivalent circuit model," *IEEE Trans. Microw. Theory Tech.*, vol. 51, no. 6, pp. 1640–1649, 2003.

[17] B. Li and S. Prasad, "Basic expressions and approximations in small-signal parameter extraction for HBT's," *IEEE Trans. Microw. Theory Tech.*, vol. 47, no. 5, pp. 534–539, 1999.

[18] G. Massobrio and P. Antognetti, "Semiconductor device modeling with SPICE," New York: McGraw-Hill, 1993.

[19] K. Morizuka, R. Katoh, K. Tsuda, M. Asaka, N. Iizuka, and M. Obara, "Electron space-charge effects on high-frequency performance of AlGaAs/GaAs HBT's under high-current-density operation," *IEEE Electron Device Lett.*, vol. 9, no. 11, Nov. 1988.

[20] R. van der Toorn, J. Paasschens, and R. J. Havens, "Physically based analytical modeling of base-collector charge, capacitance and transit time of III–V HBT's," in *IEEE CSIC Dig.*, 2004.

[21] L. H. Camnitz, S. Kofol, and T. Low, "An accurate large signal, high frequency model for GaAs HBTs," in *IEEE GaAs IC Symp.*, 1996.

[22] D. Costa, W. Liu, and J. S. Harris Jr., "A new direct method for determining the heterojunction bipolar transistor equivalent circuit model," *IEEE 1990 Bipolar Circuits and Technology Meeting*, pp. 118–121.

[23] S. Bousnina, P. Mandeville, A. B. Kouki, R. Surridge, and F. M. Ghannouchi, "Direct parameter-extraction method for HBT small-signal model," *IEEE Trans. Microw. Theory Tech.*, vol. 50, no. 2, pp. 529–536, Feb. 2002.

[24] C. C. McAndrew, J. A. Seitchik, D. F. Bowers, M. Dunn, M. Foisy, I. Getreu, M. McSwain, S. Moinian, J. Parker, D. J. Roulston, M. Schroter, P. van Wijnen, and L. F. Wagner, "VBIC95, the vertical bipolar inter-company model," *IEEE J. Solid-State Circuits*, vol. 31, no. 10, pp. 1476–1483, 1996.

[25] "HBT model equations," http://hbt.ucsd.edu.

[26] "HICUM documentation," http://www.iee.et.tu-dresden.de/iee/eb/hic_new/hic_doc.html

[27] "The Mextram bipolar transistor model," http://www.nxp.com/acrobat_download/other/philipsmodels/nlur2000811.pdf

[28] "Documentation of the FBH HBT model," http://www.designers-guide.org/VerilogAMS/semiconductors/fbh-hbt/fbh-hbt_2_1.pdf

[29] "Agilent heterojunction bipolar transistor model," Agilent Advanced Design System Documentation 2006A, *Nonlinear Devices*, Chapter 2, pp. 4–33.

[30] Y. Gobert, P. J. Tasker, and K. H. Bachem, "A physical, yet simple, small-signal equivalent circuit for the heterojunction bipolar transistor," *IEEE Trans. Microw. Theory Tech.*, vol. 45, no. 1, pp. 149–153, Jan. 1997.

[31] L. J. Giacolleto, "Measurement of emitter and collector series resistances," *IEEE Trans. Microw. Theory Tech.*, May 1972.

[32] M. Rudolph, "Introduction to modeling HBTs," Artech House, 2006.

[33] C. J. Wei, J. C. M. Hwang, "Direct extraction of equivalent circuit parameters for heterojunction bipolar transistors," *IEEE Trans. Microw. Theory Techs.*, vol. 43, No. 5, Sept. 1995.

[34] S. A.Maas and D. Tait, "Parameter-extraction method for heterojunction bipolar transistors," *IEEE Trans. Microw. Guid. Wave Lett.*, vol. 2, no. 12, Dec. 1992.

[35] E. Arnold, M. Golio, M. Miller, and B. Beckwith, "Direct extraction of GaAs MESFET intrinsic element and parasitic inductance values," *IEEE Microw. Theory Tech. Symp. Dig.*, pp. 359–362, May 1990.

[36] M. Berroth and R. Bosch, "High-frequency equivalent circuit of GaAs FET's for large-signal applications," *IEEE Trans. Microw. Theory Tech.*, vol. 39, pp. 224–229, Feb. 1991.

[37] G. Dambrine, A. Cappy, F. Heliodore, and E. Playez, "A new method for determining FET small-signal equivalent circuits," *IEEE Trans. Microw. Theory Tech.*, vol. 36, pp. 1151–1159, July 1988.

[38] R. Tayrani, J. E. Gerber, T. Daniel, R. S. Pengelly, and U. L. Rohde, "A new and reliable direct parasitic extraction method for MESFETs and HEMTs," in *Proc. 23rd Eur. Microwave Conf.*, 1993, pp. 451–453.

[39] R. Vogel, "Application of RF wafer probing to MESFET modeling," *Microw. J.*, pp. 153–162, Nov. 1988.

[40] L. Yang and S. Long, "New method to measure the source and drain resistance of the GaAs MESFET," *IEEE Electron. Devices Lett.*, vol. EDL-7, Feb. 1986.

[41] P. M. White and R. M. Healy, "Improved equivalent circuit for determination of MESFET and HEMT parasitic capacitances from 'cold FET measurements,'" *IEEE Microw. Guided Weave Lett.*, vol. 3, no. 12, pp. 453–454, Dec. 1993.

[42] F. Diamand and M. Laviron, "Measurement of extrinsic series elements of a microwave MESFET under zero current conditions," in *Proc. 12th Eur. Microwave Conf.*, (Finland), Sept. 1982, pp. 451–465.

[43] A. Jarndal and G. Kompa, "A new small-signal modeling approach applied to GaN devices," *IEEE Trans. Microw. Theory Tech.*, vol. 53, no. 11, pp. 3440–3448, Nov. 2005.

[44] E. S. Mengistu, "Large-signal modeling of GaN HEMTs for linear power amplifier design," Ph.D. dissertation, Dept. High Freq. Eng., Univ. Kassel, Germany, 2008.

[45] A. Zárate-de Landa, J. E. Zúñiga-Juárez, J. A. Reynoso-Hernández, M. C. Maya-Sánchez, E. L. Piner, and K. J. Linthicum, "A new and better method for extracting the parasitic elements of the on-wafer GaN transistors," in *IEEE MTT-S Symp. Dig.*, Honolulu, Hawaii, June 3–8, 2007, pp. 791–794.

[46] W. Kuang, R. J. Trew, G. I. Bilbro, Y. Liu, and H. Yin, "Impedance anamolies and RF performance limitations in AlGaN/GaN HFET's," WAMICON 2006.

[47] F. Kharabi, M. Poulton, D. Halchin and D. Green, "A classic nonlinear FET model for GaN HEMT devices," *Compound Semiconductor Integrated Circuit Symp.*, 2007. CSIC 2007. IEEE, Oct. 14–17, 2007, pp. 1–4.

[48] D. Resca, A. Santarelli, A. Raffo, R. Cignani, G. Vannini, and F. Filicori, "Scalable equivalent circuit pHEMT modeling using an EM-based parasitic network description," *Proc. 2nd Eur. Microw. Integrated Circuits Conf.*, Oct. 2007, pp. 60–63.

[49] D. Resca, A. Santarelli, A. Raffo, R. Cignani, G. Vannini, F. Filicori, and A. Cidronali, "A distributed approach for millimeter-wave electron device modeling," *Proc. 1st Eur. Microwave Integrated Circuits Conf.*, Sept. 2006, pp. 257–260.

[50] A. Cidronali, G. Colloddi, A. Santarelli, G. Vannini, and G. Manes, "Millimeter-wave FET modeling using on-wafer measurements and EM simulation," *IEEE Trans. Microw. Theory Tech.*, vol. 50, no. 2, pp. 425–432, Feb. 2002.

[51] S. Nedeljkovic, J. McMacken, J. Gering, and D. Halchin, "A scalable compact model for III-V heterojunction bipolar transistors," *2008 IEEE MTT-S Int. Microw. Symp. Dig.*, pp. 479–491, June 2008.

4 Uncertainties in small-signal equivalent circuit modeling

Christian Fager, Kristoffer Andersson, and Matthias Ferndahl

Microwave Electronics Laboratory, Chalmers University of Technology

4.1 Introduction

Almost all computer-aided RF and microwave circuit design work relies on small-signal equivalent circuit models [1] (see Chapters 3 and 5). These models are used to replicate, as accurately as possible, the linearized electrical characteristics of the device under test. However, despite all efforts spent on deriving physically correct model topologies, surprisingly little research has been reported on how to extract the parameters of these models accurately.

There are several sources of uncertainty in the modeling process that contribute to inaccuracies in the models and their parameter estimates. By using a stochastic approach that acknowledges the fact that microwave and RF measurements are, indeed, associated with uncertainties and incorporating this approach with existing model extraction methods, it becomes possible to quantify the uncertainties in the models obtained. The statistical approach adopted in this chapter will be used as a framework for the application of statistical parameter estimation methods that are often used in other applications but not in the context of microwave and RF small-signal modeling. As a consequence, the accuracy of the models obtained is significantly improved when compared to traditionally used methods.

Different simple modeling examples will be used to illustrate how measurement uncertainties propagate to uncertainties in the models and their parameters when different extraction methods are applied. The application to small-signal transistor models will be of particular focus, since they form the basis for large signal transistor modeling and thus indirectly influence the performance when designing complex active circuits (see Chapter 5). The parameters of the small-signal models are, in themselves, also used by semiconductor foundries for monitoring of process variations [2], since they give detailed information of the transistor performance (see Chapter 9).

Before going into the details of different statistical methods for model parameter estimation, it is important to address the sources of uncertainty in the modeling process.

4.1.1 Sources of uncertainty in modeling

There are several sources of uncertainty that contribute to uncertainty in the modeling results; see Figure 4.1. The measurement uncertainties, the model assumptions, and the parameter estimation method all affect the accuracy of the model obtained.

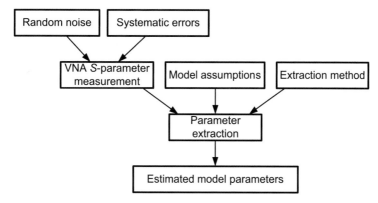

Figure 4.1 Overview of various sources of uncertainty that contribute to the model extraction results.

The techniques presented later in this chapter use a statistical approach to minimize the uncertainty contribution from the parameter estimation algorithm. However, they all assume that the statistical properties of the measurement noise are known.

4.1.2   Measurement uncertainty

The uncertainties in the measured S-parameters are ultimately determined by the performance of the VNA used. Approximate specifications are given by the instrument manufacturers, but the uncertainties depend heavily on the calibrations performed. The measurement uncertainty is also affected during the measurements, e.g., from uncertainties in the measurement reference plane. It is therefore not easy to predict the uncertainty in S-parameter measurements for the general case. As a prerequisite for the development of statistical extraction methods and assessment of the model uncertainties, it is, however, necessary to quantify the S-parameter uncertainties. For this purpose, stochastic metrics such as variances and covariances will be used and all S-parameter measurement data needs to be accompanied with corresponding uncertainties.

Figure 4.2 shows an example of the S-parameter uncertainties specified for an Agilent 8510C VNA [3], showing the uncertainty both in reflection and transmission measurements. It is interesting to note that the relative magnitude and absolute phase uncertainties are specified equally. It can also be noted that the relative uncertainties grow for low reflection coefficients as well as for low port-to-port transmission as expected.

The uncertainties in Figure 4.2 are given as worst-case, which according to the specifications should be interpreted as a 99.7% confidence interval, thus corresponding to $\pm 3\sigma$ if normal distributions are assumed for the measurement noise. The specifications can thus be used to predict the uncertainties for given measured S-parameters in terms of variances and covariances for the magnitude and phase:

$$\mathrm{Var}\left[\Delta \left|S\right| / S\right] = f_{|S|}\left(\left|S\right|\right) \tag{4.1}$$

$$\mathrm{Var}\left[\Delta \angle S\right] = f_{\angle S}\left(\left|S\right|\right). \tag{4.2}$$

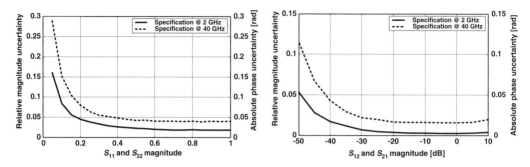

Figure 4.2 Example of S-parameter uncertainties obtained from VNA specifications.

There are many factors that contribute to the overall uncertainty in a specific measurement scenario. It is therefore, in practice, rather difficult to quantify the uncertainties very accurately. Although this may lead to over- or underestimation of the model parameter uncertainties, it will normally not have a severe effect on the quality of the model parameter estimates in themselves.

The results in the following sections are based on empirical uncertainty models approximating the instrument specifications shown in Figure 4.2. These can easily be replaced by more accurate uncertainty estimations obtained, e.g., by using the statistical calibration method presented by Williams *et al.* [4]. Additional calibration standards are then used to form an overdetermined set of equations for the calibration coefficients. The additional information is used to assess the uncertainty in the measurements. The algorithm is implemented in the software StatistiCAL [5] and the results can be used directly with the methods in this chapter to further improve the quality of the uncertainty estimations obtained.

4.2   Uncertainties in direct extraction methods

Direct extraction methods rely on closed form expressions for the model parameters in terms of the measurement data. Such expressions are available for specific model topologies including standard FET small-signal models [6] and are known to provide fast and reliable extraction results (see Chapter 3). This section describes how a statistical approach can be applied in the context of direct extraction methods.

4.2.1   Simple direct extraction example

A simple $R-C$ circuit is used in this section to illustrate the statistical model parameter estimation method.[1] Artificial measurement data with added normal distributed noise is used. The true parameter values are then known and the quality of the parameter estimate

---

[1] Parameter *estimation* will hereafter be used to emphasize the fact that the exact parameter value can never be extracted from practical measurements.

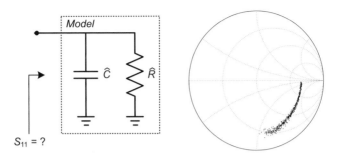

Figure 4.3 Example $R-C$ circuit model and S_{11} measurement.

can be evaluated. Although an idealized case, it illustrates the basic ideas behind the statistical model parameter estimation technique for equivalent circuit models.

4.2.1.1   Example circuit and measurements

Suppose that noisy S-parameter measurements are available over a wide frequency range for a device modeled by the simple $R-C$ circuit in Figure 4.3. This could represent, for example, measurements of the output reflection coefficient S_{22} on the combination of C_{ds} and g_{ds} in a FET model (see, e.g., Figure 4.9).

Intuitively, the resistance should be estimated from low-frequency measurements, since it is short-circuited at high frequencies by the capacitance. The capacitance, on the other hand, has high reactance at low frequency and is therefore masked by the resistance in parallel. At high frequency, the measurement uncertainties become larger and the capacitance approaches a short circuit. The capacitance should, therefore, be estimated at an intermediate frequency range.

The model parameters, R and C, are easily calculated from the real and imaginary parts of Y_{11} (which is easily calculated from S_{11} in the measurements):

$$R = 1/\text{Re } Y_{11} \tag{4.3}$$

$$C = \text{Im } Y_{11}/2\pi f. \tag{4.4}$$

Figure 4.4 shows the calculated model parameter values versus frequency. Apparently, $R \approx 250$ Ohm and $C \approx 50$ fF in this example. Their true values cannot, however, be uniquely determined from the measurement, but only estimated with some uncertainty. These uncertainties should be quantified by, e.g., confidence intervals, and presented together with the estimated values for R and C. For this purpose, an uncertainty analysis is needed as described in the following section.

4.2.1.2   Uncertainty analysis

To allow for the following analysis, we need to consider the true model parameter values as being known. The variations observed in Figure 4.4 are stochastic deviations from the true R and C values.

The model parameter deviations are transformed from deviations in the measured S-parameters. This transformation is nonlinear, but, since the deviations are usually small,

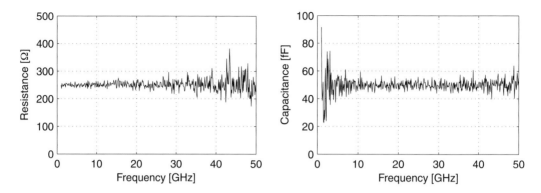

Figure 4.4 Calculated model parameters versus frequency.

a linearization may be used. A first-order sensitivity analysis may therefore be used to relate the S_{11} deviations to the deviations in R and C.

Sensitivity analysis is often included in electrical computer aided design (CAD) programs, where it is used to quantify how sensitive an output signal is to deviations in internal circuit parameters [7]. However, generally, these sensitivity analyses cannot be used for our intended purpose. We are interested in the opposite situation, i.e., how sensitive the internal circuit parameters (R and C) are to deviations in an output signal (S_{11}).

The relative sensitivity (K),[2] which will be used here, is based on a linearization of the direct extraction equations and normalized to give the percentage change in, e.g., R for a 1% change in $|S_{11}|$. Mathematically, this is expressed by:

$$\frac{\Delta R}{R} \cong \frac{\partial R}{\partial |S_{11}|} \frac{1/R}{1/|S_{11}|} \frac{\Delta |S_{11}|}{|S_{11}|} = K_{|S_{11}|}^{R} \frac{\Delta |S_{11}|}{|S_{11}|}, \tag{4.5}$$

which also gives the definition of the sensitivity in R to S_{11} magnitude deviations, $K_{|S_{11}|}^{R}$. A similar expression is used for phase sensitivities,

$$\frac{\Delta R}{R} \cong \frac{\partial R}{\partial \angle S_{11}} \cdot \frac{1}{R} \cdot \Delta \angle S_{11} = K_{\angle S_{11}}^{R} \Delta \angle S_{11}. \tag{4.6}$$

Absolute phase deviations are used, since the phase is already a relative measure of the arc-length to the radius.

Figure 4.5 illustrates how the sensitivities are used to calculate the model parameter deviations from the S-parameter deviations. The Y-parameters are used as an intermediate step to ease the sensitivity calculations as illustrated in the figure. The resulting model parameter sensitivities to deviations in the S_{11} magnitude and phase are shown versus frequency in Figure 4.6. Figure 4.6 shows that R is most sensitive to $|S_{11}|$ deviations, whereas C is more sensitive to deviations in $\angle S_{11}$, as expected. In order for

---

[2] Hereafter, only relative sensitivities will be treated. The word relative will therefore be omitted.

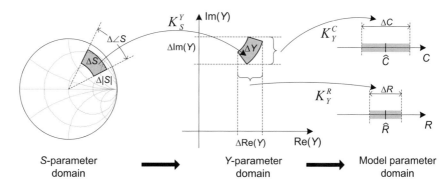

Figure 4.5 Illustration of how the sensitivities (K) are used to calculate the model parameter deviations from deviations in the S-parameter measurement. The Y-parameter domain is used intermediately to ease the sensitivity calculations.

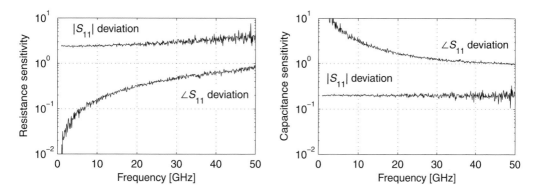

Figure 4.6 Frequency dependence of the relative model parameter sensitivities to deviations in the S-parameter magnitude and phase.

a parameter to be well defined, its sensitivities should be small, since the sensitivity to measurement noise will then also be low.

In Figure 4.6, the resistance has smaller sensitivity at low frequency, although it is quite low over the entire frequency range. The capacitance, on the other hand, has a sensitivity that decreases at higher frequencies, indicating that it is more well defined there. However, to know where it is most well defined, the measurement uncertainties must also be considered.

The S-parameter deviations arise from measurement uncertainties that are random and thus described by probability distributions. The exact distributions are usually unknown. In practice, however, they are often assumed to be normal distributed. This may be partially justified by the central limit theorem, since each of them is composed of a large number of small and reasonably independent uncertainty contributions [8].

The S-parameter magnitude and phase deviations are normal distributed with zero mean and known variances (σ^2) in this example. It is also assumed that these deviations are independent. The model parameter variances may then be calculated using the

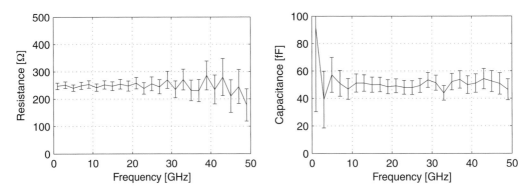

Figure 4.7 Calculated model parameters and their estimated uncertainties versus frequency. The error bars display a 2σ (95%) confidence interval.

sensitivities in (4.5) and (4.6) together with well-known variance formulas [8]:

$$\sigma_R^2 = \left(K_{|S_{11}|}^R\right)^2 \sigma_{|S_{11}|}^2 + \left(K_{\angle S_{11}}^R\right)^2 \sigma_{\angle S_{11}}^2 \tag{4.7}$$

$$\sigma_C^2 = \left(K_{|S_{11}|}^C\right)^2 \sigma_{|S_{11}|}^2 + \left(K_{\angle S_{11}}^C\right)^2 \sigma_{\angle S_{11}}^2. \tag{4.8}$$

Figure 4.7 shows the resulting model parameter uncertainties versus frequency. The uncertainties are indicated by $\pm 2\sigma$ error bars, corresponding to 95% confidence intervals for the normal distribution.

4.2.1.3   Parameter estimation

In traditional estimation methods, the calculated model parameter values are simply averaged over a predetermined frequency range. However, with knowledge about the parameter uncertainties, a weighted average may be performed to improve the estimate. The estimation uncertainty is minimized by assigning less weight to more uncertain extractions and vice versa. The appendix of reference [9] describes how this is done.

As previously mentioned, the uncertainty analysis needs the true model parameter values. Initially, we use the frequency-dependent values in Figure 4.4 for the uncertainty analysis. However, as new estimates are made, the uncertainty analysis may be refined. The model parameter estimates are not very sensitive to the uncertainty analysis results and three iterations are normally sufficient.

The resulting parameter estimates for this example become:

$$\begin{cases} \hat{R} = 249.72 \pm 0.91 \text{ [Ohm]} \\ \hat{C} = 49.82 \pm 0.28 \text{ [fF]}, \end{cases} \tag{4.9}$$

where a 95% confidence interval is also given. Since the S-parameters were normal distributed, and only linear operations have been used, the estimated model parameters will also be normal distributed with mean and variances given from (4.9) above.

The model parameter estimates correspond well to the true values used for generating the artificial measurement data: $R = 250$ Ohm and $C = 50$ fF.

4.2.1.4 Parameter correlations

Although the confidence intervals presented for $\hat{R}$ and $\hat{C}$ in (4.9) give a good understanding for the parameter estimation uncertainties, it is not a complete statistical representation. In fact, their correlations need also to be considered. Correlations between the estimated parameters can be determined by generalizing the previous analysis into matrix form. The extension of (4.5) is given by

$$
\begin{bmatrix} \Delta R/R \\ \Delta C/C \end{bmatrix} = \begin{bmatrix} K_{|S|}^R & K_{\angle S}^R \\ K_{|S|}^C & K_{\angle S}^C \end{bmatrix} \begin{bmatrix} \Delta|S|/|S| \\ \Delta\angle S \end{bmatrix}
$$

$$
\Delta\mathbf{x} = \mathbf{K_S^x}\Delta\mathbf{S}, \tag{4.10}
$$

where $\Delta\mathbf{x}$ is a vector of the model parameter deviations, $\mathbf{K_S^x}$ is a sensitivity matrix, and $\Delta\mathbf{S}$ represents S-parameter deviations. The S-parameter measurement uncertainties are then quantified, at each measurement frequency, by a covariance matrix:

$$
\mathbf{C_S} = \begin{bmatrix} \mathrm{var}[\Delta S/S] & \mathrm{cov}[\Delta S/S, \Delta\angle S] \\ \mathrm{cov}[\Delta S/S, \Delta\angle S] & \mathrm{var}[\Delta\angle S] \end{bmatrix} = \begin{bmatrix} \sigma_{|S|}^2 & 0 \\ 0 & \sigma_{\angle S}^2 \end{bmatrix}, \tag{4.11}
$$

where in the final step it was assumed that the magnitude and phase uncertainties are uncorrelated, which is normally a good approximation.

The sensitivities allow the measurement covariance matrix to be propagated to the model parameters [10],

$$
\mathbf{C_{\hat{x}}} = \begin{bmatrix} \sigma_R^2 & \mathrm{cov}[\Delta R/R, \Delta C/C] \\ \mathrm{cov}[\Delta R/R, \Delta C/C] & \sigma_C^2 \end{bmatrix} = \mathbf{K_S^x}\mathbf{C_S}\left(\mathbf{K_S^x}\right)^{\mathrm{T}}, \tag{4.12}
$$

where the generalized expression in (4.12) allows a complete statistical representation of the model parameter uncertainties and their covariances.

4.2.2 Results using transistor measurements

The same method used for estimating the $R-C$ model parameters in the example above can be applied to estimate the FET model parameters and their uncertainties [9]. Similar analyses have later been presented also for HBT models [11, 12].

4.2.2.1 Uncertainty contributions

Direct extraction of intrinsic transistor model parameters is typically divided into two steps, where in the first step the parasitic- and bias-independent parts of the model are extracted from a separate set of measurements. For FET devices, reverse and forward bias measurements at $V_{DS} = 0$ V are normally used [6]. Similar methods are also available for HBT transistors [13].

Once determined, the parasitic elements are fixed and de-embedded from subsequent measurements in direct extraction of the intrinsic bias-dependent model parameters. The propagation of uncertainties in direct extraction of transistor model parameters is outlined in Figure 4.8.

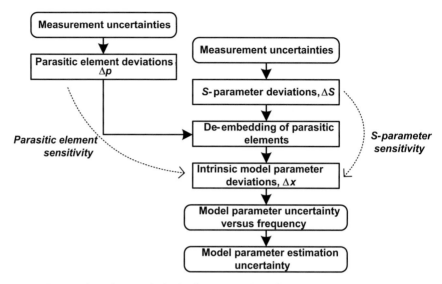

Figure 4.8 Propagation of uncertainties in direct extraction of intrinsic transistor model parameters.

In many practical cases, it has been found that the uncertainty contributions from the parasitic elements have a minor effect and may be neglected [11]. This is true in particular for on-wafer measurements where the influence of parasitics are typically small. Hence, for the analysis presented here, we will focus on the intrinsic model parameters and their uncertainties.[3] Measurements and extraction of intrinsic FET model parameters will be used to illustrate typical results.

4.2.2.2   Intrinsic model parameter sensitivities

As a first step of the uncertainty calculations, the relative sensitivities need to be derived. Similar to the expressions in (4.5) and (4.6) it is possible to derive analytical expressions for the intrinsic model parameter sensitivities to S-parameter deviations. For FET transistors, such expressions are presented in reference [9] and for HBTs similar expressions are presented in reference [12].

Small-signal modeling of on-wafer GaAs pHEMT transistors from a commercial foundry will be used to exemplify typical behavior. The transistors were characterized by on-wafer S-parameter measurements obtained using an Agilent 8510C vector network analyzer. The intrinsic small-signal model shown in Figure 4.9 has been used. This model, and variations thereof, is shown to predict measured S-parameters up to very high frequencies and for a variety of devices [6, 14, 15]. These references also provide detailed information about direct extraction methods used to calculate the intrinsic model parameters versus frequency.

---

[3] A complete analysis taking the effect of parasitic element uncertainties into consideration is included for the optimizer-based extractions in the next section.

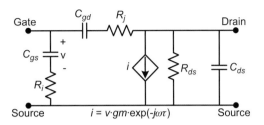

Figure 4.9 Small-signal intrinsic transistor model.

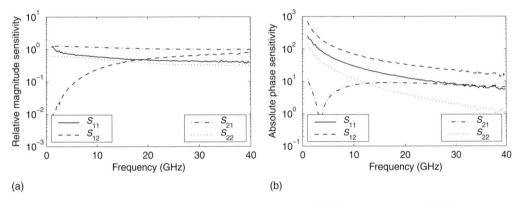

(a)

(b)

Figure 4.10 Calculated intrinsic model parameter sensitivities to S-parameter deviations: (a) Transconductance sensitivity, $K_{|S|}^{g_m}$; (b) R_j sensitivity, $K_{|S|}^{R_j}$.

Figure 4.10 shows an example of the sensitivities obtained for two of the intrinsic model parameters used in the small-signal FET model in Figure 4.9: the transconductance, g_m, and the gate-drain series resistance, R_j. The former is known to be well defined and easy to extract from measurements, while the latter is typically difficult to extract accurately [14]. This fact is also evident by observing the calculated sensitivities. The sensitivities for g_m are far smaller than for R_j, which means that g_m is less affected by deviations in the S-parameter measurements, i.e., less sensitive to measurement noise. The highest g_m sensitivity is with respect to $|S_{21}|$, which indicates that the magnitude of $|S_{21}|$ is most important for the extraction of its value. R_j, on the other hand, is most sensitive to $\angle S_{12}$. At low frequencies, the sensitivities are almost 100 times larger than for g_m which indicates a large sensitivity to measurement noise. The sensitivities decrease at higher frequencies, indicating that R_j is mostly accurately extracted from high-frequency $\angle S_{12}$ measurements. However, the measurement uncertainties also increase at high frequencies and without the uncertainty analysis presented in the following section, it is not easy to guess the optimum frequency range where the parameter estimation should be made.

4.2.2.3 Intrinsic model parameter uncertainties

The uncertainties in the small-signal model parameters are calculated by combining the sensitivities and a measurement uncertainty model, as explained in Section 4.2.1. In the

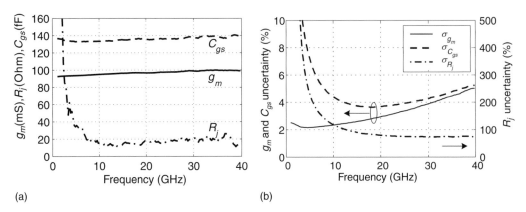

Figure 4.11 Intrinsic small-signal model parameter values extracted versus frequency and their estimated uncertainties: (a) extracted model parameter values; (b) estimated model parameter uncertainties.

intrinsic FET model parameter case, this means that the intrinsic model sensitivities illustrated above will be combined with the uncertainty models described in Section 4.1.2. As a result, it is possible to estimate the parameter uncertainties versus frequency.

Figure 4.11 shows both the parameter values calculated using regular direct extraction techniques, and the associated uncertainties calculated for three of the model parameters: g_m, C_{gs}, and R_j. Please note that the uncertainty in R_j is almost 50 times larger than for the other two parameters and therefore is displayed using a separate vertical axis. This fact is also clearly visible in the estimated parameter values shown in Figure 4.11(a) where the values at low frequency deviate from the final high-frequency asymptote. The results also clearly confirm that g_m has least uncertainty at low frequencies and should be extracted from low-frequency measurements. In the case of C_{gs}, it should generally be estimated from high-frequency measurements, but as the measurement uncertainties grow there, the resulting total uncertainty is minimized at an intermediate frequency range corresponding to approximately 20 GHz in this case. This information is difficult to assess from the direct extracted values in themselves.

4.2.2.4   Multibias extraction results

The transistor example above illustrates how the uncertainty calculation can be used to determine an optimum frequency range for reliable and accurate estimation of the intrinsic model parameters. The entire process for calculation of the model parameter uncertainties was analytically derived in reference [9]. Hence, the uncertainties can be easily calculated along with the model parameter estimation itself. This makes it ideally suitable for inclusion in automated multibias model extraction schemes. The uncertainties will certainly depend on the device and bias point used, but the aforementioned technique allows the parameters always to be estimated under optimum, minimum uncertainty conditions, thus providing reliable model extraction results that can be used, for example, for large-signal modeling (see Chapter 5). The knowledge of the uncertainties

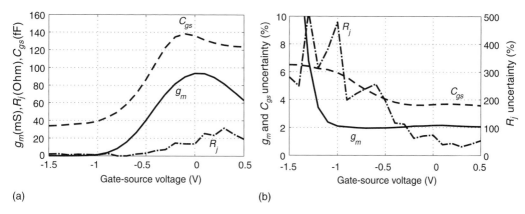

(a) (b)

Figure 4.12 Intrinsic small-signal model parameter values extracted versus bias and their estimated uncertainties: (a) extracted model parameter values; (b) estimated model parameter uncertainties.

is then obtained as an added value of this method, but can also be directly incorporated, e.g., in statistical and Monte-Carlo simulations, thus taking the uncertainties originating from the model extraction into consideration.

Figure 4.12 shows the model parameters and their uncertainties automatically estimated versus bias using the uncertainty-based method outlined above. The results illustrate that a smooth and well-behaved extraction can be performed without any a-priori knowledge about the device characteristics. It is also interesting to study the bias dependencies of the model parameter uncertainties obtained. It should be noted, however, that the relative uncertainties can be misleading when the model parameter values tend to zero, as is the case for g_m below the pinch-off voltage. Although the relative uncertainty grows rapidly there, the absolute uncertainty will still be finite and well behaved.

4.3 Optimizer-based estimation techniques

Finding an estimator with the direct extraction method was successful for the simple R–C example in Section 4.2.1 and under certain circumstances also for transistor modeling [6, 15]. However, for larger problems involving tens of parameters and where the model cannot be put in a closed form, it is normally impossible to find such an estimator and even more difficult to assess its performance.

Optimizer-based extraction methods find the parameters by minimizing an error function, which is typically the difference between measured and modeled S-parameters. Optimizer-based methods are very flexible; the model topology can easily be changed to accommodate unusual device structures. Another advantage is that microwave CAD software often includes an optimizer and thus, all work can be carried out in a single software environment. Main drawbacks include difficulty in finding a global minimum of the error function, which makes the result sensitive to starting values. Another disadvantage is that the curvature of the minimum is often rather shallow, i.e., the problem is

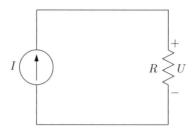

Figure 4.13 Simple resistive circuit.

ill-conditioned. Several methods for addressing both ill-conditioning and robustness to variations in starting values have been proposed [16–19].

These optimizer-based methods use a deterministic approach, where all measurements are assumed to be ideal. Similarly to the commonly used direct extraction methods, they therefore offer no possibilities for assessment of the uncertainty in the results obtained.

In this section we will show how a maximum likelihood estimator (MLE) can be used to implement a generic statistical optimizer-based method for RF and microwave modeling applications. The objective of the method is to find the model parameter estimates that give the least possible modeling uncertainties. The method will be demonstrated with both transistor modeling and de-embedding applications.

4.3.1   Maximum likelihood estimation

The MLE is a commonly used and well-established statistical estimator that is asymptotically unbiased and also asymptotically efficient [10]. Asymptotic efficiency means that, as the number of measurement values increases, the variance of the estimate approaches the theoretical limit given by the Cramer–Rao lower bound [10, section 3.4].

4.3.1.1   Simple example
To illustrate the MLE method for our modeling applications, consider an example where we are able to measure the dissipated power of a resistive circuit; see Figure 4.13.

A single measurement of the dissipated power, P_n, can then be described by the following model:

$$P_n = RI^2 + w_n, \qquad (4.13)$$

where R is the circuit resistance, I is the current through the circuit, and w_n is the noise picked up by the measurement equipment. To complete the model we assume that the measurement noise has a zero mean normal distribution and that consecutive measurements are independent. The probability density function (pdf) for w_n is therefore given by:

$$p(w_n) = \frac{1}{\sqrt{2\pi\sigma^2}} \exp\left[-\frac{1}{2\sigma^2}w_n^2\right], \qquad (4.14)$$

where σ^2 is the noise variance and is directly proportional to the noise power.

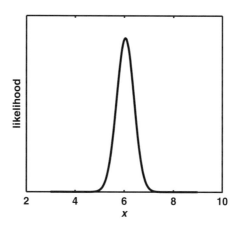

Figure 4.14 Likelihood function plotted for an unknown parameter, x.

In MLE, a complete set of measurements is considered jointly, for our example denoted by $\mathbf{P} = [P_0, P_1, \ldots, P_{N-1}]^{\mathrm{T}}$. The *joint pdf* for N measurements is then expressed as a function of the unknown parameter, x. The notation $p(\mathbf{P}) = p(\mathbf{P}; x)$ is used to acknowledge this dependence and is called the *likelihood function*. For independent measurements, which we normally can assume, the joint pdf is the product of the individual pdfs:

$$p(\mathbf{P}; x) = \prod_{n=0}^{N-1} p(P_n; x). \tag{4.15}$$

The MLE of x is then found by maximizing the likelihood function with x as an argument. The estimate is denoted with $\hat{x}$:

$$\hat{x} = \arg\max_{x} p(\mathbf{P}; x). \tag{4.16}$$

This means that we are seeking the value of x that maximizes the probability that we observe the actual set of measurements. In Figure 4.14, a pdf with fixed data is plotted versus the parameter x. It is not very likely that the true value of x is 2 or 10. From the plot, we therefore conclude that the most likely value is near 6, thus we choose $\hat{x} = 6$ as the MLE.

Returning to the measurement example, the likelihood function of the recorded measurements is:

$$p(\mathbf{P}; \mathbf{x}) = \prod_{n=0}^{N-1} \frac{1}{\sqrt{2\pi\sigma^2}} \exp\left[-\frac{1}{2\sigma^2}(P_n - RI^2)^2\right], \tag{4.17}$$

where $\mathbf{x} = [R, I]^{\mathrm{T}}$ are the parameters. Now consider the natural logarithm of the likelihood function:

$$\ln p(\mathbf{P}; \mathbf{x}) = -\frac{n}{2}\ln\left(2\pi\sigma^2\right) - \sum_{n=0}^{N-1} \frac{1}{2\sigma^2}\left(P_n - RI^2\right)^2. \tag{4.18}$$

Maximizing the log-likelihood function is equivalent to maximizing the likelihood function, since the logarithm is a strictly monotone function. Therefore, by assuming a known current, I, the maximum is found by setting the partial derivative (with respect to R) equal to zero:

$$\frac{\partial \ln p(\mathbf{P}; R)}{\partial R} = \sum_{n=0}^{N-1} \frac{I^2}{\sigma^2} \left(P_n - RI^2 \right) = 0. \tag{4.19}$$

This yields the MLE estimate of R:

$$\hat{R} = \frac{1}{I^2 N} \sum_{n=0}^{N-1} P_n. \tag{4.20}$$

For this simple example, the solution is simple and intuitive. It is the mean power measured for the sample divided by the squared current. Repeating the procedure but with unknown current gives the following MLE of I:

$$\hat{I} = \pm \sqrt{\frac{1}{R} \frac{1}{N} \sum_{n=0}^{N-1} P_n}, \tag{4.21}$$

which also in this case is a very intuitive estimator.

4.3.1.2   MLE uncertainty

The accuracy of the estimate, e.g., $\hat{R}$, is directly related to the curvature of the likelihood function and the asymptotic variance of the MLE is given by [10, section 7.5 and 7.8]:

$$\text{var}[\hat{x}] = \frac{1}{-\frac{\partial^2 \ln p(\mathbf{P};x)}{\partial x^2}}. \tag{4.22}$$

This motivates us to calculate the negative log-likelihood function instead of the likelihood function; and, as mentioned above, the maximum of the log-likelihood function is identical to that of the likelihood function. In Figure 4.15 the negative log-likelihood function for our example is plotted for three different current magnitudes. A higher current results in a sharper likelihood function, meaning that a high current gives a more accurate estimate than a low current. A higher current gives a larger dissipated power and thus improved dynamic range.

Using (4.22) to calculate the variance of $\hat{R}$ yields:

$$\text{var}\,\hat{R} = \frac{1}{-\frac{\partial^2 \ln p(\mathbf{P};R)}{\partial R^2}} = \frac{\sigma^2}{I^4}. \tag{4.23}$$

For vector estimators, this is generalized so that the covariance matrix of the MLE is equal to the Hessian of the negative log-likelihood function.

4.3.2   MLE of small-signal transistor model parameters

The use of MLE for transistor modeling applications is described below. In contrast to the simplified example above, the estimation of small-signal transistor models is normally

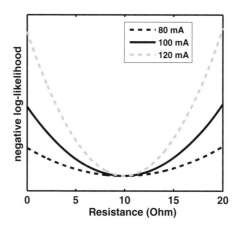

Figure 4.15 Negative log-likelihood functions for the dissipated power measurement. Different curves correspond to different values of the current source. The true value of the resistance is 10 Ohm.

a two-step process. First, the parasitic elements (and their covariance) are estimated, then this information is used to estimate the intrinsic model parameters. Normally this is done by de-embedding the parasitic elements from the measured data and then using some fitting algorithm on the de-embedded data set; unfortunately this also means that the intrinsic elements will depend on the parasitics (see Figure 4.1).

In this work the parasitic elements are treated as random variables associated with a pdf, and estimated using the MLE method. The likelihood function used to determine the intrinsic elements will therefore depend also on the parasitic elements, as will be described later.

4.3.2.1 Parasitic parameter estimation

The parasitic estimation is typically based on measurements in two or more *cold* bias points (see Chapter 3). From the S-parameter measurement at each bias point k, a stacked column vector, $\mathbf{s}_k$, is formed:

$$
\mathbf{s}_k =
\begin{bmatrix}
\mathrm{Re}\{\mathbf{s}_{k,11}\} \\
\mathrm{Im}\{\mathbf{s}_{k,11}\} \\
\mathrm{Re}\{\mathbf{s}_{k,21}\} \\
\mathrm{Im}\{\mathbf{s}_{k,21}\} \\
\mathrm{Re}\{\mathbf{s}_{k,12}\} \\
\mathrm{Im}\{\mathbf{s}_{k,12}\} \\
\mathrm{Re}\{\mathbf{s}_{k,22}\} \\
\mathrm{Im}\{\mathbf{s}_{k,22}\}
\end{bmatrix},
\tag{4.24}
$$

where each of the elements $\mathbf{s}_{k,ij}$ is a column vector containing measured data at all frequencies. In the same way, the model response at bias point k is collected in $\mathbf{g}_k(\mathbf{p})$, where $\mathbf{p}$ is the vector of parasitic model S-parameter parameters.

The residual error at bias point k, $\mathbf{h}_k(\mathbf{p})$, is now defined as the vector difference between the modeled and measured S-parameters:

$$\mathbf{h}_k(\mathbf{p}) = \mathbf{s}_k - \mathbf{g}_k(\mathbf{p}) + \mathbf{w}_k, \qquad (4.25)$$

where $\mathbf{w}_k$ is a vector containing the unknown measurement noise. The *joint* residual of the K bias points used for the parasitic extraction may then be expressed as:

$$\mathbf{h}(\mathbf{p}) = \begin{bmatrix} \mathbf{h}_1(\mathbf{p}) \\ \mathbf{h}_2(\mathbf{p}) \\ \vdots \\ \mathbf{h}_K(\mathbf{p}) \end{bmatrix} = \mathbf{s} - \mathbf{g}(\mathbf{p}) + \mathbf{w}. \qquad (4.26)$$

Assuming that the measurement noise, $\mathbf{w}$, has zero-mean normal distribution with covariance matrix $\mathbf{C}_w$, the joint likelihood function becomes [10]:

$$p(\mathbf{h}; \mathbf{p}) = e^{-\frac{1}{2}\mathbf{h}(\mathbf{p})^T \mathbf{C}_w^{-1} \mathbf{h}(\mathbf{p})}. \qquad (4.27)$$

The maximum likelihood estimate of the model parameters, $\hat{\mathbf{p}}$, is then given by the maximum of the logarithm of the likelihood function:

$$\hat{\mathbf{p}} = \arg\max_{\mathbf{p}} -\frac{1}{2}\mathbf{h}(\mathbf{p})^T \mathbf{C}_w^{-1} \mathbf{h}(\mathbf{p}). \qquad (4.28)$$

The scalar valued maximization problem (4.28) is identical to a weighted nonlinear least-squares fit to the measured data. This can be exploited so that efficient Newton methods can be used. The Jacobian required for the Newton method is formed by using sensitivities calculated using the technique of adjoint networks [20]. This scheme is very efficient and the estimate is usually found within < 20 iterations. The uncertainty in the MLE parameter estimate is approximated by [10]:

$$\mathbf{C}_{\hat{\mathbf{p}}} \approx \mathbf{H}^{-1}, \qquad (4.29)$$

where $\mathbf{H}$ is the Hessian of the log-likelihood function, whose elements are given by

$$H_{k,l} = \frac{\partial^2}{\partial p_k \partial p_l}\left(\frac{1}{2}\mathbf{h}(\mathbf{p})^T \mathbf{C}_w^{-1} \mathbf{h}(\mathbf{p})\right) \qquad (4.30)$$

and should be evaluated at $\mathbf{p} = \hat{\mathbf{p}}$. This Hessian is valid near the solution of the maximization problem [17].

4.3.2.2   Application to parasitic FET model extraction

For a typical on-wafer measurement of a common source coupled FET, pads and access regions can be described by the equivalent circuit in Figure 4.16. We refer to this model as the package model. This model can readily be modified to fit other configurations such as in-fixture measurements of packaged devices, etc.

Embedded in the package is the intrinsic FET. For zero drain-source voltage, the intrinsic FET can be described by the equivalent circuits in Figure 4.17. For a gate voltage corresponding to pinch-off, the intrinsic FET behaves as a capacitive Π-network. For gate voltages larger than zero volts, the intrinsic FET can be viewed as a Schottky

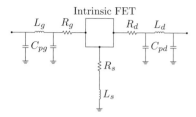

Figure 4.16 Small-signal FET parasitics.

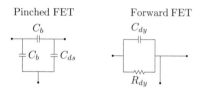

Figure 4.17 Small-signal intrinsic FET models for zero drain-source voltage.

diode. The fact that these two conditions have very different S-parameters but share the parasitic elements can be utilized in an estimator. Figure 4.18 shows typical S-parameters for pinched and forward conditions.

Simultaneously estimating parasitic, pinched, and forward model parameter values yields the following parasitic model parameter vector:

$$\mathbf{p} = \begin{bmatrix} C_{pg} & L_g & R_g & R_s & L_s & C_{pd} & L_d & R_d & C_b & C_{ds} & C_{dy} & R_{dy} \end{bmatrix}^{\mathrm{T}}. \quad (4.31)$$

This means that in total, 12 parameters are now estimated from two sets of measurements, which improves accuracy and robustness of the estimator compared to the standard direct method.

The accuracy of the MLE solution in (4.28) is verified using Monte Carlo simulations. Parasitic model parameters are estimated from a set of measured transistor S-parameter data in pinch-off and forward bias. S-parameters were measured on a transistor at 201 frequencies from 45 MHz to 50 GHz. The estimated parameters are then used as a template for generating 1000 sets of synthetic measurements. The synthetic measurements are generated by adding noise to the calculated model response. The noise is generated according to an empirical VNA uncertainty model similar to the one presented in Section 4.1.2. For each set of synthetic data, a MLE is performed – resulting in a total of 1000 sets of parasitic parameter estimates and corresponding parameter uncertainties. From the synthetic estimates, histograms and scatter plots are constructed to evaluate the validity of the MLE results.

In Figure 4.19, histograms for the parasitic parameter estimates are shown. The estimates are clearly normal distributed and the agreement between simulated and estimated distributions is excellent. In general, the parasitic elements are easy to extract, i.e., the uncertainty is small. However, for R_g the uncertainty is relatively large, as expected from normal modeling situations.

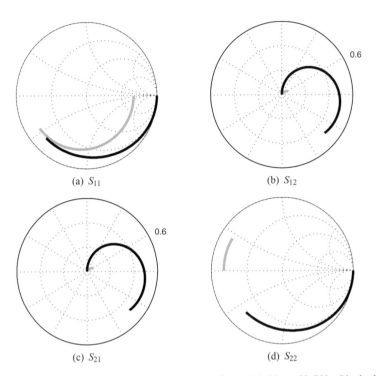

(a) S_{11} (b) S_{12}

(c) S_{21} (d) S_{22}

Figure 4.18 Typical example of cold FET S-parameters from 45 MHz to 50 GHz. Pinched bias –
black lines, forward bias – gray lines.

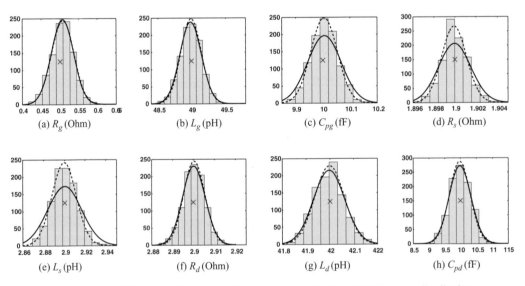

(a) R_g (Ohm)   (b) L_g (pH)   (c) C_{pg} (fF)   (d) R_s (Ohm)

(e) L_s (pH)   (f) R_d (Ohm)   (g) L_d (pH)   (h) C_{pd} (fF)

Figure 4.19 Histograms for the parasitic parameter estimates. Solid lines are distributions
predicted from the MLE. Dashed lines are normal distributions fitted to the histograms. Crosses
indicate the true value. The vertical axis is the number of counts (out of 1000) in each box.

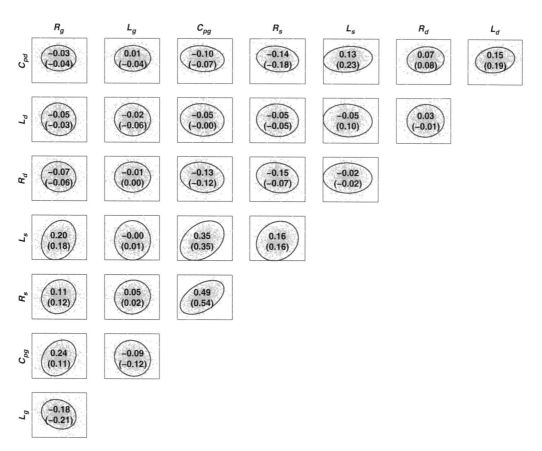

Figure 4.20 Pairwise scatter plots of parasitic estimates. Ellipses corresponds to 95% confidence regions. The number indicates the estimated correlation factor, the number within parenthesis is the true correlation factor.

To visualize the correlation between parameter estimates, pairwise scatter plots are used. In these plots all estimated parameters are plotted versus each other. Thereby it is possible to see any correlation between parameters. In each scatter plot the predicted and simulated correlation coefficients are given. Ellipses corresponding to 95% confidence regions are superimposed on the scatter plots. As can be seen in Figure 4.20, most estimates are uncorrelated, but some correlation can be seen for the pairs R_s–C_{pg} and L_s–C_{pg}. Both correlation factors are predicted by the MLE using the expression in (4.29).

4.3.2.3 MLE of intrinsic model parameters

The intrinsic FET model (Figure 4.9) is embedded in a package (Figure 4.16). The measured S-parameters will therefore not only depend on the intrinsic elements but also on the parasitic elements. However, the parasitic parameter values estimated using the MLE method above are not deterministic but associated with some uncertainty, $\mathbf{C_{\hat{p}}}$. It is therefore not possible to de-embed them directly from an active measurement and derive an estimator for the intrinsic parameters. Instead, a joint likelihood function, $p(\mathbf{h}; \mathbf{x}, \mathbf{p})$, is formed where both the intrinsic, $\mathbf{x}$, and parasitic, $\mathbf{p}$, parameters are unknown. Since the distributions of the parasitic parameters are available from the previous extraction step, it is possible to eliminate them from the likelihood function by considering the *expected* likelihood function:

$$p(\mathbf{h}; \mathbf{x}) = E[p(\mathbf{h}; \mathbf{x}, \mathbf{p})], \tag{4.32}$$

where the expectation is taken with respect to the parasitic parameters, $\mathbf{p}$. The new likelihood function can be thought of as a weighted average over all possible parasitic element values.

Using the same notation as in the parasitic MLE, the joint likelihood function can be written:

$$p(\mathbf{h}; \mathbf{x}, \mathbf{p}) = e^{-\frac{1}{2}\mathbf{h}(\mathbf{x},\mathbf{p})^T \mathbf{C}_w^{-1} \mathbf{h}(\mathbf{x},\mathbf{p})}. \tag{4.33}$$

As outlined above, the dependence of $\mathbf{p}$ can be eliminated, since an estimate of the parasitics is already available, by integrating (4.33) over all possible values of $\mathbf{p}$ and weighting with the pdf. The expected likelihood function then becomes:

$$p(\mathbf{h}; \mathbf{x}) = \int_{\Omega} e^{-\frac{1}{2}\mathbf{h}(\mathbf{x},\mathbf{p})^T \mathbf{C}_w^{-1} \mathbf{h}(\mathbf{x},\mathbf{p})} f(\mathbf{p}) d\mathbf{p}, \tag{4.34}$$

where $f(\mathbf{p})$ is the pdf of the parasitics and Ω is the domain of real numbers with dimension equal to the number of parasitic parameters. This equation may be analytically solved under the assumption that the parasitic parameter uncertainties are relatively small and normal distributed [21]. The result is used to derive the following expression for the MLE solution for the intrinsic model parameters:

$$\hat{\mathbf{x}} = \arg\max_{\mathbf{x}} \frac{e^{-\frac{1}{2}v(\mathbf{x})}}{\sqrt{|\mathbf{A}(\mathbf{x})|}} = \arg\max_{\mathbf{x}} \left(-\frac{v(\mathbf{x})}{2} - \frac{1}{2}\ln|\mathbf{A}(\mathbf{x})| \right), \tag{4.35}$$

where

$$\mathbf{A}(\mathbf{x}) = \mathbf{J}_p(\mathbf{x})^T \mathbf{C}_w^{-1} \mathbf{J}_p(\mathbf{x}) + \mathbf{C}_{\hat{\mathbf{p}}}^{-1} \tag{4.36}$$

$$v(\mathbf{x}) = \mathbf{h}_0(\mathbf{x})^T \mathbf{C}_w^{-1} \mathbf{h}_0(\mathbf{x}) - \left(\mathbf{J}_p(\mathbf{x})^T \mathbf{C}_w^{-1} \mathbf{h}_0(\mathbf{x}) \right)^T \mathbf{A}(\mathbf{x})^{-1} \mathbf{J}_p(\mathbf{x})^T \mathbf{C}_w^{-1} \mathbf{h}_0(\mathbf{x}) \tag{4.37}$$

$$\mathbf{h}_0(\mathbf{x}) = \mathbf{h}(\mathbf{x}, \hat{\mathbf{p}}) \tag{4.38}$$

$$\mathbf{J}_p(\mathbf{x}) = \frac{\partial \mathbf{h}(\mathbf{x}, \mathbf{p})}{\partial \mathbf{p}}. \tag{4.39}$$

As for the parasitics, an asymptotic covariance matrix is found as the inverse Hessian of the negative log likelihood function.

The intrinsic parameters are then estimated by solving the maximization problem (4.35) using a Levenberg–Marquardt algorithm. Both the required Jacobian and the final Hessian are calculated using adjoint techniques. To improve efficiency in finding the solution, starting values of the intrinsic parameters are calculated using the direct method by Dambrine *et al.* [6].

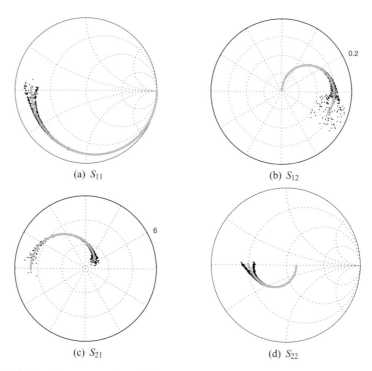

(a) S_{11} (b) S_{12}

(c) S_{21} (d) S_{22}

Figure 4.21 Active S-parameters from 45 MHz to 50 GHz. True values – gray lines, worst case – black lines.

4.3.2.4 Application to intrinsic FET model extraction

The accuracy of the intrinsic MLE is verified using Monte Carlo simulations in the same way as for the parasitic MLE above. The resulting S-parameters for an active bias point are shown in Figure 4.21. For each set of synthetic data, an intrinsic MLE is performed – resulting in a total of 1000 sets of intrinsic parameter estimates.

For the intrinsic parameters (Figure 4.22), the fit is equally good as for the parasitic elements. Although some of the distributions are a bit skewed, they can still be well approximated by a normal distribution. Interesting to note is that the two parameters R_i and R_j are associated with the largest relative error. It is well known (and verified in Section 4.2) that these two parameters are hard to estimate accurately; see Figure 4.11. With the proposed MLE method, the amount of uncertainty is calculated directly and thus it is easy to identify model parameters that have a weak influence on the S-parameters.

Figure 4.23 shows scatter plots for the intrinsic parameters. For the intrinsic parameter estimates, the two most correlated parameter pairs are τ–R_i and C_{gd}–C_{gs}. These correlations are predicted by the MLE.

4.3.3 Comparison between MLE and the direct extraction method

A head-to-head comparison between the MLE and the direct extraction method [6] is made. Histograms of both methods are shown in Figure 4.24. For all model parameters,

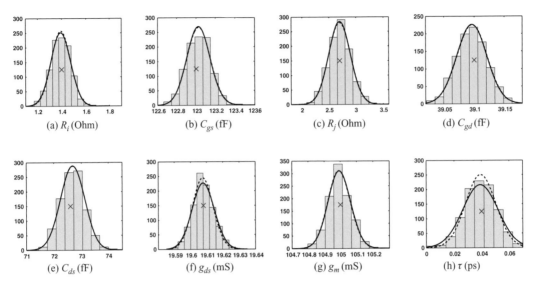

Figure 4.22 Histograms for the intrinsic parameter estimates. Solid lines are distributions predicted from the MLE. Dashed lines are normal distributions fitted to the histograms. Crosses indicate the true value. The vertical axis is the number of counts (out of 1000) in each box.

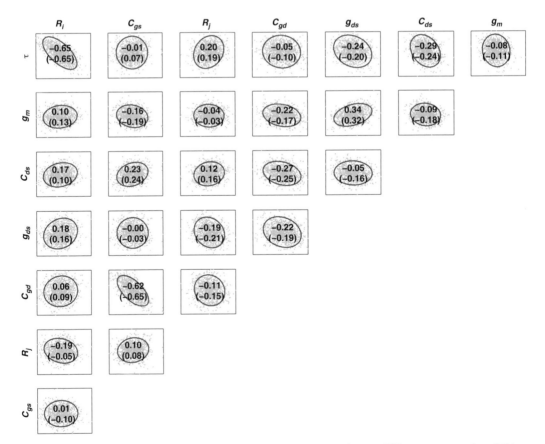

Figure 4.23 Pairwise scatter plots of intrinsic parameter estimates. Ellipses corresponds to 95% confidence regions. The number indicates the estimated correlation factor, the number within parenthesis is the true correlation factor.

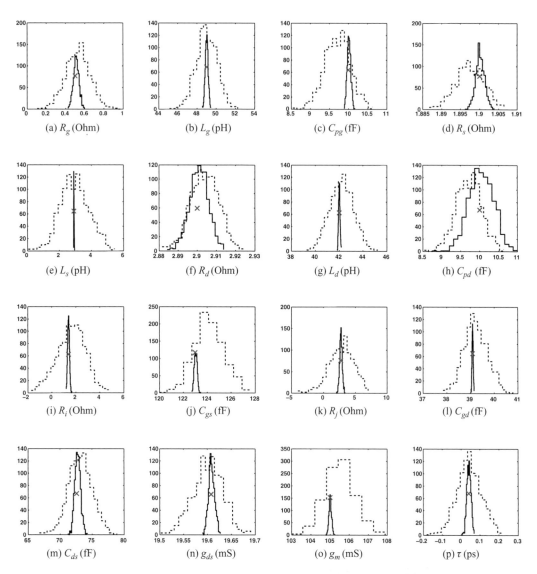

Figure 4.24 Comparison of estimates using the MLE and the direct method. The vertical axis is the number of counts (out of 1000). Dashed lines – direct method, solid lines – MLE method. Crosses indicate the true value.

the MLE is more accurate compared to the direct method and in the direct method there is also some bias in the estimates. For the two parasitic parameters R_d and C_{pd}, both methods perform equally well though. Interesting to note is that for the parameters R_i and R_j, the direct method sometimes ends up with negative values, whereas the MLE method does not.

Clearly, the modeling uncertainties are significantly reduced when using the MLE method described compared to the direct extraction method conventionally used. One

Table 4.1 Comparison of f_T and f_{max} using MLE and direct methods. The level of confidence is 95%.

	f_T [GHz]	f_{max} [GHz]
true value	103	107
MLE method	103 ± 0.2	107 ± 0.6
direct method	103 ± 1	107 ± 9

example where the modeling accuracy is very critical is in the estimation of figures of merit derived from the model parameters themselves. Using the estimated parameter covariance matrix, confidence intervals for figures of merit can be obtained. Thus, it would be possible for a process engineer to assess whether or not certain processing treatments significantly affect the maximum frequency of oscillation. As an example, the intrinsic f_T and f_{max} were therefore calculated for each extracted parameter set above. Table 4.1 shows estimated means and calculated confidence intervals. For f_{max}, the uncertainty when using the direct extraction method is almost 10%, whereas with the MLE, it is less than 1%.

4.3.4   Application of MLE in RF-CMOS de-embedding

Accurate measurements are a prerequisite for any microwave design and modeling work, irrespective of the transistor technology used. For integrated microwave circuit design, these measurements are usually performed on-wafer using coplanar probes connecting the instrumentation to the actual device under test. At DC this normally poses no problem other than de-embedding resistive losses in cabling and interconnects. For high-frequency characterization, i.e., S-parameters, noise parameters, and load-pull measurements, however, several issues can be identified. In particular for RF-CMOS applications, there are a number of issues that make it very challenging to find accurate measurement data at the transistor reference plane. These issues can be divided into four different parts; see also Figure 4.25.

1. The pad capacitance. The capacitance between the silicon substrate and the pad will tend to mask the device performance. This is especially true for transistors operating above 100 GHz where C_{pad} can be a factor of 10 times higher than the intrinsic capacitances, C_{gs} and C_{ds}.
2. The contact resistance. As the contact pads are made of aluminum, a hard layer of AlO_2 can be formed on top of the pads, which will drastically increase the contact resistance.
3. The interconnect parasitic components. These are mainly inductive and resistive contributions from the wiring between the pad and the actual device, but also include fringing capacitances.
4. The difference in dielectric constant. As the calibration substrate and the substrate of the device under test usually differ, the reference plane will move away from the probe tip [22].

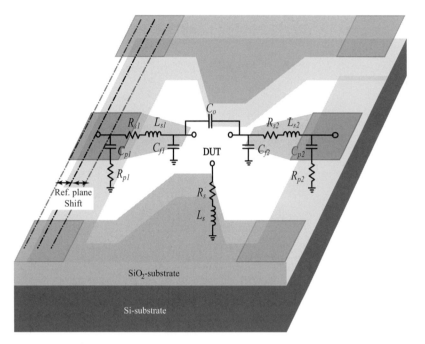

Figure 4.25 Layout of transistor test structure and a typical package network. Not shown are the contact resistances between the probe and the pads.

Thus, the effects from the embedding structure and the difference in effective dielectric constants must be dealt with by proper de-embedding and/or calibration procedures so that accurate intrinsic data can be acquired. De-embedding is in this context, therefore, closely related to the de-embedding of transistor parasitics within the device as discussed in Section 4.3.2.

4.3.4.1 Method description

The most common de-embedding methods used for RF-CMOS measurements, 'open-short' [23] or 'improved three step' [24], are performed as a post-measurement process. These methods use a set of dedicated standards, beside the device test structure, to find and de-embed a fixed network describing the electrical environment. These methods, however, have several disadvantages and a new statistical equivalent circuit-based approach based on the statistical MLE methods presented in Section 4.3.1 has therefore been developed [25].

The de-embedding method in reference [25] uses an equivalent circuit model, *the package model*, to describe the embedding electrical environment; see Figure 4.25. The MLE algorithm described in Section 4.3.1 is then used with measurement data of several de-embedding standards to estimate the package model parameters with minimum uncertainty. The procedure is very similar to the one presented for the FET parasitic extraction in Section 4.3.2 replacing the multiple cold-FET measurements with

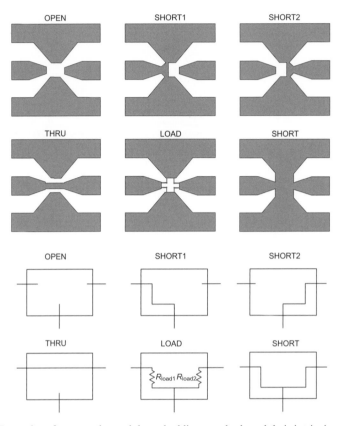

Figure 4.26 Examples of commonly used de-embedding standards and their intrinsic models.

measurements of various on-wafer de-embedding standards each being represented by a simple intrinsic model as shown in Figure 4.26.

As discussed further in Section 4.4, a too-simple package model cannot represent the behavior of the package, as expected, while a too-complex model including many parameters leads to very uncertain parameter values, and so-called overfitting. The model should thus be chosen carefully to be complex enough to capture the behavior of the package but not too complex as this will lead to very uncertain parameter estimates.

The uncertainties in the package model parameters will also depend on the number of standards used. Few standards result in large package parameter uncertainties and a simplified model must be used, while a large number of diverse standards allows more complex package models to be identified. There is, hence, a tradeoff between the number of standards used (and thus the wafer area consumed) and the maximum complexity and accuracy of the package model.

4.3.4.2   Example using 130 nm RF-CMOS measurements

Measured data from transistor and de-embedding structures on 130 nm CMOS process has been used to study the effects of the package in a practical case. The package model

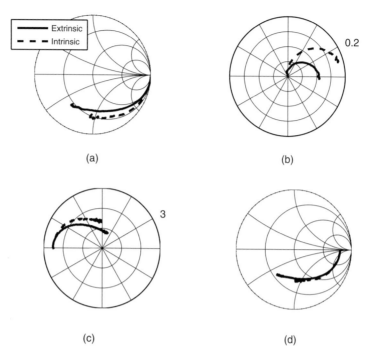

Figure 4.27 Measured extrinsic and de-embedded intrinsic data for 130 nm CMOS (40 MHz to 50 GHz, active bias point) using the package model in Figure 4.25.

shown in Figure 4.25 has been used and its parameters have been estimated using the MLE method using measurements of the on-wafer standards in Figure 4.26. The resulting *S*-parameter data before and after de-embedding the package is shown in Figure 4.27. Clearly, the package parasitics have a large effect on the data obtained, and accurate estimation of the package parameters is essential.

Figure 4.28 shows the uncertainties in some of the estimated package parameters when different combinations of on-wafer standards are included in the estimation process. Clearly, the uncertainties in the package parameters improve when more standards are used.

4.3.4.3   Comparison between different de-embedding methods

The statistical equivalent circuit-based de-embedding method is difficult to compare directly to other methods using measured data of two reasons: (a) the "true" intrinsic data is not known, (b) the de-embedding methods require different standards.

In order to compare the methods, synthetic data has therefore been used. Synthetic "measurement" data has been created for both an embedded transistor and a number of different standards and allows the true result to be known and thus the de-embedding error to be calculated. As can be seen from Figure 4.29 and Figure 4.30, the open-short method, which is something of an industry standard, is correct up to about 30 GHz, whereas the improved three step is fairly accurate all the way up to 100 GHz. The statistical MLE-based equivalent circuit method is the most accurate method.

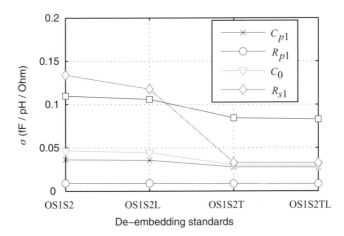

Figure 4.28 Standard deviations of package model parameter estimates for different combinations of de-embedding standards. Measurement data from 130 nm CMOS devices was used.

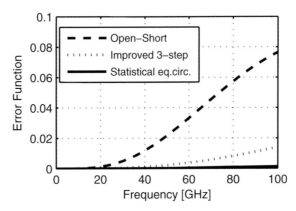

Figure 4.29 Residual sum-of-squares error as function of frequency for different de-embedding methods using synthetic data.

4.3.5   Discussion

The MLE-based model extraction method described in this section provides both accurate parameter estimates and their covariance. The estimated covariance can also be of direct use. Compared to the direct extraction methods discussed in the preceding section, the optimizer-based MLE method is general in nature and may, with little modification, be used to estimate model parameters for spiral inductors or any other passive elements.

The flexibility of the MLE method was also demonstrated with application to an on-wafer de-embedding problem. Accurate de-embedding of the device interconnects is of prime importance in such applications, in particular for high-frequency RF-CMOS applications where the effects of the parasitics often dominate over the intrinsic device

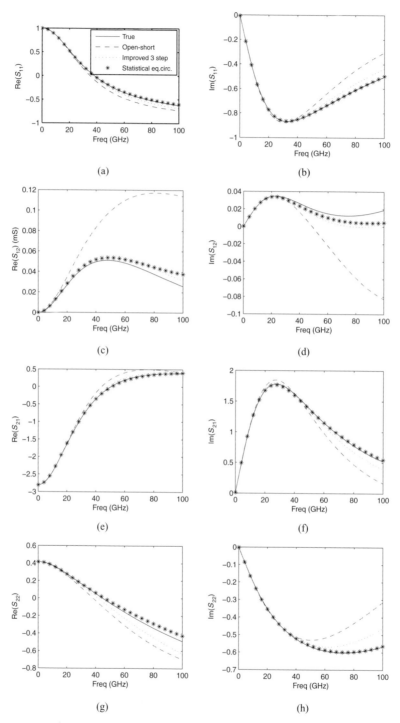

Figure 4.30 De-embedding using the MLE method [25], open-short [23], and improved three step [24] based on synthetic/simulated data.

characteristics. The importance is expected to increase, since these device technologies now begin to be used in the millimeter wave frequency range.

The MLE-based model extraction method presented here is also ideally suited to the study of model topologies. For instance, the influence of R_i and R_j may be negligible for some devices. This will be manifested by a large variance. The elements R_i and R_j may then be removed and the influence on the remaining estimates may be checked. The MLE-based extraction method will therefore play a key role when we, in the following section, introduce a formalized method for finding an optimum model topology.

4.4   Complexity versus uncertainty in equivalent circuit modeling

The statistical methods presented in this chapter allow the parameters of a given small-signal device model to be estimated with high accuracy and even give an estimation of the uncertainty in the parameter values obtained. It is, in practice, however, not easy to know which model topology to consider. A variety of models with varying complexities and topologies are, for example, available for modeling of FETs and bipolar transistors, and it is normally not easy to know which one to choose that gives the best performance and actually describes the device physics best.

It will be shown in this section that the results of the statistical model parameter estimations presented in this chapter provide a foundation that can be used to carefully evaluate different models against each other and thereby identify the optimum model topology that gives the best tradeoff between complexity and accuracy from a given set of measurements.

4.4.1   Finding an optimum model topology

The quality of the models will be evaluated by their ability to reproduce the measured S-parameters and the mean-square error (MSE) metric will be used to quantify the model agreement. Special care will then be taken as to the fact that the estimated model parameters are indeed uncertain and are therefore stochastic variables rather than fixed numbers. Having this statistical approach means that the expression for the MSE will be given by:

$$\mathrm{MSE}(\hat{\mathbf{x}}) = \mathrm{E}\left[|S_{mod}(\hat{\mathbf{x}}) - S_{meas}|^2\right]$$

$$= \underbrace{\mathrm{var}\left[S_{mod}(\hat{\mathbf{x}})\right]}_{\text{variance}} + \underbrace{|\mathrm{E}\left[S_{mod}(\hat{\mathbf{x}}) - S_{meas}\right]|^2}_{\text{(bias)}^2}. \tag{4.40}$$

The first part of the MSE expression above corresponds to the contribution from the model parameter uncertainties and will dominate if the parameter uncertainties are very large. This is typically the case if the model complexity is too high for given measurement uncertainties. The second term denotes the model bias. This term dominates if the model is too simple or if the model topology is not correct. Hence, with the use of the statistical MSE expression in (4.40), it is possible to compare models of different complexities and topologies and find the one that gives the lowest MSE value, thus providing the

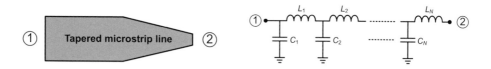

(a) Tapered transmission line DUT.   (b) Lumped element model.

Figure 4.31 DUT: tapered transmission line. Ideal S-parameter data available for 45 MHz $< f <$ 110 GHz.

optimum complexity versus accuracy tradeoff. This concept of *bias-variance tradeoff* is well established in a wide range of statistical modeling applications [26, Chapter 13].

4.4.2 An illustrative example

The use of the statistical MSE metric for finding an optimal complexity/accuracy compromise in a microwave application will be illustrated by a simple example where measurements on a tapered transmission line will be approximated by a lumped element equivalent circuit model; see Figure 4.31.

Small-signal S-parameter simulations are used as a basis for the example. Similarly to the direct extraction example in Section 4.2.1, normal distributed noise with known covariance matrix, $\mathbf{C_w}$, will be added to represent measurement noise.

The lumped element model used to approximate the tapered line DUT is shown in Figure 4.31(b). The model consists of multiple L–C sections that are used to approximate the transmission line in the DUT. As indicated in the figure, it is not clear how many sections, N, of the model are needed, though. Clearly, adding more sections in the model will allow the tapered line S-parameters to be better approximated, but it also becomes more and more difficult to estimate all the L and C parameters accurately. Hence, according to the discussion above, the objective is to find the optimum number of model sections and the corresponding model parameters that minimize the MSE for the model.

4.4.2.1 MSE estimation procedure

In order to calculate the MSE in (4.40) it is necessary to estimate the model parameters, $\hat{\mathbf{x}}$ and their uncertainties, $\mathbf{C_{\hat{x}}}$ for all models being evaluated. Depending on the model topology, either the stochastic direct extraction method in Section 4.2 or the optimizer-based method in Section 4.3 can be used. For the example presented here, the MLE method will be used.

First, the model bias term in (4.40) is estimated from the measurements and the model parameters using the following expression:

$$(\text{bias})^2 \approx \sum |S_{mod}(\hat{\mathbf{x}}) - S_{ideal}|^2, \tag{4.41}$$

where the summation should extend over all four S-parameters and frequency points in the measurement data. Note that measured S-parameters can be used to replace the ideal S-parameters, S_{ideal}, in practical applications where the true value is unknown.

The second step is to estimate the model variance in (4.40), i.e., $\mathrm{var}\,[S_{mod}(\hat{\mathbf{x}})]$. This means to transform the covariance in the estimated model parameters to corresponding variance of the modeled S-parameters. This transformation is nonlinear, and special care needs to be taken to get accurate variance estimation. We have used a Monte Carlo-based approach here, but other numerically efficient methods could also be used, e.g., the unscented transform method [27].

The Monte Carlo method is based on K realizations of the model parameters according to their uncertainties. Each realization is denoted, $\mathbf{x}_k$, and for the usual situation where the uncertainties may be considered normal distributed, we have $\mathbf{x}_k \sim N(\hat{\mathbf{x}}, \mathbf{C}_{\hat{\mathbf{x}}})$. The corresponding modeled S-parameters are denoted $S_{mod}(\mathbf{x}_k)$. The MSE variance contribution may then be approximated by:

$$\mathrm{var}\,[S_{mod}(\mathbf{x})] \approx \frac{1}{K-1} \sum_{k=1}^{K} \left| S_{mod}(\hat{\mathbf{x}}_k) - \overline{S_{mod}} \right|^2, \tag{4.42}$$

where

$$\overline{S_{mod}} = \frac{1}{K} \sum_{k=1}^{K} S_{mod}(\mathbf{x}_k). \tag{4.43}$$

The total MSE for each model is then easily found by adding the results in (4.41) and (4.42) according to (4.40).

4.4.2.2   Results

The procedure above has been applied for the modeling example presented above. Initially, it is assumed that measurements are available in the 45 MHz $< f <$ 110 GHz frequency range. Measurement noise has been generated and added to the idealized simulation data using the uncertainty model presented in Section 4.1.2. Models with $N = 1$ to $N = 10$ sections (see Figure 4.31b) have been estimated using the MLE method in Section 4.3, which also returns the uncertainties of the model parameters obtained.

The results of the MSE calculation are shown in Figure 4.32. The figure shows that the minimum MSE occurs at the $N = 5$ model order, corresponding to a model with five elements. For smaller complexities, i.e., simpler models, the bias term dominates, while for more complex models with more parameters, the model variance starts to dominate. The optimum model and its parameter estimates are shown in Figure 4.33.

The model in Figure 4.33 was found optimum for the measurements available. However, it is also interesting to investigate the MSE performance and the optimum model when a smaller frequency range is used. The same procedure as above is therefore repeated when truncating the upper frequency limit to 50 GHz and 20 GHz, respectively. The MSE results obtained are shown in Figure 4.34.

The results shown in Figure 4.34 and Figure 4.32 clearly illustrate that the optimum model complexity depends on the frequency range of the measurements and the model with three sections is actually optimum when the measurements only cover up to 20 GHz. It is also interesting to note that the minimum MSE value decreases at lower frequencies.

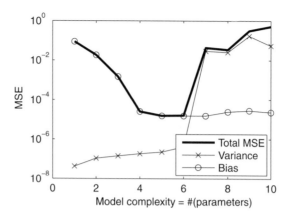

Figure 4.32 Estimated MSE versus model complexity.

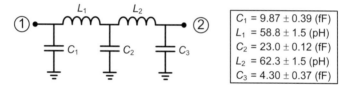

| $C_1 = 9.87 \pm 0.39$ (fF) |
| $L_1 = 58.8 \pm 1.5$ (pH) |
| $C_2 = 23.0 \pm 0.12$ (fF) |
| $L_2 = 62.3 \pm 1.5$ (pH) |
| $C_3 = 4.30 \pm 0.37$ (fF) |

Figure 4.33 Optimum model complexity and estimated parameter values with their 95% confidence intervals.

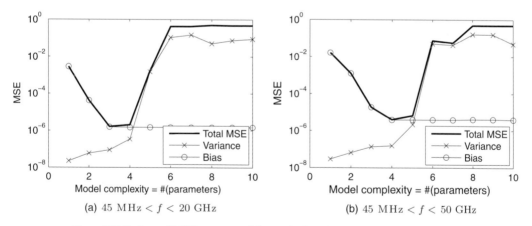

(a) $45\ \mathrm{MHz} < f < 20\ \mathrm{GHz}$ (b) $45\ \mathrm{MHz} < f < 50\ \mathrm{GHz}$

Figure 4.34 Estimated MSE versus model complexity when using different measurement frequency ranges.

This indicates that the overall accuracy of the model improves despite the fact that a simpler model topology can be used.

For the idealized example presented above, it may seem obvious that a simpler model can be used if the upper frequency limit is reduced. However, for more complicated models, such as the ones used in transistor modeling, there is no way to assess intuitively which topology to choose. The statistical framework presented here and in references

[9, 21, 25], establishes a formal means of quantifying which model is actually better. It can therefore be used as a general tool for computer-aided development of models for existing and new devices of all kinds.

4.5   Summary and discussion

This chapter has introduced stochastic methods suitable for use in equivalent circuit modeling of microwave devices. These methods extend the commonly used direct extraction and optimizer-based extraction methods by incorporating knowledge of the fact that microwave measurements are, indeed, associated with uncertainties. A variety of typical microwave device modeling examples have been included to illustrate that the statistical methods also result in improved modeling accuracy compared to the methods typically used.

A major difference compared to the traditional methods is, however, that these statistical model extraction techniques offer an inherent possibility to quantify the uncertainty in the models obtained. This opens up new insights into microwave device modeling where these are no longer considered as ideal, and a variety of new applications emerge.

One typical example of where the modeling uncertainty is important is where the parameters of the models obtained are used to interpret the physical properties of a device. Knowing the uncertainties means that it will now be possible to assess whether variations observed are due to real changes of the device properties, or if they are only caused by stochastic variations originating from measurement uncertainties. Furthermore, many commercial semiconductor foundries offer statistical models allowing the MMIC designers to perform Monte Carlo simulations to evaluate the influence of stochastic processing variations on their circuit performances. Similarly, using the uncertainty information automatically obtained from the stochastic modeling methods, it would be possible to take the modeling uncertainties to be considered at the design stage using the same simulation framework.

Finally, it was also demonstrated that the uncertainties can be very useful when comparing different models. Most often the model topology is not known, and the model designer has to evaluate models with different topologies and complexities against each other. From engineering experience, it is well known that very complex models will fit the measurements better, but at the cost of increased parameter uncertainties, while too simplified models may not be sufficiently accurate. The formal stochastic framework presented for the evaluation of models allow the user easily to find the optimum topology that gives the most favorable tradeoff between highest accuracy and minimum complexity.

References

[1] S. A. Maas, *Nonlinear Microwave Circuits*. Norwood, MA: Artech House, 1988.

[2] R. Anholt, R. Worley, and R. Neidhard, "Statistical analysis of GaAs MESFET S-parameter equivalent-circuit models," *Int. J. Microw. Millimeter-Wave Comput. Aided Eng.*, vol. 1, pp. 263–270, Mar., 1991.

[3] "8510C Network Analyzer Data Sheet," Agilent Technologies, Tech. Rep., 1999.

[4] D. Williams, J. Wang, and U. Arz, "An optimal vector-network-analyzer calibration algorithm," *IEEE Trans. Microw. Theory Techn.*, vol. 51, no. 12, pp. 2391–2401, 2003.

[5] StatistiCAL VNA calibration software package. [Online]. http://www.nist.gov/eeel/electromagnetics/related-software.cfm

[6] G. Dambrine, A. Cappy, F. Heliodore, and E. Playez, "A new method for determining the FET small-signal equivalent circuit," *IEEE Trans. Microw. Theory Tech.*, vol. 36, pp. 1151–1159, July, 1988.

[7] J. Bandler, Q. Zhang, R. Biernacki, O. Inc, and O. Dundas, "A unified theory for frequency-domain simulation and sensitivity analysis of linear and nonlinear circuits," *IEEE Trans. Microw. Theory Tech.*, vol. 36, no. 12, pp. 1661–1669, 1988.

[8] J. A. Rice, *Mathematical Statistics and Data Analysis*, 2nd ed. Belmont: Duxbury Press, 1993.

[9] C. Fager, P. Linnér, and J. Pedro, "Optimal parameter extraction and uncertainty estimation in intrinsic FET small-signal models," *IEEE Trans. Microw. Theory Tech.*, vol. 50, pp. 2797–2803, Dec., 2002.

[10] S. M. Kay, *Fundamentals of Statistical Signal Processing: Estimation Theory*. Englewood Cliffs, NJ, Prentice-Hall, 1993.

[11] D. Schreurs, H. Hussain, H. Taher, and B. Nauwelaers, "Influence of RF measurement uncertainties on model uncertainties: practical case of a SiGe HBT," in *64th ARFTG Microw. Measurements Conf.*, 2004, pp. 33–39.

[12] S. Masood, T. Johansen, J. Vidkjaer, and V. Krozer, "Uncertainty estimation in SiGe HBT small-signal modeling," in *Eur. Gallium Arsenide and Other Semiconductor Applicat. Symp. (GAAS)*, 2005, pp. 393–396.

[13] M. Rudolph, R. Doerner, and P. Heymann, "Direct extraction of HBT equivalent-circuit elements," *IEEE Trans. Microw. Theory Tech.*, vol. 47, no. 1, pp. 82–84, 1999.

[14] N. Rorsman, M. Garcia, C. Karlsson, and H. Zirath, "Accurate small-signal modeling of HFET's for millimeter-wave applications," *IEEE Trans. Microw. Theory Tech.*, vol. 44, pp. 432–437, Mar., 1996.

[15] M. Berroth and R. Bosch, "Broad-band determination of the FET small-signal equivalent circuit," *IEEE Trans. Microw. Theory Tech.*, vol. 38, pp. 891–895, July, 1990.

[16] H. Kondoh, "An accurate FET modelling from measured S-parameters," in *IEEE – MTT-S Int. Microw. Symp. Dig.*, 1986, pp. 377–380.

[17] A. D. Patterson, V. F. Fusco, J. J. McKeown, and J. A. C. Stewart, "A systematic optimization strategy for microwave device modelling," *IEEE Trans. Microw. Theory Tech.*, vol. 41, no. 3, p. 395, 1993.

[18] C. van Niekerk, P. Meyer, D. P. Schreurs, and P. B. Winson, "A robust integrated multibias parameter-extraction method for MESFET and HEMT models," *IEEE Trans. Microw. Theory Tech.*, vol. 48, no. 5, pp. 777–786, 2000.

[19] C. van Niekerk, J. A. Du Preez, and D. M.-P. Schreurs, "A new hybrid multibias analytical/decomposition-based FET parameter extraction algorithm with intelligent bias point selection," *IEEE Trans. Microw. Theory Tech.*, vol. 51, no. 3, pp. 893–902, 2003.

[20] J. Vlach and K. Singhal, *Computer Methods for Circuit Analysis and Design*. Van Nostrand, 1994.

[21] K. Andersson, C. Fager, P. Linnér, and H. Zirath, "Statistical estimation of small signal FET model parameters and their covariance," in *IEEE MTT-S Int. Microw. Symp.*, Fort Worth, USA, 2004, pp. 695–698.

[22] D. F. Williams and R. B. Marks, "Compensation for substrate permittivity in probe-tip calibration," in *44th ARFTG Conf. Dig.*, vol. 26, 1994, pp. 20–30.

[23] M. C. A. M. Koolen, J. A. M. Geelen, and M. P. J. G. Versleijen, "An improved de-embedding technique for on-wafer high-frequency characterization," in *Bipolar Circuits and Tech. Meeting, Proc.*, 1991, pp. 188–191.

[24] E. P. Vandamme, D. P. Schreurs, and G. Van Dinther, "Improved three-step de-embedding method to accurately account for the influence of pad parasitics in silicon on-wafer RF test-structures," *IEEE Trans. Electron Devices*, vol. 48, no. 4, pp. 737–42, 2001.

[25] M. Ferndahl, C. Fager, K. Andersson, L. Linner, H.-O. Vickes, and H. Zirath, "A general statistical equivalent-circuit-based de-embedding procedure for high-frequency measurements," *IEEE Trans. Microw. Theory Tech.*, vol. 56, pp. 2692–2700, Dec., 2008.

[26] J. Spall, *Introduction to Stochastic Search and Optimization: Estimation, Simulation, and Control*. John Wiley & Sons, 2003.

[27] S. Julier and J. Uhlmann, "The scaled unscented transformation," *Proc. American Control Conf.* vol. 6, Citeseer, 2002, pp. 4555–4559.

5 The large-signal model: theoretical foundations, practical considerations, and recent trends

David E. Root, Jianjun Xu, Jason Horn, and Masaya Iwamoto

Agilent Technologies, Santa Rosa, California, USA

5.1 Introduction

This chapter presents a survey of selected theoretical foundations of large-signal device modeling for nonlinear circuit simulation. Topics covered include conditions for well-defined nonlinear constitutive relations, nonlinear charge modeling including a comprehensive discussion of terminal charge conservation, and also diffusion charge, transit time, and capacitance cancelation modeling in III–V HBTs. Practical considerations are presented for regularizing poorly defined constitutive relations, constructing and using nonlinear table-based models, and extrapolating measurement-based models for robust convergence. Recent advances in nonlinear measurement instrumentation, specifically the commercial availability of the nonlinear vector network analyzer (NVNA), and the growing sophistication of artificial neural networks for device modeling, are simultaneously exploited to develop an advanced electrothermal and trap-dependent III–V FET model constructed directly from large-signal data.

5.2 The equivalent circuit

5.2.1 Intrinsic and extrinsic elements

The separation of a circuit-level transistor model into intrinsic and extrinsic parts is an idealization that simplifies the treatment of an otherwise very complicated device. Equivalent circuits of a simple quasi-static III–V FET model [1] and a modern III–V HBT model [2–4] are shown in Figure 5.1 and Figure 5.2, respectively.

Conceptually, the intrinsic model describes the dominant nonlinearities of the transistor that occur in the active region, inside the feed networks, manifolds, and other parasitic particularities of the layout. For FETs, the intrinsic model includes that part of the active drain-source channel controlled by the gate and modulated by gate–source and drain–source voltages. The channel current from drain to source and charge storage between the gate and channel are the dominant phenomena, represented in the intrinsic model by nonlinear circuit elements, I_{DS}, Q_{GS}, and Q_{GD}, respectively, within the dashed box of Figure 5.1. Other elements can be added to account for gate leakage and breakdown in reverse bias and forward gate conduction at large forward bias conditions.

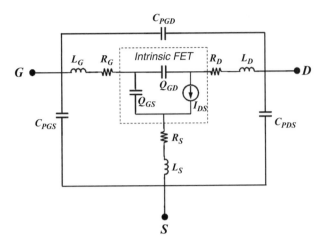

Figure 5.1 Simple FET nonlinear equivalent circuit.

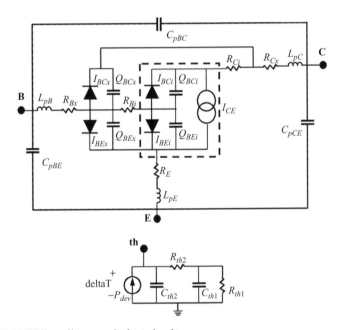

Figure 5.2 III–V HBT nonlinear equivalent circuit.

In FET models, parasitic elements are usually modeled with simple circuit elements whose parameter values do not change with bias. The parasitic elements that make up the extrinsic model are usually associated with capacitive coupling between the electrodes, and inductance and resistance of feed structures and manifold metallization, dependent on the device layout [5, 6]. Even the FET access resistances – those parts of the semiconducting channel outside the control of the gate – are usually modeled by simple resistors with fixed values, R_S and R_D, independent of the current or the voltages applied.

For III-V HBTs, the intrinsic model includes the major nonlinearities of the base–emitter and base–collector junctions. The intrinsic model incorporates the effects of the basic semiconductor physics, including junction diode currents, nonlinear charge transport characteristics, and heterojunction effects depending on the epitaxial layer structure of the semiconductor device. These effects are represented in the intrinsic model by elements for the base–emitter current I_{BE_i}, base–collector current I_{BC_i} collector–emitter current I_{CE}, base–emitter charge Q_{BE_i}, and base–collector charge Q_{BC_i} respectively, inside the dashed box of Figure 5.2.

For bipolar models, some parasitic elements are modeled by simple nonlinear elements, such as diode currents (e.g., I_{BC_x}), and depletion capacitances outside the intrinsic area of the device (e.g., Q_{BC_x}), as indicated in Figure 5.2.

Procedurally, the intrinsic model can also be defined as that part of the model that remains after the parasitic elements have been identified and removed. Failure to account properly for parasitics can therefore negatively impact the modeling of the intrinsic device. Both perspectives on the intrinsic model should be consistent for a reliable and useful overall model.

5.2.2 The intrinsic nonlinear model: dynamics, constitutive relations, and parameter values

From the FET equivalent circuit of Figure 5.1, we have the intrinsic model dynamic equations given by (5.1) and (5.2):

$$I_G(t) = \frac{dQ_{GS}(V_{GS}(t))}{dt} + \frac{dQ_{GD}(V_{GD}(t))}{dt} \tag{5.1}$$

$$I_D(t) = I_{DS}(V_{GS}(t), V_{DS}(t)) - \frac{dQ_{GD}(V_{GD}(t))}{dt}. \tag{5.2}$$

Next is the specification of the constitutive relations, namely the functional form of the (generally nonlinear) current–voltage and charge–voltage relations. For example, a classical model due to Curtice and Ettenberg [1] defines constitutive relations by equations (5.3) – (5.5).

$$I_{DS}(V_1, V_2) = \left(A_0 + A_1 V_1 + A_2 V_1^2 + A_3 V_1^3\right) \tanh(\gamma V_2) \tag{5.3}$$

$$Q_{GS}(V) = -\frac{C_{GS0}\phi}{\eta + 1}\left(1 - \frac{V}{\phi}\right)^{\eta+1} \tag{5.4}$$

$$Q_{GD}(V) = C_{GD0}V. \tag{5.5}$$

To simulate with the model, numerical values must be specified for each of the model parameters in all of the constitutive relations (5.3)–(5.5), typically by relating them to measurements on a reference device. This is the goal of parameter extraction. Obviously, the specific form of the intrinsic model constitutive relations can influence the parameter extraction strategy, the methodology by which the parameters are assigned numerical values. For the above model, $\{A_n\}$ define the gate–voltage dependence of the channel current, γ the saturation of the I–V curves with drain bias, and $C_{GS0}, C_{GD0}, \phi, \eta$ are parameters determining the charge variation of the two capacitors as functions of their respective controlling voltages. Equation (5.4) is a standard "physically based" bias-dependent junction model, while (5.5) describes a capacitance with a fixed value independent of bias. Any modification of these constitutive relations may mean a corresponding modification of the extraction process.

5.2.3   Electrothermal models

The intrinsic electrical elements of the HBT model of Figure 5.2 are specified by their constitutive relations, namely how the current or charge associated with a particular branch element depends on the controlling voltages at the intrinsic terminals and, in this case, also the junction temperature, T_j. The source element, P_{dev}, in the thermal equivalent circuit depends on the dissipated electrical power from the currents and voltages in the electrical equivalent circuit. Mathematically, this mutual interaction is represented by coupled ordinary differential equations for the time evolution of the electrical and thermal variables of the overall circuit. The simulator solves these equations simultaneously and self-consistently [2–4, 7, 8].

One can consider the thermal subcircuit of Figure 5.2 (the smaller of the two equivalent circuits in the figure) to be part of the intrinsic electrothermal model. However, if some of the thermal resistances and heat capacitors are meant to include parts of the thermal environment outside the active semiconductor device, it can be useful to consider those elements as parasitics. It is therefore critical to "thermally de-embed" when extracting model parameters. That is, the thermal environment of the device must be carefully accounted for when it is characterized for modeling. When the model is used in a simulation, it is important to re-embed the intrinsic portion of the thermal subcircuit in the thermal environment in which the device will be used, which might be quite different from that in which the device was characterized [7].

5.2.4   Scaling with frequency and geometry

The frequency dependence of the overall model depends on the types of circuit elements, their topological arrangement in the equivalent circuit, and the correct parameter values for both the intrinsic element constitutive relations and the parasitic elements. The reactive elements provide explicit time-dependence according to whether they are capacitors, inductors, or some type of distributed component. Series versus parallel combinations of resistive and reactive elements define the overall frequency dependence of the model. A good separation of the intrinsic and extrinsic model, therefore, is important for accurate broadband frequency performance of the model. An important corollary is that a well-formulated and well-extracted model with the appropriate equivalent circuit topology can simulate accurately at frequencies well beyond the upper frequency of the data used in the parameter extraction process. Such a model can also enable good scaling capabilities with geometrical variations of the layout. The ability of the model to extrapolate accurately beyond measured data is much less true of the bias dependence of the intrinsic constitutive relations, to which we now turn.

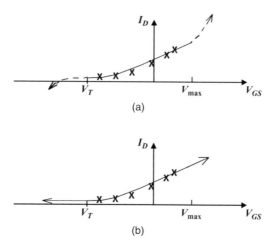

Figure 5.3 Problems created by naive parameter extraction methods for poorly defined constitutive relations.

5.3   Nonlinear model constitutive relations

5.3.1   Good parameter extraction requires proper constitutive relations

Even a simple intrinsic model constitutive relation, like that of equation (5.3), can be extracted improperly leading to disastrous results. The straightforward way to extract parameters in (5.3) is to measure I_{DS} versus V_1 (essentially V_{GS}), say at fixed V_2 (V_{DS}) for which the tanh(.) term is effectively unity, and solve for the $\{A_n\}$ coefficients using a robust least-squares fitting process. Negative consequences of this direct approach appear when the constitutive relation (5.3) is evaluated outside the range of bias used to extract the coefficients. The resulting channel current may take physically unreasonable values for reasonable values of V_1. This is illustrated in Figure 5.3 [9–11]. The polynomial model may never pinch off (even if the device does) in some cases, or become negative and therefore unphysical in other cases.[1]

5.3.2   Properties of well-defined constitutive relations

The root cause of these problems is the formulation of the model constitutive relation themselves. Care must be used to define the domain of voltages where equation (5.3) applies, and then appropriately extend this domain to define the constitutive relationship properly for all values of V_1, while imposing reasonable conditions on the global I–V relationship. Model constitutive relations must have certain mathematical properties for a robust device model. They must be well defined for all voltages, even at values far outside the range over which a real device might operate in any application. The first-order partial derivatives of the constitutive relations must be continuous everywhere. Usually, the second partial derivatives should be bounded as well. These conditions are

---

[1] The channel current should be positive for positive V_{DS}, assuming that the gate current is modeled by other elements.

imposed mainly by the underlying Newton-type algorithms used by the simulator to converge to a solution of the circuit equations.

Even for solutions within a subdomain of intrinsic terminal voltages where the constitutive relations are well behaved, the process of iterating to convergence may require the evaluation of the constitutive relations at values of the controlling variables far outside this region. If a singularity in the function evaluation of the constitutive relation is encountered, or if the derivatives point in the wrong direction, the simulator could be led astray for the next iteration, and convergence may fail altogether. It may also happen that the simulation converges to a nonphysical solution that happens to satisfy the circuit equations.

Further constraints on constitutive relations are induced by accuracy requirements for certain types of simulation. Accuracy requirements for distortion simulation, such as IM3 or IM5 at low-signal amplitudes, impose higher order continuity constraints on the model constitutive relations. That is, constitutive relations should have nonvanishing partial derivatives of sufficiently high orders. Note that the constitutive relations defined by (5.3) do not satisfy this requirement, since all fourth-order and higher partial derivatives are identically zero.

5.3.3   Regularizing poorly defined constitutive relations: an example

The solution to making (5.3) well defined can be obtained by enforcing additional constraints on the model – namely, that it pinches off and attains a maximum value – to make it physically reasonable everywhere. Specifically, the model channel current should be constrained to be zero for all values of voltages at or below a value of $V_1 = V_T$, a threshold voltage. The condition on continuous partial derivatives then requires the polynomial to have a double root at $V_1 = V_T$ so that its derivative with respect to V_1 also vanishes there. We can also assert that there is a value, $V_1 = V_{max}$, where the current attains its maximum value and is constant for all higher values of V_1. These conditions enable us to reformulate (5.3) in terms of the three new parameters V_T, V_{max}, I_{max} as given in (5.6).

$$I_D(V_1, V_2) =$$

$$\begin{cases} 0 & V_1 < V_T \\ I_{max} \frac{(V_1 - V_T)^2}{(V_{max} - V_T)^3} (V_{max} - V_T + 2(V_{max} - V_1)) \tanh(\gamma V_2) & V_T \leq V_1 \leq V_{max} \\ I_{max} & V_1 > V_{max}. \end{cases} \quad (5.6)$$

Equation (5.6) satisfies all the constraints required for a well-defined and reasonable constitutive relation [9–11],[2] provided that only $V_T < V_{max}$ and I_{max} and γ are positive. Moreover, the new parameters now have a clear interpretation in terms of minimum and maximum values of the device response. (The relationship of V_T, V_{max}, I_{max} to the original polynomial coefficients can be obtained by expanding (5.6) and collecting

---

[2] This discussion neglects the general consideration of adding some residual very small positive conductance as a further aid to convergence.

powers of V_1.) However, this strategy of parameter extraction has used up all of the original fitting degrees of freedom in equation (5.3). Distortion figures of merit, such as IM3, are completely determined by the three parameters, V_T, V_{max}, I_{max}. For real transistors, variations in the doping density can result in devices with identical values of V_T, V_{max}, I_{max}, but with different shapes of the I–V curves between zero and I_{max}. Therefore, more general and flexible constitutive relations than those of (5.6) are required for independently and accurately modeling intermodulation distortion.

It should be clear from the above discussion that the charge constitutive relation (5.4) also needs to be extended for values of V_{GS} approaching ϕ and beyond. This is easily accomplished by linearizing (5.4) at some fixed voltage, $V_0 < \phi$. The result is given by equation (5.7).

$$Q_{GS}(V) = \begin{cases} -\frac{C_{GS0}\phi}{\eta+1}\left(1-\frac{V}{\phi}\right)^{\eta+1} & V < V_0 \\ -\frac{C_{GS0}\phi}{\eta+1}\left(1-\frac{V_0}{\phi}\right)^{\eta+1} + C_{GS0}\cdot(V-V_0) & V \geq V_0 \end{cases} \tag{5.7}$$

5.3.4 Comment on polynomials for model constitutive relations

Polynomials are fast to evaluate – so models using polynomial constitutive relations can simulate quickly. Polynomials are linear in the coefficients, so they are efficient to extract (e.g., by least-squares or pseudo-inverse methods) without nonlinear optimization. However, polynomials diverge for very large magnitudes of their arguments. They have only finite orders of nonvanishing derivatives, so they can cause discontinuities of simulated distortion at low signal levels. Extending the domain of polynomial constitutive relations beyond the boundary over which they are used for extraction becomes much more difficult when the expressions depend on more than one variable (e.g., both V_1 and V_2), unlike the simple case discussed above where the polynomial part of (5.3) depends only on V_1. In general, therefore, polynomial constitutive relationships should be used with great care or avoided altogether, if possible.

5.3.5 Comments on optimization-based parameter extraction

For constitutive relations more complicated than (5.6), parameter extraction generally involves a simulation–optimization loop. An example of the general flow is given in Figure 5.4. However, such direct approaches can be slow. The model must be evaluated and parameters updated many times before a good result is obtained. Gradient-based optimization schemes may be sensitive to the initial parameter values or get stuck in local minima in the cost function (the error function between the desired value and actual value of the simulation with particular values of the model parameters). There are other techniques, such as simulated annealing [8] and genetic algorithms [12] that can help to find a global solution to the nonlinear optimization problem, but these techniques are usually much slower and more complex.

The parameters in (5.4) must be constrained not to take specific values during optimization where the constitutive relations might become singular (e.g., for $\eta = -1$) or unphysical (complex values). Modern parameter extraction software usually allows the

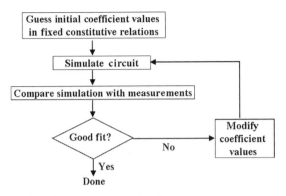

Figure 5.4 Optimization-based parameter extraction flow.

user to restrict the parameter values to specific ranges during the iterative optimization process.

Advanced nonlinear models like that of Figure 5.2 have complicated nonlinear electrical constitutive relations. Most of these electrical nonlinear constitutive relations have nonlinear thermal dependences as well. It is usually best to extract parameters of such models by using an iterative scheme where dominant electrical parameters are extracted from specific subsets of data to which those parameters exhibit high sensitivities. A good flow is necessary for good, global fits to the data and also to get physically reasonable parameter values, which can then be scaled to model devices of other sizes without the need for an additional comprehensive extraction. A general parameter extraction flow for the *Agilent HBT* model is indicated in reference [7].

Of course, no matter what parameters are used, a given model with fixed, a-priori closed-form constitutive relations may never give sufficiently accurate results. The model may just be too simple to represent the actual behavior of the device. We turn now to other approaches that are more flexible.

5.4   Table-based models

Table models are usually classified as extreme forms of empirical models, since the constitutive relations are not physically based, but depend directly on measured data. In fact, there are no fixed, a-priori model constitutive relations with parameters to be extracted at all. Table models are examples of "nonparametric" models. The *data are the constitutive relations*. The idea is simple enough – measure the I–V curves, tabulate the results, and interpolate to evaluate the constitutive relations and their derivatives as needed during simulation.

5.4.1   Nonlinear re-referencing for table-based models

The intrinsic model constitutive relations are defined on the set of intrinsic voltages, V_{GS}^{int} and V_{DS}^{int} after accounting for the voltage drop across the parasitic resistances. Measured

I–V data, on the other hand, are defined on the applied – or extrinsic – voltages.[3] The relationship between extrinsic and intrinsic voltages is simple, given the resistive parasitic element values, previously extracted, and the simple equivalent circuit topology. The equations are given in (5.8) [13]. An important issue for table models is that the extrinsic voltages at which the measurements are taken are usually defined on a grid, but the resulting intrinsic voltages, explicitly computed by substitution using (5.8), do not fall on a grid, and therefore cannot be directly tabulated. This is shown in Figure 5.5.

$$\begin{bmatrix} V_1^{int} \\ V_2^{int} \end{bmatrix} = \begin{bmatrix} V_1^{ext} \\ V_2^{ext} \end{bmatrix} - \begin{bmatrix} R_g + R_s & R_s \\ R_s & R_d + R \end{bmatrix} \cdot \begin{bmatrix} I_1^{DC} \\ I_2^{DC} \end{bmatrix}. \tag{5.8}$$

If the measured extrinsic I–V data is fit or interpolated, equation (5.8) can be interpreted as a set of implicit nonlinear equations for the extrinsic voltages, V_i^{ext}, given specified intrinsic voltages, V_i^{int} [9, 13]. Solving (5.8) in this sense enables the data to be re-gridded on the intrinsic space so that the terminal currents can be tabulated as functions of the intrinsic voltages.

Modeling the measured I–V data as functions of the intrinsic voltages reveals characteristics quite different from the model expressed in terms of extrinsic data. This is shown in Figure 5.6. In Figure 5.6a, the modeled I–V curves as functions of the applied (extrinsic) voltages V_1^{ext} and V_2^{ext} are plotted. In Figure 5.6b, intrinsic I-V modeled constitutive relations, defined on V_1^{int} and V_2^{int}, are plotted. There is a big difference between Figure 5.6a and 5.6b, especially around the knee of the curves. This process also makes clear that errors in parasitic extraction can distort the characteristics that we would otherwise attribute to the intrinsic model.

Alternatively, one can tabulate the extrinsic I–V data, as measured on the original grid, and treat the coupled equations (5.8) as additional model equations to be solved dynamically during simulation. This allows the simulator to sense the intrinsic voltages, and look up the associated interpolated values of the measured I–V curves consistent with the solution of (5.8). This saves the post-processing step of regridding during the parameter extraction, but adds the two nonlinear equations (5.8) to the model, thus increasing simulation time.

Table-based models can be both accurate and general. The same procedure and modeling infrastructure can be used to model devices in very different material systems (e.g., Si and GaAs) and manufacturing processes [13]. An example of the same table-based model applied to Si and GaAs transistors is given in Figure 5.7. For physically based models, each transistor would have very different constitutive relations requiring different parameter extraction procedures.

5.4.2 Issues with table-based models

A critical issue with table-based models is the nature of the interpolation algorithms that define the constitutive relations as differentiable functions on the continuous domain containing all the discrete data points stored in the data tables. The interpolator needs

---

[3] For the purposes of this chapter, we do not distinguish between applied and extrinsic voltages.

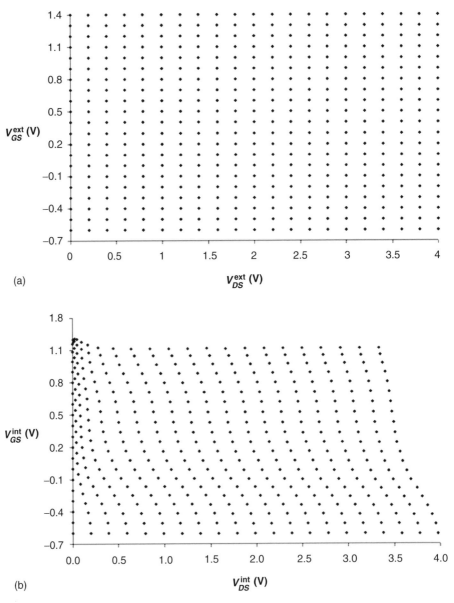

Figure 5.5 Extrinsic (gridded) and corresponding intrinsic (nongridded) voltage domain of a FET.

to define the partial derivatives continuously, and extrapolate appropriately – the same conditions that apply to any constitutive relations. At relatively small signal levels, it has been shown that simulations of harmonic distortion can become inaccurate when the amplitude (in volts) of the signal is comparable to or smaller than the distance between voltage data points at which the constitutive relations are sampled [14]. This is demonstrated in Figure 5.8. At small voltage swings associated with low power signals, simulation results depend on the mathematical properties of the interpolant between

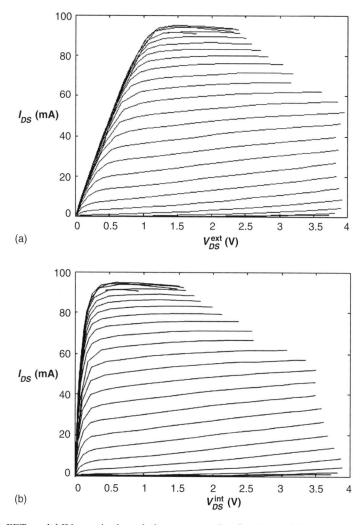

Figure 5.6 FET model IV constitutive relations expressed as functions of (a) extrinsic and (b) intrinsic voltages.

data points rather than the underlying data itself. For larger signals, the corresponding applied voltage swings average out the local characteristics of the interpolant, and the table-based model simulations become quite accurate. It can sometimes help to increase the density of the data points. For some spline schemes, however, this can make the interpolant oscillate nonphysically between data points. Ultimately, there is a practical limit on taking too many data points, leading to increased measurement time, file size, and interpolation of noise [14].

Various types of spline-based model have been explored in the literature. Methods involving B-splines have been applied in reference [15], but unphysical behavior due to spline oscillations between knots can still sometimes occur. Variation-diminishing

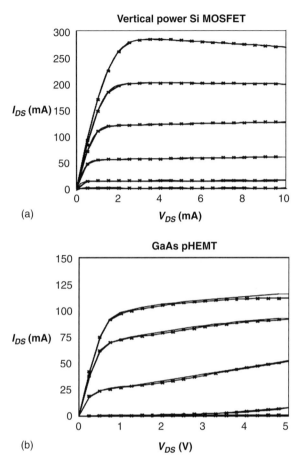

Figure 5.7 Table-based I–V models (-) and measurements (X) for Si MOSFET and GaAs pHEMT transistors.

splines [16] can tame oscillations, but their low polynomial order precludes their use for intermodulation simulation. "Smoothing splines" [17] can have a variable spline order and tradeoff accuracy for smoothing noise.

For good large-signal simulation results with table-based models, it is necessary to acquire data over the widest possible range of device-operating conditions. This range should include regions of breakdown, high power dissipation, and forward gate conduction, since these phenomena are critical to limiting the large-signal RF device performance. A portion of the domain of static measurements is shown in Figure 5.9 for a GaAs FET. The figure labels the major mechanisms that constrain the data to the interior of the boundary. The precise shape depends on the detailed device-specific characteristics and the compliance limits set on the measurement equipment [18, 19]. Covering a wide range of device operation during characterization reduces the likelihood of uncontrolled extrapolation during simulation and possible poor convergence as a consequence. Unfortunately, these extreme operating conditions can stress the device to

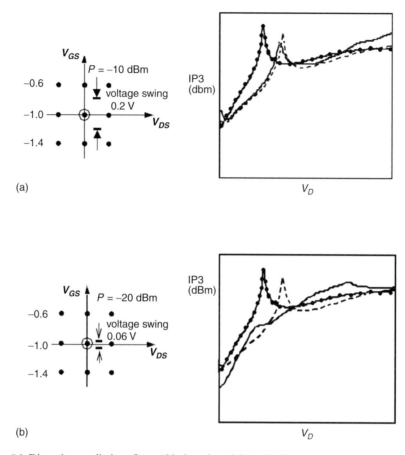

Figure 5.8 Distortion predictions from table-based models (solid lines) at (a) large and (b) small-signal amplitudes, compared to analytical model (black dots) and measured data (dashed lines).

the point of changing its characteristic during characterization [14]. This is especially true for static operating conditions at which DC I–V and S-parameters are measured. An excellent model of a degraded device can be obtained unless care is used. A delicate balance must be maintained between complete characterization and device safety. It is therefore very important to defer stressful static measurements until as late as possible in the characterization process [14].

5.5   Models based on artificial neural networks (ANNs)

Many of the problems of table models, including issues of gridding, ragged boundaries, and poor interpolation properties, can be obviated by replacing the tables with artificial neural networks (ANNs) [20–23]. An ANN is a parallel processor made up of simple, interconnected processing units, called neurons, with weighted connections that

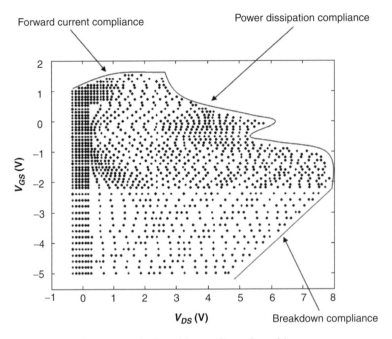

Figure 5.9 Data domain for pHEMT device with compliance boundries.

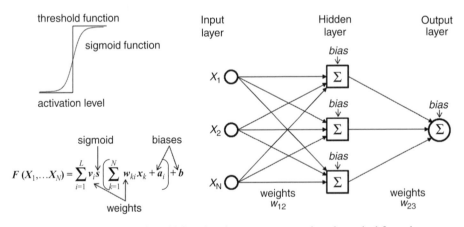

Figure 5.10 ANN illustrating sigmoid function, layer structure, and mathematical formula.

constitute the parameters [24,25]. A schematic of an ANN is given in Figure 5.10. Each neuron is a simple univariate nonlinear "sigmoid" function, with range zero to unity, monotonically increasing, and infinitely differentiable with respect to its argument. The layer structure and interconnectedness of the neurons – specified by the weights – endows the overall network with powerful mathematical properties. The universal approximation (UA) theorem states that any nonlinear function, in an arbitrary number of variables, can be approximated arbitrarily well by such a network [24].

ANNs provide a powerful and flexible way to approximate the required model multivariate constitutive relations by smooth nonlinear functions from discretely sampled scattered data. ANNs provide an alternative to using multivariate polynomials, rational functions, or other more conventional basis sets to approximate the data. There are now powerful third-party software tools [25,26] available to train the networks, that is, extract the weights and biases that form the parameters of the resulting function, such that the network approximates the nonlinear constitutive relationship well.

A key benefit of ANNs is the infinite differentiability of the resulting constitutive relations, providing a smooth approximation also for all the partial derivatives necessary for good low-level distortion simulation.[4] Another key benefit is that ANNs can be trained on scattered data. In particular, they can be trained directly on the scattered intrinsic I–V^{int} data without the need for regridding. An example of an ANN I–V constitutive relation trained on the nongridded intrinsic voltage space of a pHEMT device is presented in Figure 5.11. Hard constraints on model constitutive relations, such as required by discrete symmetry properties, can also be accommodated by ANN technology. An example of an ANN-based FET model with drain-source exchange invariance is presented in reference [21].

The mathematical form of an ANN nonlinear constitutive relation is a very complicated expression, typically involving many transcendental functions – even nested transcendental functions if there are multiple hidden layers. The expression can take many lines of mathematical symbols just to write out explicitly. However, from the point of view of the simulator, it is just a closed-form nonlinear expression like that of (5.6). The implementation of an ANN-based model in the simulator requires the values of the neural-based constitutive relations and their partial derivatives at all values of the independent variables, just like any conventional compact model. The parameters (weights and biases) can be placed in a datafile and read by the simulator for each model instance. The partial derivatives can be efficiently computed by evaluating a related neural network, called the adjoint network, obtained from the original network and weights [27].

5.6   Extrapolation of measurement-based models

Conventional parametric empirical models, when properly formulated, are well defined everywhere, even far outside the training data used to extract the parameters. As we have discussed, this can be achieved by defining the constitutive relations first in a bounded domain and then appropriately extending this domain, often by linearization.

More of a challenge is how to systematically extend the domain of table-based or ANN models. Table models based on polynomial splines can often extrapolate very poorly, causing failures in convergence of simulation. An example of a table-model extrapolation is shown in Figure 5.12a. The symbols indicate the actual data points. The

---

[4] Well-modeled partial derivatives of constitutive relations are necessary, but not sufficient, for good low-level distortion simulations. The correct dynamical model equations are also needed.

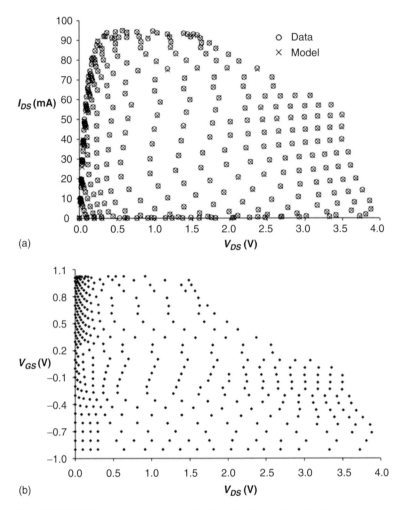

Figure 5.11 ANN FET IV model (x) and data (circles). (b) Nongridded intrinsic voltage space.

solid lines correspond to the table-based model. Within the region of actual data, the model fits extremely well. At high drain voltages, the model extrapolates and the model curves cross in a nonphysical way. Eventually, the drain current of the model becomes negative (nonphysical) and the model does not converge robustly.

ANN models do not diverge as rapidly as polynomials, but their extrapolation properties are also poor from the perspective of simulation robustness. Successful deployment of table-based or ANN models can depend on a good "guided extrapolation" to help the simulator find its way back into or near the training region should it stray far away on iteration. The method reported in reference [28] defines a compact domain containing the training region in terms of the convex hull constructed from the data points themselves [29]. Inside this region, the table or ANN is evaluated. Outside the boundary, an

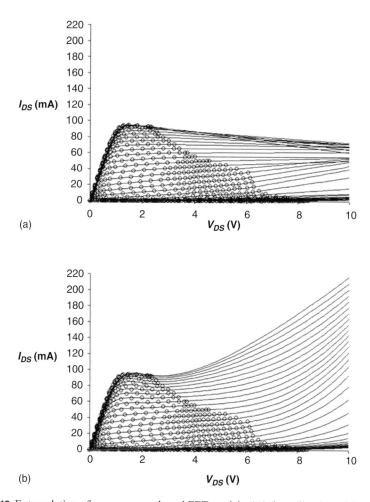

Figure 5.12 Extrapolation of measurement-based FET models, (+) data, (lines) model: (a) without guided extrapolation, (b) with guided extrapolation.

algorithm is applied to extend the current constitutive relation smoothly in a way that sharply increases the model branch conductances outside the training range. An example is shown in Figure 5.12b. This method increases the robustness of DC convergence, and also the maximum power levels at which the model converges in large-signal harmonic balance analysis. Transient analysis becomes more robust as well. Other methods of extrapolation of ANN models can be found in references [22] and [30].

5.7   Charge modeling

Nonlinear charge modeling has been shown to be critical to accurately simulating bias-dependent high-frequency S-parameters [31], intermodulation distortion and ACPR in

FETs [32, 33], and harmonic and intermodulation distortion in III–V HBTs [4, 34–36]. FET models with identical I–V relations, differing from one another only by the form of their charge models, can result in differences of 5 to 10 dB or more in simulated IM3 and differences of more than 5 dB in simulations of ACPR [32]. Moreover, conventional nonlinear charge models, based on textbook junction formulas, tend to show significant discrepancy compared to actual measured device characteristics.

The charge modeling problem can be simply stated as the specification of the nonlinear constitutive relations defining the independent terminal charges at the (intrinsic) nodes of the circuit model, as functions of the relevant independent controlling variables, usually voltages. The charge-based contribution to the current at the ith terminal is then the total time derivative of the charge function, Q_i. This is expressed in equation (5.9).

$$I_i(t) = \frac{dQ_i(V_1(t), V_2(t))}{dt}. \tag{5.9}$$

Here, V_1 and V_2 are the two independent intrinsic port voltages, which, for a FET, can be taken to be V_{GS} and V_{DS}. For an HBT, V_1 and V_2 can be taken to be V_{BE} and V_{CE}. I_1 is the gate (base) current and I_2 is the drain (collector) current for the FET (HBT) cases, respectively. Charge constitutive relations contribute to the current model through the time derivative operator in (5.9). This makes it apparent that charge plays an increasingly important role as the stimulus frequency increases.

We have seen earlier that the empirical Curtice charge model is a simple case of (5.9) with the branch elements given by (5.4) and (5.5). The model admits a representation in terms of terminal charges at the gate and drain, respectively, according to (5.10).

$$Q_G(V_{GS}(t), V_{DS}(t)) = Q_{GS}(V_{GS}(t)) + Q_{GD}(V_{GD}(t))$$
$$Q_D(V_{GS}(t), V_{DS}(t)) = -Q_{GD}(V_{GD}(t)). \tag{5.10}$$

The Shockley model [37] is a classical physically based model that also fits into the form (5.10). However, there is only one independent charge function, $Q_{GS}(V) = Q_{GD}(V) = Q_{1d}(V) = -C_0\phi\sqrt{1 - V/\varphi}$. Here, $Q_{1d}(V)$ is the standard expression for stored charge in a one-dimensional (1D) ideal constant-doped semiconductor junction [37]. The Curtice and Shockley model gate charge expressions each separate into the sum of two 1D pieces. It is a consequence of the idealized simplicity of these models that there are only two 1D functions, $Q_{GS}(V)$ and $Q_{GD}(V)$, defining the entire two-port charge model. The more general equation (5.9) admits two functions, $Q_G(V_1, V_2)$ and $Q_D(V_1, V_2)$, each dependent on two independent variables. In general, neither of these two terminal charge functions need be separable in terms of univariate functions.

5.7.1   Measurement-based approach to charge modeling

Charge cannot be measured directly like the DC I–V curves. To specify the charge model, it is most convenient to establish a relationship between the model nonlinear charge constitutive relations and simpler quantities that can be directly compared to bias-dependent S-parameter data. Mathematically, this means linearizing equations (5.9) and expressing them as admittance matrix elements. Experimentally, this means measuring S-parameters, de-embedding the effects of parasitics, and transforming to intrinsic voltages using (5.8) to arrive at the intrinsic device small-signal scattering matrix – and then transforming to admittance representation. We can use this correspondence to extract model parameters for an empirical model such as (5.4). Ultimately, we can use this correspondence to invert the problem (when possible) to construct the model constitutive relations for the Q_i functions directly from the appropriate data.

The imaginary parts of the model intrinsic admittance matrix elements are computed by taking the partial derivatives of the model charge functions with respect to their controlling voltages. Assuming a FET common source (or HBT common emitter) configuration, we obtain the following matrix equation, which relates the four nonlinear functions of bias comprising the (imaginary parts of) the model admittance matrix, to the partial derivatives of the two model nonlinear charge functions, Q_i, evaluated at the operating point. Here, indices i and j range from 1 to 2, the number of independent terminals, or ports. For devices like MOSFETs, with additional terminals (e.g., bulk), i and j range from 1 to 3. The right-hand equality defines a capacitance matrix, C_{ij}, used for notational simplicity.

$$\frac{\text{Im}\left[Y_{ij}\left(V_1, V_2, \omega\right)\right]}{\omega} = \frac{\partial Q_i(V_1, V_2)}{\partial V_j} \equiv C_{ij}(V_1, V_2) \tag{5.11}$$

The assumption that the middle and right-hand sides of (5.11) are independent of frequency is necessary for consistency with (5.9). In practice, with a good intrinsic and extrinsic equivalent circuit topology, and good parasitic extraction, (5.11) is approximately true for frequencies approaching the cutoff frequency of the device. For higher frequencies, the intrinsic model topology needs to be augmented to deal with "non quasi-static effects" [38], a topic beyond the scope of the present review.

The measured admittance parameters can be obtained by simple linear transformations of the (properly de-embedded) S-parameters according to (5.12) [39]. The measured capacitance matrix then follows by taking the imaginary part of (5.12) and dividing by angular frequency.

$$Y_{ij} = \left[(I - zS)(I + zS)^{-1}\right]_{ij}. \tag{5.12}$$

At this point, modeled and measured intrinsic admittance functions, or capacitance matrix elements, C_{ij}, can be directly compared. However, it is more customary to compare modeled and measured small-signal responses in terms of linear equivalent circuit elements. There are many different intrinsic linear equivalent circuit representations that lead to the same intrinsic capacitance matrix. Defining equivalent circuit elements therefore requires a specific choice of equivalent circuit topology. A common choice [40, 41] of linear equivalent circuit is depicted in Figure 5.13. This choice defines the linear equivalent circuit elements in terms of simple and invertible linear transformations of

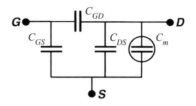

Figure 5.13 Linear equivalent circuit model of capacitive part of intrinsic FET model.

the four independent capacitance matrix elements defined by equations (5.13), where we suppress the bias dependence for convenience. When applied to the transformed *S*-parameter measurements using equation (5.12), equations (5.13) define "measured" equivalent circuit elements. When applied to the linearized model admittances, (5.13) results in "modeled" equivalent circuit elements.

$$
\begin{aligned}
C_{GS} &= C_{11} + C_{12} \\
C_{GD} &= -C_{12} \\
C_{DS} &= C_{22} + C_{12} \\
C_m &= C_{21} - C_{12}.
\end{aligned}
\tag{5.13}
$$

We note that there are four "capacitance" equivalent circuit elements defined by equation (5.13). This is not surprising, given that there are four functions, C_{ij}, corresponding to the four imaginary parts of the two-port admittance matrix. However, there are only three nodes in the equivalent circuit diagram of Figure 5.13. Historically, it was customary to place one capacitance element between each pair of nodes. This procedure neglected, entirely, the fourth element, C_m, called the transcapacitance. Its presence is very significant in the small-signal data [42]. Its existence can be roughly understood in terms of an expansion of the factor representing a time delay of the transconductance element to first order in $\omega\tau$ in a standard linear equivalent circuit [43]. This is shown in equation (5.14). For the treatment here, however, we identify the transcapacitance using equations (5.11)–(5.13).

$$
g_m e^{-j\omega\tau} \approx g_m - j\omega g_m \tau \equiv g_m + j\omega C_m.
\tag{5.14}
$$

A three-terminal (two-port) intrinsic equivalent circuit with two independent terminal charges generally leads to (at least) one transcapacitance. The equivalent circuit of Figure 5.13, or equivalently (5.13), places the transcapacitance in the device channel branch connecting drain and source, (parallel to the transconductance – not shown). It is important to note that the relationship of terminal charge partial derivatives to admittance matrices given in (5.11) is unique, and is more fundamental than the set of transformations that define linear equivalent circuit elements in equation (5.13).

Using (5.13), the measured capacitances can be compared to theory. An example of measured capacitances is given by Figure 5.14 for the elements C_{GS} and C_{GD}, and in Figure 5.15 for the measured elements C_{DS} and C_m. Several facts are immediately

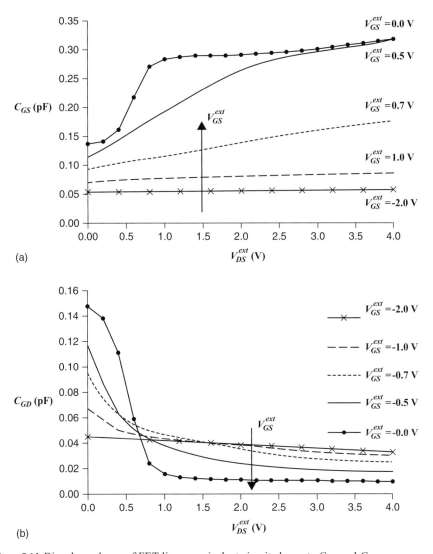

Figure 5.14 Bias-dependence of FET linear equivalent circuit elements C_{GS} and C_{GD}.

apparent from the figures. C_{GS} depends not just on V_{GS}, the voltage across the element, but also on the other independent voltage, V_{DS}. This is qualitatively different from the Shockley and Curtice models, where the model C_{GS} is completely independent of V_{DS}. This also means that C_{GS} cannot be modeled by a standard two-terminal nonlinear capacitor, no matter what the dependence on the (single) voltage across the element. The feedback capacitance, C_{GD}, depends on both V_{GS} and V_{DS}, in a more complicated way than $V_{GD} = V_{GS} - V_{DS}$. That is, C_{GD} also cannot be modeled by a standard two-terminal nonlinear capacitor, despite the familiar-looking symbol in the linear equivalent circuit. Moreover, the V_{GS} dependence of C_{GD}, for large V_{DS}, when the device is in the saturation region of operation, is exactly the opposite of

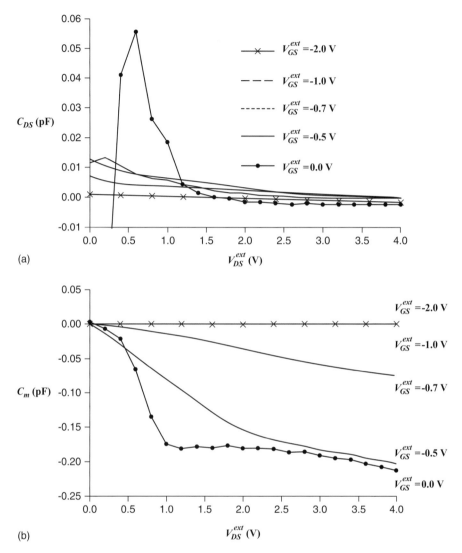

(a)

(b)

Figure 5.15 Bias-dependence of FET linear equivalent circuit elements C_{DS} and C_m.

the Shockley model's prediction. That is, the feedback capacitance is actually much larger when the device is pinched off ($V_{GS} = -1.5\,\text{V}$) than when the channel is open ($V_{GS} = 0\,\text{V}$) and conducting current. The Curtice model deals with this by assigning a constant capacitance value, independent of bias. More elaborate physical theories, which yield results closer to measured characteristics of modern FETs, lead to equations sufficiently complicated that they can be expressed, usually, only in approximate form [44]. A more recent approach, based on decomposition of the charge model into simple 1D depletion charges and two-dimensional (2D) drift charges defined in terms of voltage and current, has been proposed in reference [45] and is presented later in this chapter.

5.7.2 Constructing model nonlinear charges from small-signal data

The above development, beginning with equations (5.9), started from large-signal model equations and computed the small-signal responses that can be compared easily to measured S-parameter (Y-parameter) data. In what follows, we present a general treatment based on trying to reverse the above flow. That is, we seek to solve the inverse problem to determine the functional form of the large-signal model constitutive relations, Q_i, that enter (5.9), directly from the measured bias-dependence of the small-signal characteristics. Unfortunately, but not surprisingly, this inverse process is generally ill-posed. However, under certain specific and verifiable conditions, a practical inverse modeling process can be constructed for transistors manufactured in a variety of material systems. This enables the generation of accurate nonlinear circuit simulation models for devices of great practical utility, from simple DC and *linear* (S-parameter) measurements. This means that accurate and predictive nonlinear circuit design is possible long before detailed, efficient, first-principles physical compact models are available.

Equation (5.11), with the left-hand side referring to measured data, can be interpreted as the mathematical statement of the inverse problem. That is, the *measured* bias-dependences of the intrinsic admittance elements are equated to the partial derivatives of the respective *model* terminal charges. The mathematical problem then becomes to determine the conditions under which equations (5.11) can be solved for the model terminal charge functions, Q_i.

The necessary and sufficient conditions for the terminal charges to be recovered from bias-dependent capacitance matrix elements defined from measured data using equations (5.13) are succinctly expressed by equation (5.15) [13, 18, 46].

$$\frac{\partial C_{ij}(V_1, V_2)}{\partial V_k} = \frac{\partial C_{ik}(V_1, V_2)}{\partial V_i}. \tag{5.15}$$

Through definitions (5.11), equations (5.15) are constraints on pairs of bias-dependent measured admittances, one pair per row of the admittance matrix labeled by the index i. If equation (5.15) is satisfied, then the terminal charges can be constructed directly from the measured capacitance matrix elements by a path-independent contour (line) integration expressed by (5.16).

$$Q_i = \int_{contour} C_{i1}dV_1 + C_{i2}dV_2. \tag{5.16}$$

This result is unique up to an arbitrary constant that has no observable consequences and so can be set equal to zero. Moreover, the partial derivatives of this charge reduce exactly to the measured bias-dependent capacitance measurements. That is,

$$\frac{\partial Q_i^{\text{model}}}{\partial V_j} = C_{ij}^{meas}. \tag{5.17}$$

If (5.15) is not exactly satisfied, then, strictly speaking, there is no function, Q_i^{model}, that is consistent with equation (5.17). In this case, the line integral using measured capacitance functions in equation (5.16) produces charge functions that depend on the path chosen for the contour. Different contours produce models that fit some capacitances versus bias better than others, with no perfect fits of all capacitances possible.

5.7.3   Terminal charge conservation

Equations (5.15), expressing the equality of mixed partial derivatives with respect to voltages of different capacitance functions with the same first index, can be interpreted as meaning that pairs of capacitance functions attached to the ith node, form a conservative vector field in voltage space [18, 31, 47]. Capacitance functions that obey (5.15) are said to obey "terminal charge conservation at the ith node."[5] We use the nomenclature "terminal charge conservation" to distinguish it from the physical law of charge conservation that is embodied in circuit theory by Kirchoff's Current Law (KCL). Terminal charge conservation is a constraint that can, but need not, be imposed by the modeler to approximate the behavior of a device. Physical charge conservation is a fundamental physical law and a requirement of any circuit model that is consistent with KCL. An example of a nonlinear model not based on terminal charge conservation, and its consequences, is presented below.

Any *model* starting from equation (5.9) has model capacitance functions that conserve terminal charge at each node. This is true because the model capacitances are derived in (5.11) starting from model charges and then (5.15) follows from the derivative properties of smooth functions. However, starting from independent *measurements* and trying to go back to *model* charges via (5.17) requires the constraints (5.15) to be satisfied by the measured C_{ij}^{meas} data.

The degree to which actual bias-dependent admittance data is consistent with the modeling principle of terminal charge conservation was investigated in references [31, 47]. For III–V FETs it was found to hold extremely well at the gate, and slightly less well at the drain. Its applicability to III–V HBTs will be discussed later.

5.7.4   Practical considerations for nonlinear charge modeling

The parameterization of line-integral (5.16) for two distinct paths, shown in Figure 5.16, are written explicitly in (5.18) and (5.19), respectively. Path independence means that the same charge function can be computed from completely independent sets of bias-dependent data along the two paths of Figure 5.16.

$$Q_G(V_g, V_d) = \int_{V_{g0}}^{V_g} C_{11}(\bar{V}_1, V_{d0}) d\bar{V}_1 + \int_{V_{d0}}^{V_d} C_{12}(V_g, \bar{V}_d) d\bar{V}_d \qquad (5.18)$$

$$Q_G(V_g, V_d) = \int_{V_{g0}}^{V_g} C_{11}(\bar{V}_g, V_d) d\bar{V}_g + \int_{V_{d0}}^{V_d} C_{12}(V_{g0}, \bar{V}_d) d\bar{V}_d. \qquad (5.19)$$

---

[5] Earlier treatments did not use the "terminal" prefix and denoted this concept by "conservation of charge" [48].

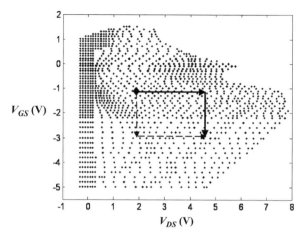

Figure 5.16 Two different paths in voltage space for line-integral calculations of terminal charges.

There are several issues with respect to implementing (5.18) and (5.19) directly on measured data. The measured capacitance data is defined only at the discrete voltages (points of Figure 5.16), so the integrals have to be done numerically. If the data is not on a rectangular grid, interpolation along some of the paths may be required (as in Figure 5.16 along the V_{GS} direction). Fundamentally, if equation (5.15) is not exactly satisfied due to measurement errors or the neglect of effects such as temperature and traps (to be considered later), the different paths effectively tradeoff model fidelity of $\mathrm{Im}Y_{12}$ and $\mathrm{Im}Y_{11}$ versus bias, respectively. Integration error accumulates for large paths, so the charge value at points far away from the starting point of integration (V_{G0} and V_{D0}) will be less accurate. The integration is also difficult to perform along the ragged nonorthogonal boundary of the data domain (see Figure 5.16).

Despite these practical difficulties, table-based models using full 2D nonlinear gate charge functions constructed directly from small-signal device data have found their way into practical commercial tools [49]. Table-based charge models can be much more accurate than closed-form empirical models, where the complex 2D nature of the Q–V constitutive relations have not received as much attention as have I–V relations, which are directly measurable.

5.7.5 Charge functions from adjoint ANN training

There are now robust methodologies to train ANNs to construct Q–V constitutive relations directly from knowledge of the desired function's partial derivatives as represented by measured capacitances [27]. This has been a major breakthrough for the practical measurement-based charge modeling of transistors.

All the practical problems described above of computing multidimensional charge functions by line integration of suitably decomposed small-signal data are ameliorated by using the adjoint ANN training approach [27]. This method directly results in a neural network that represents the $Q_i(V_{GS}, V_{DS})$ functions from information only about their

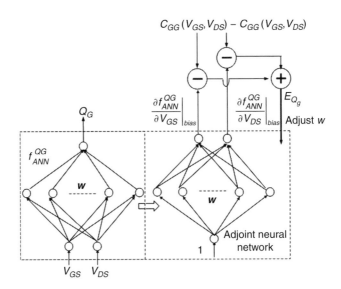

Figure 5.17 Adjoint ANN training of model gate terminal charge function from C_{11} and C_{12} data.

partial derivatives, as represented by the bias-dependent measured capacitances defined by (5.17). A diagram of the training method is given in Figure 5.17. If (5.15) is not exactly satisfied by the device data, the adjoint method still returns a charge function that generally gives a much better global compromise between the capacitances than the typical line-integral methods. The training can take place directly on the intrinsic non-gridded intrinsic bias data, and the ragged boundary presents no difficulty. Validation of the adjoint ANN approach to simultaneously fit the detailed 2D FET input capacitance behavior with bias is shown in Figure 5.18. The validation for the independent fit for the drain capacitances is shown in Figure 5.19.

With current-voltage and charge-voltage nonlinear constitutive relations modeled by ANNs, the improvement in simulation accuracy over spline-based table models can be demonstrated. A comparison is shown for the case of a GaAs pHEMT device in Figure 5.20. At moderate to high power levels, where the voltage swings are comparable or greater than the distance between discrete data points, the ANN and table-based models are nearly identical and compare well with measurements. At low power levels, the distortion simulation of the table-based model is determined by the numerical properties of the interpolating functions. Piecewise-cubic splines, used in this case, do not do a good job for high-order distortion at low signal levels, hence the ragged variation of distortion with power. However, the ANN model is very well behaved at all power levels, and has the correct asymptotic dependence as the power decreases.

5.7.6　Transcapacitances and energy conservation

The transcapacitance, C_m, clearly shows up in device small-signal data as evidenced in Figure 5.15. See also reference [42]. However, attempts to calculate stored charge

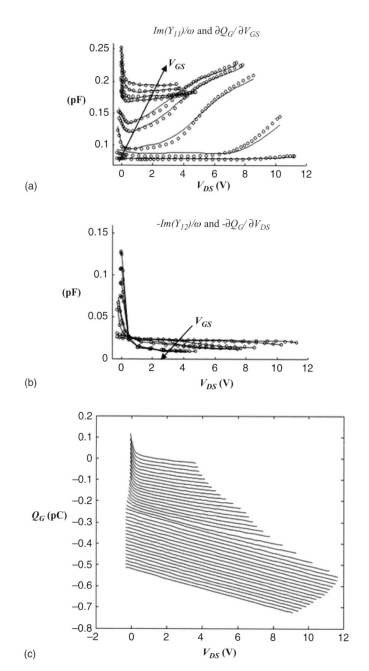

Figure 5.18 Validation of ANN-based gate charge model for fitting, (a) C_{11}, (b) C_{12}. Data (symbols), model (lines); (c) gate charge function, Q_G.

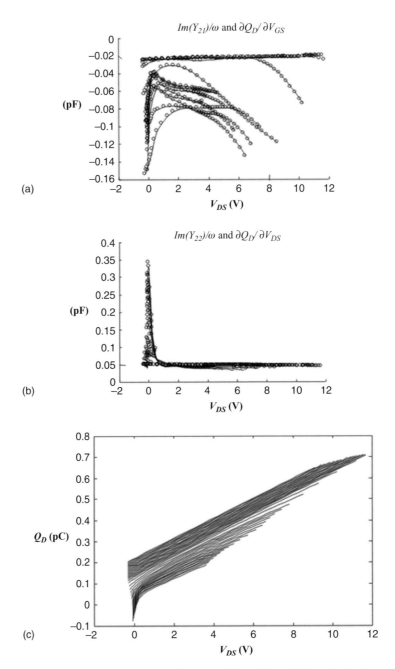

Figure 5.19 Validation of ANN-based gate charge model for fitting, (a) C_{21}, (b) C_{22}. Data (symbols), model (lines); (c) drain charge function, Q_D.

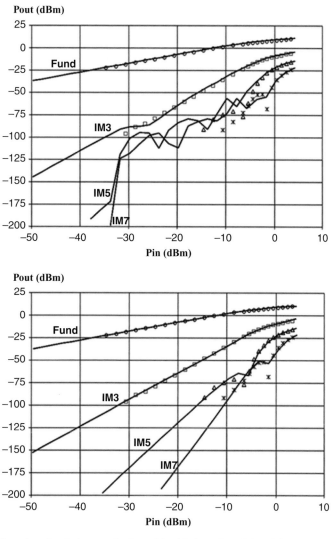

Figure 5.20 Distortion simulation results from (a) table-based model and (b) ANN-based model constructed from the same DC and small-signal data. Measured nonlinear validation data (symbols) and model predictions (lines).

from simple theoretical conditions including energy conservation lead to the conclusion that $C_m = 0$ [13, 50, 51], inconsistent with the small-signal data. The magnitude of the channel current can increase with signal frequency in models that have large-signal terminal charges that admit transcapacitances (this is the general case unless the mixed partial derivatives of the terminal charge functions are equal). This can cause an anomalously large simulated gain at high frequencies. Fortunately, the parasitic network limits the rate of intrinsic voltage variation to a maximum frequency determined by the input resistance and input capacitance product, limiting the undesirable consequences.

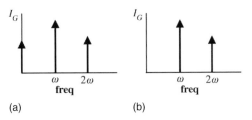

Figure 5.21 Spectra generated by (a) nongate-charge conserving capacitance models, and (b) gate-charge conserving models.

5.7.7   Capacitance-based nonlinear models and their consequences

Strictly speaking, if (5.17) is not exactly satisfied, the assumption that (5.9) models the noncurrent-source terms in the intrinsic device is not consistent with the data. An alternative is to write the time-dependent node current directly in terms of the measured two-port capacitance matrix elements [13]. That is, one can propose to replace (5.9) with more general equations according to (5.20).

$$I_i(t) = C_{i1}(V_1(t), V_2(t))\frac{dV_1(t)}{dt} + C_{i2}(V_1(t), V_2(t))\frac{dV_2(t)}{dt}. \tag{5.20}$$

In (5.20) the model functions C_{ij} can be completely independent of one another, without the constraints of (5.15). It is quite easy to implement models like (5.20) for nonterminal-charge-conserving capacitance-based large-signal equations in nonlinear circuit simulators. Equivalently, using the definitions (5.13), it is possible to rewrite (5.20) in terms of contributions from the four equivalent circuit elements of Figure 5.13 [13]. As compared to equation (5.9), which is specified by two nonlinear functions, equation (5.20) is defined by four model nonlinear functions (for a two-port device), which, if desired, can be taken to be precisely the measured (or independently fit) relations from equation (5.13). Such a model will exactly fit the measured bias-dependent small-signal dependence, by construction. However, as proved in references [10, 52, 53], such models will generally lead to spectra containing a DC component, proportional to the stimulus frequency and the square of the signal amplitude, generated from such capacitance elements when stimulated by large signals at the device terminals [9, 13, 46, 54]. The spectra of nonterminal charge conserving capacitance models and terminal charge conserving models are shown in Figures 5.21a and b, respectively. A spectrum with a DC component cannot happen for modeling the true displacement current, as is the physical origin of gate current (neglecting leakage) in reverse biased FETs, owing to modulated stored charge. Things are less clear in the channel of a FET, where current arises by a combination of transport and time-varying electric field [55]. Nevertheless, enforcing terminal charge conservation on large-signal intrinsic models results in a simpler model, with no "strange" consequences in large-signal analysis. Finally, it is possible to model the gate current using equation (5.9) (for $i = 1$) and the drain current using (5.20) (for $i = 2$). That is, terminal charge conservation can be enforced at the gate terminal but not at the drain terminal, if desired.

5.8   Terminal charge conservation, delay, and transit time for HBT models

5.8.1   Measurement-based HBT models

FETs are naturally described in terms of their controlling voltages. For bipolar devices, such as GaAs or InP based HBTs, it is more common to specify a mixed voltage-current set of independent variables, such as V_{BC} and I_C [2,34,56]. From the measurement-based perspective, it is desirable to deduce the relationship of terminal charges to these mixed current and voltage independent variables. This is done starting from a simplified three-node intrinsic topology, like the FET case, making the correspondence between gate and base nodes, drain and collector nodes, and source and emitter nodes, respectively. The computation of the common emitter admittance matrix requires application of the chain rule to convert the charge equations, defined in terms of voltages in equation (5.13) into the mixed representation.

$$
\frac{Im(Y_{int})}{\omega} = \begin{bmatrix} \frac{\partial Q_B}{\partial V_{BE}} & \frac{\partial Q_B}{\partial V_{CE}} \\ \frac{\partial Q_C}{\partial V_{BE}} & \frac{\partial Q_C}{\partial V_{CE}} \end{bmatrix}
$$

$$
= \begin{bmatrix} \frac{\partial Q_B}{\partial V_{BC}} + \frac{\partial Q_B}{\partial I_C} g_m & -\frac{\partial Q_B}{\partial V_{BC}} + \frac{\partial Q_B}{\partial I_C} g_{CE} \\ \frac{\partial Q_C}{\partial V_{BC}} + \frac{\partial Q_C}{\partial I_C} g_m & -\frac{\partial Q_C}{\partial V_{BC}} + \frac{\partial Q_C}{\partial I_C} g_{CE} \end{bmatrix} \tag{5.21}
$$

$$
= \begin{bmatrix} \tilde{C}_B + \tilde{\tau}_B \cdot g_m & -\tilde{C}_B + \tilde{\tau}_B \cdot g_{CE} \\ \tilde{C}_C + \tilde{\tau}_C \cdot g_m & -\tilde{C}_C + \tilde{\tau}_C \cdot g_{CE} \end{bmatrix}.
$$

Here, g_m and g_{CE} are the transconductance and common emitter output conductance, respectively. Solving for the time delay and capacitance functions yields the following expressions [3,56].

$$
\tilde{\tau}_i = \frac{\partial Q_i}{\partial I_C} = \frac{Im(Y_{i1} + Y_{i2})}{\omega (g_m + g_{CE})} \tag{5.22}
$$

$$
\tilde{C}_i = \frac{\partial Q_i}{\partial V_{BC}} = \frac{Im(Y_{i1})}{\omega} - \tilde{\tau}_i \cdot g_m. \tag{5.23}
$$

These equations relate, in an unambiguous way, partial derivatives of the model terminal charges to pairs of functions, each related directly and simply to specific combinations of small-signal high frequency data.

Necessary and sufficient conditions for solving equations (5.22) and (5.23) for the model charge functions are [3,56]:

$$
\frac{\partial \tilde{\tau}_i(V_{BC}, I_C)}{\partial V_{BC}} = \frac{\partial \tilde{C}_i(V_{BC}, I_C)}{\partial I_C}. \tag{5.24}
$$

Equation (5.24) is the mixed voltage-current analog of the equality of the mixed partial derivatives of FET model capacitances expressed by (5.15). That is, (5.24) is a statement of "terminal charge conservation" at the ith node. The bias-dependent small-signal data can be tested to see the degree to which this modeling constraint is satisfied, as done for FETs in reference [47]. The above considerations lead to a prescription for computing the effective model nonlinear charges from knowledge of the voltage and current-dependent transit times and capacitances. Provided that the constraints (5.24) are satisfied, the terminal charges can be constructed by path-independent line integrals according to (5.25).

$$Q_i = \int_{contour} \left[\tilde{C}_i dV_{BC} + \tilde{\tau}_i dI_C \right]. \tag{5.25}$$

Direct application of the adjoint neural network training algorithms apply similarly as in the FET case. That is, instead of numerically evaluating (5.25), an ANN function can be trained to compute Q_i given measured $\tilde{C}_i$ and $\tilde{\tau}_i$ as input functions.

5.8.2   Physical considerations for empirical HBT charge models

It is possible to use the basic relationships between capacitance, delay, and charge to help with the construction of a physically based HBT nonlinear charge model. The physics of high-injection effects in the collector of III–V HBTs is related to charge screening of the positive dopant ionic charge by the large electron concentration moving at the saturation velocity [57,58]. These considerations mean that the base–collector capacitance depends on the current through the collector, not just on the junction voltage V_{BC}. That is, the total charge associated with Q_{BC} must be a nonlinear function of both I_C and V_{BC}. The carrier velocity depends on the field in the collector. Velocity-field characteristics for majority carriers in III–V semiconductors are very different from those in Si devices. In particular, the nonmonotonic velocity-field characteristics of III–V transport means that the collector transit delay (related to the inverse of the velocity) has precisely the opposite dependence on V_{BC} compared to Si devices. Attempts to extract parameters of Si BJT models to fit III–V HBT data yield unphysical parameter values and inconsistent results over the device operating range. Distinct models with constitutive relations based on III–V physics are therefore required for GaAs single heterojunction and InP double heterojunction devices.

5.8.3   Delay and diffusion capacitance in physically based and empirical III–V HBT models

A flexible and powerful analytical charge model can be formulated by decomposing the base-collector charge into a physically distinct 1D current-independent depletion charge and a diffusion charge that depends both on current and voltage [56]. That is, we define the base–collector charge contribution to the total base charge according to:

$$Q_{BC}(V_{BC}, I_C) = Q_{BC}^{dep}(V_{BC}) + Q_{BC}^{dif}(V_{BC}, I_C) = \int_0^{V_{BC}} C_{BC}^{dep}(V)dV + \int_0^{I_C} \tau(V_{BC}, I)dI.$$

$$\tag{5.26}$$

The last two terms of (5.26) correspond to be a particular parameterization of (5.25) (segments of the contour taken along the line $I = 0$ and then $V = V_{BC}$) in terms of a 1D depletion capacitance, $C_{BC}^{dep}(V_{BC})$, and 2D transit delay, $\tau(V_{BC}, I_C)$. The total bias and

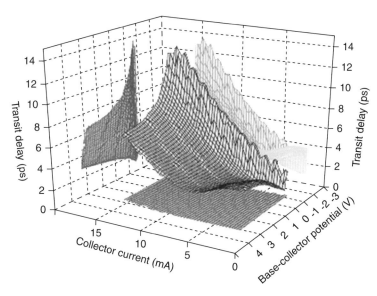

Figure 5.22 Total base-collector delay of a III–V HBT device as a function of collector current and base-collector voltage.

current dependent feedback capacitance, $C_{BC}(V_{BC}, I_C)$, is then defined by the partial derivative of (5.26) with respect to voltage evaluated at the operating point (V_{BC}, I_C).

$$C_{BC}(V_{BC}, I_C) = \left. \frac{\partial Q_{BC}}{\partial V} \right|_{V=V_{BC}; I=I_C}. \tag{5.27}$$

The current dependence of $C_{BC}(V_{BC}, I_C)$, therefore, comes entirely from the transit-time dependence on current in (5.26).

An advantage of this approach is that the 1D dependence of the depletion charge, $Q_{BC}^{dep}(V_{BC})$, can be easily formulated from physical principles or identified from simple measurements when the collector current is zero. The remaining current-dependent contribution to the charge involves integrating the time delay with respect to current. The time delay is inversely related to the velocity of the carriers in the device, which in turn depends upon the applied field. These relationships correspond to the well-known negative differential velocity-field characteristics of III–V semiconductors. Therefore, the basic physics of transport in these devices, through the relations (5.26), can be used to derive a detailed nonlinear charge model for the transistor starting from velocity-field and depletion charge considerations [2, 34, 56].

Another benefit of this approach is the identification of the diffusion charge with an *integral* of the transit time with respect to the current. This is the proper way to define diffusion charge, rather than the inconsistent (but still popular) method obtained by simply multiplying the transit time by the current. This resolved a long-standing issue with the mathematical consistency of the UCSD HBT model implementation [2, 56, 58].

The measured characteristics of the (total) base-collector delay function for a GaAs HBT is shown in Figure 5.22. It is noted that, for low current densities, the delay increases with increasing V_{CB}. This is the *opposite* behavior from what is observed in Si transistors. The shape of the device cutoff frequency, f_T, has been demonstrated to be related to the distortion characteristics of the device in large-signal conditions [35]. Since the device f_T is essentially inversely related to the transit time, the above considerations imply that unless the charge model is formulated consistently with (5.25), it is impossible to fit well, over the full range of current and voltage, the device f_T and capacitances simultaneously. This implies that a good charge model is necessary for accurate distortion simulation, just as in the FET case.

A key result that emerges from this formulation is the relationship between the current dependence of the total feedback capacitance, $C_{BC}(V_{BC}, I_C)$, and the voltage dependence of the transit delay $\tau(V_{BC}, I_C)$. Since the delay decreases with decreasing field (as V_{BC} becomes less negative, the velocity increases), equation (5.24) predicts that the capacitance C_{BC} must decrease with increasing collector current. This is the well-known "capacitance cancellation" effect [57]. An example of a commercially available III–V HBT charge model fitting both the f_T and C_{BC} bias-dependence is shown in Figure 5.23 [3, 4]. The collector voltage, V_{CE}, ranges from 0.5 V to 3 V in 0.5 V steps. Similar results were obtained for InP double heterojunction devices.

5.9   FET modeling in terms of a drift charge concept

An interesting application of the diffusion charge concept applied above for HBTs is to re-examine the FET charge model from this perspective. This approach postulates that the total FET charge model can be constructed from a current-independent depletion model of separable two-terminal nonlinear charge-based capacitors, with an additional voltage and current-dependent "drift charge" term added to the gate and drain terminals, respectively. This model is specified by (5.28). This approach, while strictly a subset of the considerations of Section 5.6, decomposes the charge model into simpler constituents involving 1D depletion-type charge storage elements plus bi-variate voltage and current-dependent drift charges (transport-related) in order to isolate the independent physical mechanisms contributing to the detailed charge storage. The drift charge in a unipolar FET device plays the role of the diffusion charge of a bipolar HBT transistor.

$$Q_G(t) = Q_{GS}(V_{GS}(t)) + Q_{GD}(V_{GD}(t)) + Q^{drift}(V_{GD}(t), I_D(t))$$
$$Q_D(t) = -Q_{GD}(V_{GD}(t)) + Q_{DS}(V_{DS}(t)) - \lambda Q^{drift}(V_{GD}(t), I_D(t)).$$

(5.28)

In (5.28), λ is a parameter that determines how much of the gate diffusion charge, Q^{drift}, is partitioned between the drain and the source nodes of the equivalent circuit. A value of $\lambda = 1$ means that all the gate diffusion charge (with a negative sign) is associated with the drain and the effect can be modeled by a branch element between gate and drain. A value of $\lambda = 0$ means that all the gate diffusion charge is associated with the source. Anything in between can be chosen to best fit the data. The interpretation of

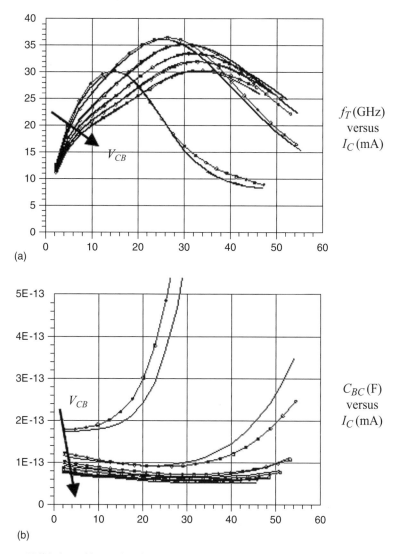

Figure 5.23 Validation of large-signal III–V HBT model f_T and C_{BC} dependence on collector current and base–collector voltage. Data (points), model (lines).

Q^{drift} with a drift charge requires that it vanish when no current flows. This is expressed by equation (5.29).

$$Q^{drift}(V_{GD}, I_D = 0) = 0. \qquad (5.29)$$

An appealing aspect of (5.28) with condition (5.29) is that the 1D charge functions Q_{GS} and Q_{GD} can be obtained from simple considerations of semiconductor depletion physics or, alternatively, from measurements via gate-bias-dependent S-parameters at $V_{DS} = 0$, where there is no drain current. Adjoint training then produces the full current and voltage-dependent drift charge function $Q^{drift}(V_{GD}, I_D)$ from the measured intrinsic admittance parameter bias-dependence after subtracting the contributions from the linearized depletion capacitances [45]. For GaAs pHEMT devices, a value of $\lambda = 1$ in (5.28) was found to fit the measured capacitance measurements accurately over all bias space. The current dependence of the drift charge can be shown to be related to a delay associated with the transport of carriers through the channel, and hence establishes a connection between the charge storage and transport phenomena. More details can be found in reference [45].

The approach of (5.28) is strictly a subset of the considerations of Section 5.6. However, the complexity of the multidimensional voltage-based formalism of Section 5.6 can obscure the fundamentally simple physical relationship between III–V carrier transport and charge storage mechanisms that is responsible for the measured characteristics of Figures 5.14 and 5.15. The approach of Section 5.9 has the potential for enabling intuitive and accurate bottom-up analytical expressions for empirical and physically based FET models, while preserving the simple principles of device operation.

5.10   Parameter extraction of compact models from large-signal data

A NVNA or large-signal network analyzer (LSNA) measures the magnitudes and relative phases of incident and scattered spectral components on each DUT port [59–61]. The device may be stimulated by one or more large- and small-signal sinusoidal inputs at one or more DUT ports simultaneously. An illustration is shown in Figure 5.24. Because the cross-frequency relative phases of the spectral components are measured, it is possible to convert the complex spectra into time-domain waveforms. Thus, the NVNA provides fully calibrated waveform measurements under high drive levels at microwave frequencies at input and output ports of the device.

Over the past 15 years, there has been significant research applying large-signal measurement systems to the field of nonlinear device modeling. Much of this work has focused on the validation of compact models under realistic large-signal operating conditions, the modifications of parameter extraction methodologies, tuning model parameter values to get better agreement with large-signal performance, and the ability to evaluate the limitations of a particular nonlinear model and suggest improvements [62]. Using NVNA data, one can get the optimal parameter set for a given compact model. Such data enables the modeling engineer to manage tradeoffs between fits to DC and S-parameters on one hand and distortion and large-signal waveforms on the other. Applications of NVNA data as an alternative to small-signal data for the computation of nonlinear model functions were addressed in references [63,64]. NVNA data extends the range of device characterization beyond that possible under DC or static operating-point conditions. This is especially important for high-power devices and when operating the device into limiting regions of operation such as breakdown. An example is given in Figure 5.25 for the case of a GaAs pHEMT [23]. The measured load-line extends well beyond the range over which DC and S-parameter data can be taken. The model responsible for the accurate prediction will be discussed now.

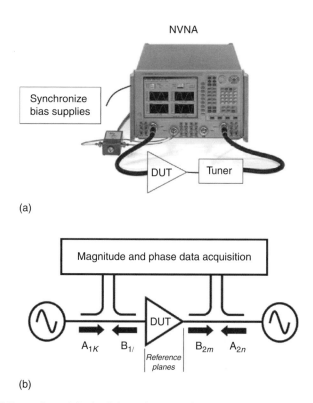

(a)

(b)

Figure 5.24 NVNA configured for load-dependent waveform measurements.

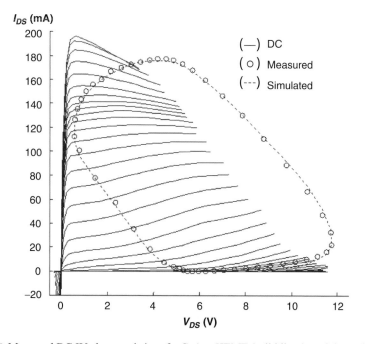

Figure 5.25 Measured DC IV characteristics of a GaAs pHEMT (solid lines), and dynamic loadline measurement (circles) and simulation (dashed line) for a reactive load condition.

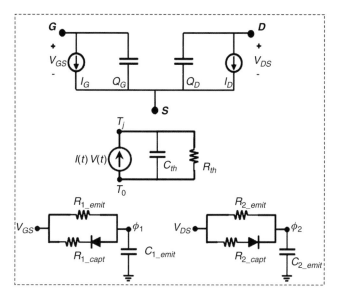

Figure 5.26 Equivalent circuit of advanced FET model with dynamic self-heating and two types of trap capture and emission process.

5.10.1   Identification of advanced FET models from large-signal NVNA data

Detailed characterization of III–V FETs in GaAs and, more recently GaN material systems, show evidence of additional dynamic effects for which the intrinsic model represented by the equivalent circuit in Figure 5.1 or equivalently, by the dynamical equations (5.1) and (5.2), are inadequate – even when dynamic self-heating effects are added (like those in the HBT model considered earlier). Frequency dispersion of small-signal behavior – differences in the equivalent circuit conductances at high frequencies compared to their values at DC – can only partially be accounted for by self-heating mechanisms. Pulsed transient phenomena like gate-lag and drain-lag, and related effects of "power slump" and knee walk-out with increasing RF power, require more complex explanations [65] and correspondingly more complicated dynamical models [23, 66].

A more complete equivalent circuit intrinsic model for III–V FETs is presented in Figure 5.26 [23]. The top subcircuit is the conventional lumped electrical topology, in common source configuration, for the currents and charges. The middle subcircuit is a simple one-pole thermal equivalent circuit that models the self-heating. The two remaining subcircuits model dynamic charge trapping and emission controlled by the gate and drain potentials, respectively [66]. The electrical constitutive relations for current and charge now depend on five state variables: the instantaneous gate and drain voltages V_{GS} and V_{DS}, the time-varying junction temperature, T_j (as in described in reference [7]), and two time-varying voltages, ϕ_1 and ϕ_2, associated with trap mechanisms [23]. More formally, we can write the dynamic equations of this intrinsic model as:

$$I_G(t) = I_G(V_{GS}(t),\ V_{DS}(t),\ T_j(t)) + \frac{d}{dt} Q_G(V_{GS}(t),\ V_{DS}(t),\ T_j(t)) \quad (5.30)$$

$$I_D(t) = I_D(V_{GS}(t),\ V_{DS}(t),\ T_j(t),\ \phi_1(t),\ \phi_2(t))$$

$$+ \frac{d}{dt} Q_D(V_{GS}(t),\ V_{DS}(t),\ T_j(t),\ \phi_1(t),\ \phi_2(t)) \quad (5.31)$$

$$\dot{T} = \frac{T_0 - T(t)}{\tau_{th}} + \frac{1}{C_{th}} \langle I(t)\ V(t) \rangle \quad (5.32)$$

$$\dot{\phi}_1 = f_1(V_{GS}(t) - \phi_1(t)) + \frac{V_{GS}(t) - \phi_1(t)}{\tau_{1\_emit}} \quad (5.33)$$

$$\dot{\phi}_2 = f_2(V_{DS}(t) - \phi_2(t)) + \frac{V_{DS}(t) - \phi_2(t)}{\tau_{2\_emit}}. \quad (5.34)$$

Equations (5.32)–(5.34) are state equations – first-order differential equations for the evolution of key dynamical (state) variables that are arguments of the electrical constitutive relations appearing in equations (5.30) and (5.31). The functions f_1 and f_2 appearing in (5.33) and (5.34), respectively, are diode-like nonlinearities that account for preferential trapping rates when the instantaneous gate (drain) voltage becomes more negative (positive) than the values of ϕ_1 and ϕ_2. The parameters τ_1 and τ_2 are characteristic emission times, typically assumed to be very long compared to the RF timescales.

The model problem becomes defining the detailed functional form of the terminal current and terminal charge functions at the gate and the drain, as functions of the many internal controlling state variables.

Rather than try to infer these complicated functional dependences from small-signal data, or pulsed I–V and pulsed S-parameter data [13, 67], more modern approaches use *large-signal* RF and microwave data, such as can be easily obtained from modern NVNAs [23, 59–61, 68, 69]. An NVNA configured with a passive tuner or using the second source as an active vector load-pull configuration system (see Figure 5.24) can measure the current and voltage waveforms and the large-signal dynamic load lines at microwave frequencies. Samples of measured load-lines are shown in Figure 5.27. This is the experimental counterpart of large-signal harmonic balance simulations.

Data is taken at several different ambient temperatures, power levels, RF frequencies, and (complex) output impedances, all at several different quiescent bias points. A great advantage of NVNA data is that the extreme regions of the device operation can be characterized with much less degradation of the transistor. This is because the instantaneous voltages only enter the high-stress regions for subnanosecond periods as the device is stimulated with signals at one GHz or higher frequency. The device also dissipates much less energy in the device at high instantaneous power regions than under DC conditions. The larger domain of device operation means that the need for the final model to extrapolate during large-signal simulation is dramatically reduced or even eliminated completely. Actual nonlinear data obtained under realistic operating conditions means that the modeling process does not have to "extrapolate" from

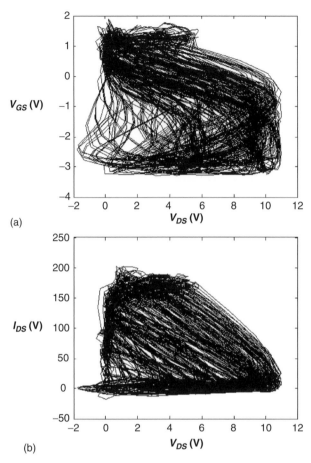

(a)

(b)

Figure 5.27 Dynamic trajectories measured as functions of load impedance on a NVNA for a GaAs pHEMT: (a) voltage space representation, and (b) drain current versus drain voltage.

linear and DC data to predict nonlinear RF behavior, as do the approaches based on small-signal data parameterized by DC bias conditions. Moreover, the NVNA data provides detailed waveforms for comprehensive nonlinear model validation without the need for additional instruments, such as spectrum analyzers, that only give the magnitude of the generated spectrum – the NVNA measures the magnitudes and the phases of the distortion products.

The modeling identification process is depicted in Figure 5.28. Details of the trap dynamics – fast capture and slow emission – mean that the trap states ϕ_1 and ϕ_2 take constant values on a given trajectory, values that can be identified with the minium V_{GS} and maximum V_{DS} values on the given contour [23]. Similarly, the junction temperature, T_j, is constant for a given load line, and can be calculated in terms of the average power dissipated along the load line and the thermal resistance extracted as in reference [7]. Prior to the availability of large-signal NVNA data, the behavior of the traps and self-heating effects were often separated by careful pulsed measurements [66, 67]. The

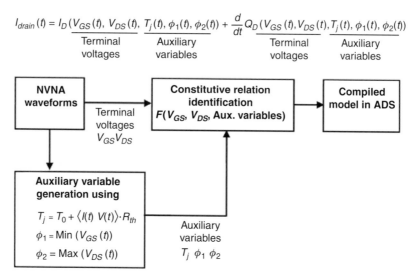

$$I_{drain}(t) = \underbrace{I_D(V_{GS}(t),\ V_{DS}(t),}_{\text{Terminal}\atop\text{voltages}}\ \underbrace{T_j(t),\ \phi_1(t),\ \phi_2(t))}_{\text{Auxiliary}\atop\text{variables}} + \frac{d}{dt}\underbrace{Q_D(V_{GS}(t),V_{DS}(t),}_{\text{Terminal}\atop\text{voltages}}\ \underbrace{T_j(t),\ \phi_1(t),\ \phi_2(t))}_{\text{Auxiliary}\atop\text{variables}}$$

Figure 5.28 Model indentification process for advance FET model from waveform measurements.

NVNA data probes the device at timescales several orders of magnitude faster than typical pulsed I–V systems. That is, NVNA data is often more indicative of the actual operating conditions of devices manufactured to operate at frequencies in the tens of GHz.

The functional form of the current and charge constitutive relations are determined using the machinery of ANNs. It would be highly impractical to determine complicated nonlinear functional dependence on five independent variables without a powerful mathematical infrastructure for approximating multivariate functions. The final model can be compiled into a conventional nonlinear circuit simulator as with any compact model. Previously, specific and simplified assumptions about how the trap state values affect the shape of the current and charge-storage characteristics had to be assumed, typically by modifying the intrinsic terminal voltages or parameters in the constitutive relations such as threshold voltage [66].

There is great insight that can be obtained by looking at the constructed constitutive relations based on large-signal steady-state waveforms. Two examples of generated intrinsic constitutive relations for different sets of trap states are shown in Figure 5.29. The model current constitutive relation corresponding to extreme trap states (Figure 5.29a) bears a striking resemblance to pulsed bias characterization from quiescent bias points associated with the trap state biases [13, 67]. The advantage of the NVNA approach is that the model characteristics are inferred from DUT responses to signals typically three or more orders of magnitude faster than what can be measured with most pulsed systems that are typically limited to 0.1 to $1\,\mu s$.

The complete model solves for the trap states, junction temperature, and currents self-consistently during simulation. When embedded back into the parasitic model, final comparison can be made to measured data. Figure 5.30 shows the validation with

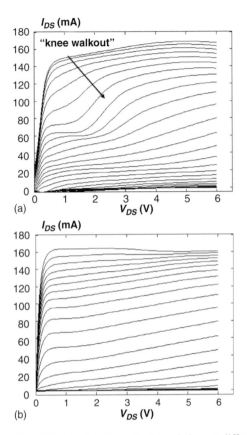

Figure 5.29 Advanced FET model intrinsic IV constitutive relations at different trap state values: (a) fixed trap states corresponding to the extremes of a large-signal trajectory; and (b) trap states following the DC bias conditions.

measured DC I–V curves. Note how different the static nonisothermal I–V curves are from the intrinsic model constitutive relations under the conditions of Figure 5.30. The dependence of the key constitutive relations on five state variables, some depending on slow dynamics such as the junction temperature and trap states, provides sufficient degrees of freedom to fit the bias dependence of the small-signal model over the entire bias space at both DC and high frequencies. That is, frequency dispersion phenomena are predicted accurately under small-signal and large-signal conditions by having a model with both dynamic trapping and electrothermal effects. Models with just electrothermal effects are not capable of such good fits to both DC and high-frequency behavior at all biases.

Figure 5.31 shows the nonlinear validation results for the advanced FET model for power-dependent gain and bias current versus power. The distinctive car-shaped gain compression characteristic and significant nonmonotonic dependence of the bias current with power is a result of the dynamics of drain-lag and the detailed constitutive

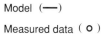

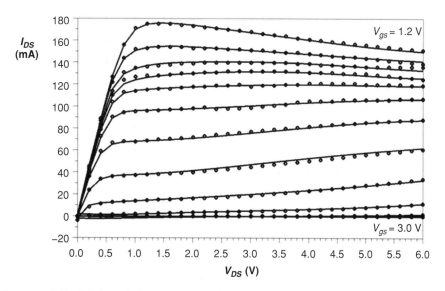

Figure 5.30 DC validation of advanced FET model.

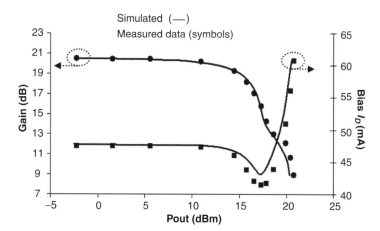

Figure 5.31 Large-signal validation of advanced FET model. Gain and drain current versus output power.

relation obtained with the ANN training. Figure 5.32 shows the model validation of distortion versus power for this device, validating both the dynamical description and accuracy and robustness of the ANN approach to modeling the complicated constitutive relations.

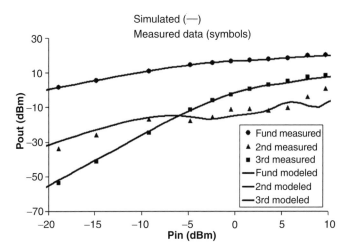

Figure 5.32 Large-signal validation of advanced FET model. Output power at the fundamental, second, and third harmonics versus input power.

5.11   Conclusions

A survey of theoretical foundations of large-signal device modeling for nonlinear circuit simulation has been presented. Requirements for well-defined nonlinear model constitutive relations have been described. Nonlinear charge modeling, including the principle of terminal charge conservation, has been reviewed in some detail, where tradeoffs and their consequences were enumerated. Applications were developed to depletion and diffusion charge modeling, and to a consistent treatment of transit delay for III–V HBTs. Several practical considerations for large-signal modeling have been described, including tradeoffs associated with constructing nonlinear charge models from bias-dependent linear data for table-based and ANN models. The importance of a good parasitic topology and good parameter extraction methods has been highlighted. Recent advances in ANN modeling techniques, and the commercial availability of powerful large-signal microwave measurement instrumentation – specifically the NVNA – are important trends that are likely to have a large positive impact on the field. Advantages of these new developments compared to earlier methodologies, specifically table-based models with splines and pulsed measurements, respectively, have been stressed throughout. In particular, we reviewed a recently developed electrothermal and trap-dependent III–V FET model constructed directly from nonlinear NVNA data using advanced ANN modeling techniques.

References

[1] W. R. Curtice and M. Ettenberg, "A nonlinear GaAs FET model for sse in the design of output circuits for power amplifiers," *IEEE Trans. Microw. Theory Tech*, vol. 33, pp. 1383–1394, Dec. 1985.

[2] Agilent heterojunction bipolar transistor model (AHBT), *Agilent Advanced Design System Manual*, nonlinear devices, ch. 2.

[3] M. Iwamoto and D. E. Root, "Large-signal III–V HBT model with improved collector transit time formulations, dynamic self-heating, and thermal coupling," *Int. Workshop on Nonlinear Microw. and Millimeter Wave Integrated Circuits (INMMIC)*, Rome, Nov. 2004.

[4] M. Iwamoto and D. E. Root, "Agilent HBT model overview," *Compact Model Council Meeting*, San Francisco, CA, Dec. 2006. Available: http://www.eigroup.org/cmc/minutes/4q06_presentations/agilent_hbt_model_overview_cmc.pdf

[5] S. Nedeljkovic, J. Gering, F. Kharabi, J. McMacken, B. Clausen, P. Partyka, and S. Parker, "Extrinsic parameter and parasitic elements in III–V HBT and HEMT modeling," *Nonlinear Transistor Parameter Extraction Techniques*, M. Rudolph, D. E. Root, C. Fager, Eds., Cambridge Univ. Press, ch. 3.

[6] J. Wood and D. E. Root, "Bias-dependent linear scalable millimeter-wave FET model," *IEEE Trans. Microw. Theory Tech.*, vol. 48, pp. 2352–2360, Dec. 2000.

[7] M. Iwamoto, J. Xu, and D. E. Root, "DC and thermal modeling for III–V FETs and HBTs," in *Nonlinear Transistor Parameter Extraction Techniques*, M. Rudolph, C. Fager, D. E. Root, Eds., Cambridge Univ. Press, ch. 2.

[8] S. Kirkpatrick, C. D. Gelett, and M. P. Vecchi, "Optimization by simulated annealing," *Sci.* vol. 220. pp. 621–680, May 1983.

[9] D. E. Root "Overview of microwave FET modeling for MMIC design, charge modeling and conservation laws, and advanced topics," *1999 Asia Pacific Microw. Conf. Workshop Short Course on Modeling and Characterization of Microw. Devices and Packages*, Singapore, Nov. 1999.

[10] D. E. Root, "Principles and procedures for successful large-signal measurement-based FET modeling for power amplifier design," Nov. 2000. Available: http://cp.literature.agilent.com/litweb/pdf/5989-9099EN.pdf

[11] S. Maas, "Fixing the Curtice FET model," *Microw. J.*, Mar. 2001.

[12] G. Antoun, M. El-Nozahi, and W. Fikry, "A hybrid genetic algorithm for MOSFET parameter extraction," *IEEE CCECE*, vol. 2, May 2003, pp. 1111–1114.

[13] D. E. Root "Measurement-based mathematical active device modeling for high frequency circuit simulation," *IEICE Trans. Electron.* vol. E82-C, pp. 924–936, June 1999.

[14] D. J. McGinty, D. E. Root, and J. Perdomo, "A production FET modeling and library generation system," in *IEEE GaAs MANTECH Conf. Tech. Dig.*, San Francisco, CA, July 1997 pp. 145–148.

[15] S. Akhtar, P. Roblin, S. Lee, X. Ding, S. Yu, J. Kasick, and J. Strahler, "RF electro-thermal modeling of LDMOSFETs for power-amplifier design," *IEEE Trans. Microw. Theory Tech.*, vol. 50, pp. 1561–1570, Jun., 2002.

[16] W. M. Coughran, W. Fichtner, and E. Grosse, "Extracting transistor charges from device simulations by gradient fitting," *IEEE Trans. Electron Devices*, vol. 8 pp. 380–394, 1989.

[17] V. Cuoco, M. P. van den Heijden, and L. C. N de Vreede, "The 'Smoothie' data base model for the correct modeling of non-linear distortion in FET devices," *IEEE Int. Microw. Symp. Dig.*, vol. 3, pp. 2149–2152, 2002.

[18] D. E. Root, S. Fan, and J. Meyer, "Technology independent non quasi-static FET models by direct construction from automatically characterized device data," *21st Eur. Microw. Conf. Proc.*, Stuttgart, Germany, pp. 927–932, Sept. 1991.

[19] *Agilent 85190A IC-CAP Manual*, nonlinear device models, vol. 2, ch. 1.

[20] J. Xu, D. Gunyan, M. Iwamoto, A. Cognata, and D. E. Root, "Measurement-based non-quasi-static large-signal FET model using artificial neural networks," *IEEE Int. Microw. Symp. Dig.*, pp. 469–472, June 2006.

[21] J. Xu, D. Gunyan, M. Iwamoto, J. Horn, A. Cognata, and D. E. Root, "Drain-source symmetric artificial neural network-based FET model with robust extrapolation beyond training data," *IEEE Int. Microw. Symp. Dig.*, June 2007.

[22] J. Wood, P. H. Aaen, D. Bridges, D. Lamey, M. Guyonnet, D. S. Chan, and N. Monsauret, "A nonlinear electro-thermal scalable model for high-power RF LDMOS transistors," *IEEE Trans. Microw. Theory Tech.*, vol. 57, pp. 282–292, Feb. 2009.

[23] J. Xu, J. Horn, M. Iwamoto, and D. E. Root, "Large-signal FET model with multiple time scale dynamics from nonlinear vector network analyzer data," *IEEE Int. Microw. Symp. Dig.*, May, 2010.

[24] S. Haykin, *Neural Networks: A Comprehensive Foundation* (2nd ed.). Prentice Hall, 1998.

[25] Q. J. Zhang and K. C. Gupta, *Neural Networks for RF and Microwave Design.* Artech House, 2000.

[26] Matlab Neural Network Toolbox™

[27] J. Xu, M. C. E. Yagoub, D. Runtao, and Q. J. Zhang, "Exact adjoint sensitivity analysis for neural-based microwave modeling and design," *IEEE Trans. Microw. Theory Tech.*, vol. 51, pp. 226–237 Jan. 2003.

[28] A. Pekker, D. E. Root, and J. Wood, "Simulating operation of an electronic circuit," US patent application #20050251376 A1, May 10, 2004.

[29] C. B. Barber, D. P. Dobkin, and H. T. Huhdanpaa, "The Quickhull algorithm for convex hulls," *ACM Trans. on Math. Software*, vol. 22, pp. 469–483, Dec. 1996.

[30] L. Zhang and Q. J. Zhang, "Simple and effective extrapolation technique for neural-based microwave modeling," *IEEE Microw. and Wireless Components Lett.*, vol 20, pp. 301–303, June 2010.

[31] D. E. Root, "Measurement-based active device modeling for circuit simulation," *Eur. Microw. Conf. Advanced Microw. Devices, Characterization, and Modeling Workshop*, Madrid, Sept. 1993 (available from author).

[32] J. Staudinger, M. C. De Baca, and R. Vaitkus, "An examination of several large signal capacitance models to predict GaAs HEMT linear power amplifier performance," *IEEE Radio and Wireless Conf.*, Aug. 1998, pp. 343–346.

[33] D. E. Root, "Nonlinear charge modeling for FET large-signal simulation and its importance for IP3 and ACPR in communication circuits," *Proc. 44th IEEE Midwest Symp. on Circuits and Sys.*, Dayton OH, Aug. 2001, pp. 768–772 (corrected version available from author)

[34] M. Rudolph, *Introduction to Modeling HBTs*. Norwood, MA: Artech House, 2006.

[35] M. Iwamoto, P. M. Asbeck, T. S. Low, C. P. Hutchinson, J. B. Scott, A. Cognata, X. Qin, L. H. Camnitz, and D. C. D'Avanzo, "Linearity characteristics of GaAs HBTs and the influence of collector design," *IEEE Trans. Microw. Theory Tech.*, vol. 48, pp. 2377–2388, 2000.

[36] M. Rudolph, R. Doerner, K. Beilenhoff, and P. Heymann, "Unified model for collector charge in heterojunction bipolar transistors," *IEEE Trans. Microw. Theory Tech.*, vol. 50, pp. 1747–1751, July 2002.

[37] W. Shockley, "A unipolar 'Field-Effect' transistor," *Proc. IRE*, vol. 40, Nov. 1952, pp. 1365–1376.

[38] M. Fernandez-Barciela, P. J. Tasker, M. Demmler, and E. Sanchez, "A simplified non quasi-static table based FET model," 26th *Eur. Microw. Conf. Dig.*, vol. 1, pp. 20–23, 1996.

[39] G. Gonzalez, *Microwave Transistor Amplifiers* (2nd ed.). Prentice Hall, 1984, p. 61.

[40] B. Hughes and P. J. Tasker, "Bias-dependence of the MODFET intrinsic model element values at microwave frequencies," *IEEE Trans. Electron Devices*, vol. 36, pp. 2267–2273, 1989.

[41] G. Dambrine, A. Cappy, F. Heliodore, and E. Playez, "A new method for determining the FET small-signal equivalent circuit," *IEEE Trans. Microw. Theory Tech.*, vol. 36, pp. 1151–1159, July 1988.

[42] A. E. Parker and S. J. Mahon, "Robust extraction of access elements for broadband small-signal FET models," *IEEE Int. Microw. Symp. Dig.*, pp. 783–786, 2007.

[43] A. D. Snider, "Charge conservation and the transcapacitance element: an exposition," *IEEE Trans. Edu.*, vol. 38, pp. 376–379, Nov. 1995.

[44] R. van der Toorn, J. C. J. Paasschens, and R. J. Havens, "A physically based analytical model of the collector charge of III–V heterojunction bipolar transistors," *IEEE Gallium Arsenide Integrated Circuit (GaAs IC) Symp.*, pp. 111–114, Nov. 2003.

[45] M. Iwamoto, J. Xu, J. Horn, and D. E. Root, "III–V FET high frequency model with drift and depletion charges," *IEEE Int. Microw. Symp.*, Baltimore, MD, June, 2011.

[46] D. E. Root, J. Xu, D. Gunyan, J. Horn, and M. Iwamoto, "The large-signal model: theoretical and practical considerations, trade-offs, and trends," *IEEE Int. Microw. Symp. parameter extraction strategies for compact transistor models workshop (WMB)*, Boston, 2009.

[47] D. E. Root and S. Fan, "Experimental evaluation of large-signal modeling assumptions based on vector analysis of bias-dependent *S*-parameter data from MESFETs and HEMTs," *IEEE Int. Microwave Symp. Dig.*, pp. 255–259, 1992.

[48] D. Ward and R. Dutton, "A charge-oriented model for MOS transistor capacitances," *IEEE J. Solid-State Circuits*, vol. 13, pp. 703–708, Oct. 1978

[49] ADS Root FET, *Agilent Advanced Design System Manual*, nonlinear devices, ch. 3.

[50] H. Statz, P. Newman, I. W. Smith, R. A. Pucel, and H. A. Haus, "GaAs FET device and circuit simulation in SPICE," *IEEE Trans. Electron Devices*, vol. 34, pp. 160–169, Feb. 1987.

[51] I. W. Smith, H. Statz, H. A. Haus, and R. A. Pucel, "On charge nonconservation in FETs," *IEEE Trans. Electron Devices*, vol. 34, pp. 2565–2568, Dec. 1987.

[52] D. E. Root "Elements of measurement-based large-signal device modeling," *IEEE Radio and Wireless Conf. (RAWCON) Workshop on Modeling and Simulation of Devices and Circuits for Wireless Commun. Syst.*, Colorado Springs, Aug. 1998.

[53] D. E. Root, ISCAS tutorial/short course and special session on high-speed devices and modeling," Sydney, pp. 2.71–2.78, May, 2001.

[54] D. E. Root, M. Iwamoto, and J. Wood, "Device modeling for III–V semiconductors: an overview," *IEEE Compound Semiconductor IC Symp.*, Oct. 2004.

[55] A. C. T. Aarts, R. van der Hout; J. C. J. Paasschens, A. J. Scholten, M. Willemsen, and D. B. M. Klaassen, "Capacitance modeling of laterally non-uniform MOS devices," *IEEE IEDM Tech. Dig.*, pp. 751–754, Dec. 2004.

[56] M. Iwamoto, D. E. Root, J. B. Scott, A. Cognata, P. M. Asbeck, B. Hughes, and D. C. D'Avanzo, "Large-signal HBT model with improved collector transit time formulation for GaAs and InP technologies," *IEEE Int. Microw. Symp. Dig.*, Philadelphia, PA, pp. 635–638, June 2003.

[57] L. H. Camnitz, S. Kofol, T. S. Low, and S. R. Bahl, "An accurate, large signal, high frequency model for GaAs HBT's," *IEEE GaAs IC Tech. Dig.*, pp. 303–306, Nov. 1996.

[58] UCSD HBT Model. Available: http://hbt.ucsd.edu

[59] Agilent Technologies. Available: http://www.agilent.com/find/nvna

[60] P. Blockley D. Gunyan, and J. B. Scott, "Mixer-based, vector-corrected, vector signal/network analyzer offering 300kHz–20GHz bandwidth and traceable phase response," *IEEE Int. Microw. Symp. Dig.*, Long Beach, pp. 1497–1500, June 2005.

[61] J. Verspecht, "Calibration of a measurement system for high frequency nonlinear devices," Ph. D. Dissertation, Dept. ELEC, Vrije Universiteit Brussel, Nov. 1995.

[62] E. P. Vandamme, W. Grabinski, and D. Schreurs, "Large-signal network analyzer measurements and their use in device modeling," *Proc. 9th Int. Conf. Mixed Design of Integrated Circuits and Systems (MIXDES)*, Wroclaw, 2002.

[63] D. Schreurs, J. Verspecht, B. Nauwelaers, A. Van de Capelle, and M. Rossum, "Direct extraction of the non-linear model for two-port devices from vectorial nonlinear network analyzer measurements," *27th Eur. Microw. Conf. Proc.*, pp. 921–926, 1997.

[64] M. C. Curras-Francos, P. J. Tasker, M. Fernandez-Barciela, Y. Campos-Roca, and E. Sanchez, "Direct extraction of nonlinear FET Q-V functions from time domain large signal measurements," *IEEE Microw. and Guided Wave Lett.*, vol. 10, pp. 531–533, 2000.

[65] A. M. Conway and P. M. Asbeck, "Virtual gate large-signal model of GaN HFETs," *IEEE Int. Microw. Symp. Dig.*, pp. 605–608, June 2007.

[66] O. Jardel, F. DeGroote, T. Reveyrand, J. C. Jacquet, C. Charbonniaud, J. P. Teyssier, D. Floriot, and R. Quere, "An electrothermal model for AlGaN/GaN power HEMTs including trapping effects to improve large-signal simulation results on high VSWR," *IEEE Trans. Microw. Theory Tech.*, vol. 55, pp. 2660–2669, Dec. 2007.

[67] A. E. Parker and D. E. Root, "Pulse measurements quantify dispersion in pHEMTs," *URSI Int. Symp. on Signals, Systems, and Electronics (ISSSE)*, Pisa, Sept. 1998, pp. 444–449.

[68] D. E. Root, J. Xu, J. Horn, M. Iwamoto, and G. Simpson, "Device modeling with NVNAs and X-parameters," *IEEE Integrated Nonlinear Microw. and Millimeter-Wave Circuits (INMMIC) Conf.*, Gotenborg, Apr. 2010.

[69] P. J. Tasker, M. Demmler, M. Schlechtweg, and M. Fernandez- Barciela, "Novel approach to the extraction of transistor parameter from large signa measurements," *24th Eur. Microw. Conf.*, pp. 1301–1306, Sept. 1994.

6 Large and packaged transistors

Jens Engelmann,[1] Franz-Josef Schmückle,[1] and Matthias Rudolph[2]

[1] Ferdinand-Braun-Institut, Leibniz-Institut für Höchstfrequenztechnik
[2] Brandenburg University of Technology

6.1 Introduction

Microwave power transistors commonly internally consist of a number of smaller transistor cells combined together in order to reach the desired performance. The individual transistor cells are positioned side by side, sometimes repeated in two dimensions. But often, only one single line of parallel cells is used, since power splitting and combining is less challenging compared to other configurations. Relying on an array of small transistors instead of only one large power transistor allows higher power at high frequencies to be realized. Reaching high frequencies calls for small and fast transistors. Inherently, reducing the physical size of a transistor will reduce the power-handling capabilities. Increasing the size of a single transistor with just one emitter or drain connection, on the other hand, is no option in the microwave regime, since unequal current or heat distribution within the device will rapidly degrade performance. Proper combination of many small transistors within a package to get one power device is therefore the only option. In addition to the advantages regarding electrical behavior, thermal management of the power transistor can be significantly simplified.

Various configurations of transistors in packages have emerged in recent years. Common to most of these solutions is the arrangement in bars, as single or multiline (i.e., 2D). However, 2D configurations of bars restrict the power transistor to lower frequencies where line lengths in general (e.g., bondwires) are small compared to the wavelength.

Besides protecting the transistor from any negative environmental factors such as humidity, and besides enabling easy insertion of the package into a circuit during production, an electrically good package ensures that all the individual internal cells operate as if it was a single lumped transistor. This requires:

- combining the output power of all cells in-phase. Any phase difference will degrade the maximum achievable output power at the additional cost of higher power dissipation;
- guaranteeing equal power handling, and current concentration, among the individual cells. Typically, the current tends to concentrate either at the edges of the transistor, or somewhere in the center. Thus, a few internal cells carry most of the load, which degrades the power-handling capability and also reliability and device lifetime. Proper splitting and combining is required to minimize these effects;
- providing good thermal management. In order to achieve equal electrical performance, it is required that all the individual cells operate at the same temperature.

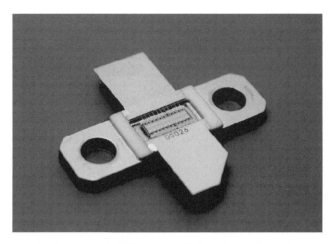

Figure 6.1 Photograph of a packaged GaN transistor, before closing the package.

In this chapter, we will focus on a generic type of transistor package, see Figure 6.1. The transistor consists of a bar of cells that are directly connected to the package leads through bondwires. Depending on the type of transistor and on the application, prematching is provided within many commercial packaged transistors. The prematching is achieved by proper design of the bondwire connections and chip capacitors that are mounted inside the package together with the transistor. But even in the simpler case without the prematching, we will be able to show, e.g., how the bondwires interact through inductive cross-coupling, and how all electrical effects taking place within the package can be characterized and transformed into an equivalent circuit model.

What makes the characterization of a transistor package highly complicated is the fact that the package has not only the external connectors where measurement equipment is easily attached. Most of the ports are the interface between package and transistor chip. Not only are these ports inside the package and therefore not accessible for measurement, there are also quite a lot of them. Modeling the packaged transistor is somewhat involved and it is wise to study the literature that discusses the topic from various points of view [1–8].

Ultimately, the package model should look like that shown in Figure 6.2. A lumped equivalent circuit that describes the overall performance of the many parallel power cells, of course, reduces the numerical effort required in circuit simulation. It improves numerical robustness, convergence, and simulation speed. However, the lumped model provides almost no insight into the properties of the packaged transistors, owing to:

- inductance of the individual bondwires;
- mutual inductance between the different bondwires, on the gate side and drain side, as well as gate-drain inductive feedback;
- phase differences between power cells located at the center or the edges;
- unequal heating, possibly resulting in the formation of hot spots.

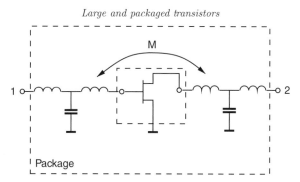

Figure 6.2 Generic lumped equivalent circuit of a transistor package.

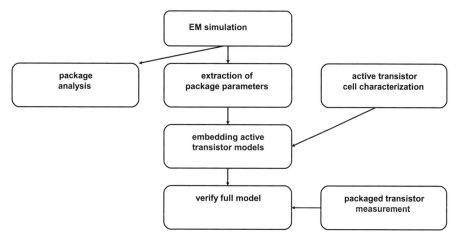

Figure 6.3 Workflow for extraction of a packaged power transistor model.

Some of the effects, such as phase difference between cells and hot spot formation, basically mean that the packaged transistor can no longer be treated as a lumped element. For example, a 10-cell transistor would behave differently than expected when extrapolated from the performance of a single transistor cell. Within certain bounds, differences can be tolerated and accounted for by tuning parameters of the packaged transistor model. In the case of commercial transistors, especially, it can be expected that these effects are controllable.

Nevertheless, when deriving a package model, it is advantageous to study the electromagnetic and thermal coupling effects in detail. Only after the characterization of the distributed structure one can safely attempt to derive a lumped model for actual circuit design.

Taking these arguments into account, it is clear that electromagnetic simulation, and thermal simulation, of the package plays an important role when determining the package properties. Figure 6.3 shows the general workflow for the electrical part of the characterization.

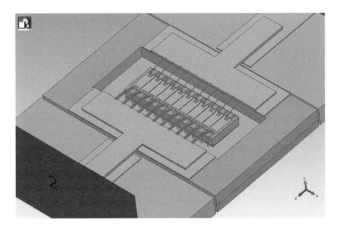

Figure 6.4 Example of the EM-simulation structure for an 11-cell transistor in a package.

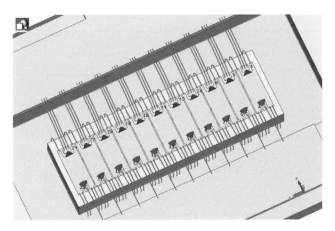

Figure 6.5 Example of the EM-simulation structure for an 11-cell transistor in a package. Closeup showing chip and bondwires.

The starting point is an electromagnetic (EM) simulation of the full package. A simulation structure is shown in Figures 6.4 and 6.5. It is important to simulate a geometry that approximates the real package as closely as possible. This requires that the transistor chip is replaced by a dielectric block, and that lumped ports are inserted at all places where gate and drain connections of the active devices would be. The following section will detail this task.

Once the EM simulation is performed, it can be used directly to investigate physical phenomena such as current flow on the leads, or magnetic fields. Interpreting these results allows, e.g., geometrical shortcomings of the package that might lead to current concentration at edges, and the like to be identified. It also provides insight into the EM effects that need to be considered in the package model, such as cross-coupling of bondwires. While these investigations are hugely helpful to understand and possibly optimize the package, they do not allow any quantitative characterization.

Therefore, the EM simulation is utilized to get an S-parameter matrix of the structure. While the S-parameters no longer provide much insight into the 3D fields and currents, they are suitable for developing an equivalent circuit and determining its parameter values. How the equivalent-circuit elements are determined unambiguously is discussed in Section 6.4 following the discussion of the EM simulation.

Once the equivalent circuit of the package is known, it is time to embed the active devices. In this chapter, we take it for granted that the large-signal model of the transistor power cells has been determined already. How to do so is discussed in detail elsewhere in the book. The final step is now to verify the model, i.e., to compare simulations with measurements of the complete packaged transistor. This is the first time that measurements come into play. Due to the complexity of the package model, it is not easy to tweak the model parameters in order to improve accuracy. Therefore, this step is more an assessment of how careful the package model was determined than a further step in model characterization. But since the package is a purely passive structure, EM simulation will be able to provide us with a very good basis for highly accurate model determination.

The final step is to transform the distributed model into a lumped model. This is usually necessary in order to relax the numerical burden of the simulator. This final step is addressed last and concludes this chapter.

6.2 Thermal modeling

The larger a device gets, the higher the risk that the temperature is not uniform over the entire area. Depending on the type of device, there can be either a positive or negative feedback mechanism between current and temperature. Positive feedback, as in the case of HBTs, can result in thermal breakdown phenomena where the warmer part of the transistor takes over more or less all the current, resulting in a hot spot.

The impact of temperature on device performance, and how to determine the respective parameters, has been discussed already in Chapter 2. However, when dealing with large and packaged transistors, one can always assume that the device is prone to unequal temperatures in the individual transistor fingers.

This case is hard to determine from measurement, and it is also not easily described through one compact model – the hot spot is a distributed effect, which is in contradiction to a lumped model approach.

Hot-spot formation is a very undesirable effect. It leads to degradation of device performance and to heavy stress in certain regions of the transistor, leading to a significantly reduced reliability. Therefore, it is the goal of all device manufacturers to offer only stable devices. But if the device needs to be analyzed in detail, a distributed thermal model is required.

Direct measurement of the mutual thermal coupling of the individual transistor fingers within a device or within a package is basically impossible. There are some ways to determine hot spots, e.g., by infrared imaging, but it is really hard to obtain any model parameters out of such measurements.

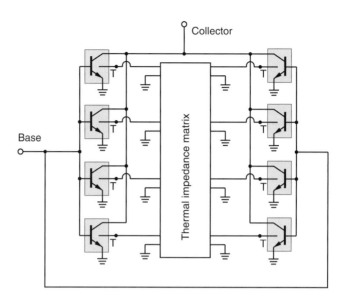

Figure 6.6 Distributed thermal model of a multifinger or multicell transistor.

The only way is therefore through numerical thermal simulation. These simulators today are commercially available, and the mathematics of heat flow are much easier compared to the electromagnetic effects that will be discussed in the following sections. The following points need to be considered when investigating the mutual heating:

1. The relevant structure needs to be discretized, including any backside heatsink, including any solder layer, etc. Even very thin layers can be of significant impact if the thermal conductivity is low.
2. The area where the heat is generated within the transistor – under the gate, or at the base–emitter junction, or at the base–collector junction – needs to be known in order to define the heat sources.
3. The thermal conductivity of the semiconductor material is, in general, a function of temperature. Therefore, it is a good idea to simulate the thermal properties around the typical range of dissipated powers. Now it can be assumed that the temperature level in simulation is close to the temperatures during operation.
4. In order to determine self- and mutual heating, apply an increment of power at one of the basic transistor structures. Depending on the application, it can be a single transistor finger within a larger device, up to one of the power transistors within a package. Determine the difference in temperature at all the other basic transistor structures. The thermal resistance R_{ji} is defined by the ratio $\Delta T_j / \Delta P_i$.
5. If, also, the thermal time constants need to be known, transient simulations are required, and how the temperature responds over time when the power is altered needs to be simulated.

Finally, a whole matrix of thermal resistances or impedances is known. For large-signal simulation, the individual transistors are now interconnected thermally, as shown in Figure 6.6 for the example of an HBT. The nonlinear transistor models need a thermal

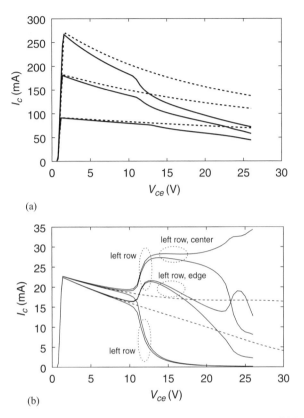

Figure 6.7 Simulation results of an eight-finger fishbone-type GaAs HBT. Solid lines: thermally unstable device; dashed lines: stabilized version. (a): total collector current, (b): emitter current of the individual cells.

port in addition to the three electrical ones. This thermal port provides a current that is equivalent to the dissipated power of the device, while the port voltage is equivalent to the temperature. Provided that the nonlinear model is at hand, the further distributed thermal simulation can be carried out using any circuit simulator.

Figure 6.7 gives an example of a large-signal simulation using a distributed thermal model. The device under test is a GaAs HBT [9]. It consists of eight emitter fingers. Four fingers are in one row, and the two rows are facing each other, a so-called fishbone-type layout. This example is used because HBTs are prone to hot-spot effects and need considerable thermal stabilization. On the other hand, the packaged GaN transistor was stable from the outset, and therefore not suitable to give an impressive example. Two cases were investigated. The solid lines in Figure 6.7 show the performance of a device where the emitter fingers were basically only electrically connected, without much care of the thermal performance. The device shows the typical thermal breakdown (Figure 6.7a), and the currents concentrate first in one row, and eventually in one center finger of this row (Figure 6.7b). The same device was then thermally stabilized through an emitter-ballasting resistor and thick airbridges that keep the temperature more or less on the same level (dashed lines). In this case, the emitter fingers at the edges of the transistor are slightly cooler, but the device is stable overall.

6.3   EM simulation

The EM simulation code needs to be able to simulate the complete 3D structure of the package. In this example, we use Microwave Studio by CST which is based on the finite differences method in time domain. Finite difference in time domain starts with subdividing the space under investigation into cubic subregions, although other types of subgrid are also possible, e.g., tetraeders. Two cubic systems are used, one for the electric field and one for the magnetic field. The two cubic systems are shifted with respect to each other in such a way that the corners of cubes defining the magnetic field are located in the center of the cubes defining the electric field and vice versa. These so-called Yee cells [10] help to minimize the error inherent in the mathematical differences or derivatives. By assigning the electric and magnetic field components in the center of the cubic sides, one can approximate Maxwell's equations as for, e.g., in (6.1) for one component of the electric field as shown in Figure 6.8.

$$\frac{\partial}{\partial t} E = \frac{1}{\varepsilon} \text{rot } \vec{H}$$

$$\downarrow$$

$$E_x^{n+1}(i, j, k) = E_x^n(i, j, k) + \frac{\Delta t}{\varepsilon \cdot h} \left(H_z^{n+1/2}(i, j, k) - H_z^{n+1/2}(i, j-1, k) - \right.$$

$$\left. - H_y^{n+1/2}(i, j, k) + H_y^{n+1/2}(i, j, k-1) \right), \tag{6.1}$$

with i, j, k being the discrete x, y, z coordinates, and n denoting the time step.

In time domain the final result is achieved by a so-called leapfrog algorithm in which the fields are calculated step by step starting from an initial set of known fields (e.g., no fields except at the ports where the structure is excited), and calculation moves on in time, one time step after the other.

In the following sections, the implications of this discretization will be further addressed. One issue is, for example, to choose a grid that approximates the rather thin bondwires and their shapes well enough. Another important topic is that of the boundary conditions. In our case, there are three different types to take into account: the gate and the drain side will be excited through a microstrip connection, i.e., fields of a microstrip line mode. The entire simulation region, on the other hand, needs to be large enough for the bounds of the simulated world not to interfere with the fields inside the package. This would obviously deter simulation accuracy, but might not be observed immediately when extracting the package equivalent circuit. The third issue, finally, concerns the internal ports. Traditional EM simulation codes use basically 2D boundaries in order to describe the ports of the simulated structure. These are placed far away from the structure, e.g., connected through microstrip lines. In this way, if the field

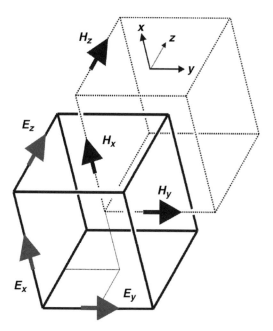

Figure 6.8 Subdivided region discretized in basic Yee cells for electric fields (solid cube) and magnetic fields (dashed line cube).

perturbations due to the simulation structure did fade away over the distance to the gates, the modes that can propagate along the lines are known. Inside the package, on the other hand, we need closely spaced 50 Ω ports at each chip–package interface (gate, drain, or source connections). Some kind of interference from these ports is to be expected.

The aim of the entire effort is to express the electrical performance of the package through an equivalent circuit. Besides being compatible with circuit simulation software, this approach allows package behavior to be understood analytically, as the complex electromagnetic interaction is broken down basically in capacitances, and mutual and self-inductances. But on the basis of the equivalent circuit, it is, in general, not possible to predict how changes in the package geometry will impact the electrical performance. This would require the different elements of the package model to be independent of each other. However, the package is a complex structure, and changes impact 3D currents and voltages that can impact parts of the model that are described through more than one equivalent circuit element.

6.3.1   Geometry of package and simulated structure

The basic geometry of the transistor packages under consideration is shown in Figure 6.3. It is mainly a bar of transistor cells inside a package, with bondwires as electrical connections. The number of transistor cells varies in our case between five and eleven. Also, the lead can be wider, as in Figure 6.3, or narrow, as in Figure 6.4. The shape of the bondwires might be different. The source of the chip is connected to the package flange, either via holes through the chip, or through bondwires, as shown in the figures. In general, we focus on a generic structure with a few variations.

The transistor chip is first soldered onto the flange, which is connected to the source. As already stated, this is achieved either via holes or bond wires. The sidewalls of the package consist of a dielectric frame. The gate and drain connections are placed on top of the frame. These gold leads fit to a connection of an external microstrip line and distribute the current inside the package, providing enough space to attach bondwires. These bondwires finally connect gate and drain on the chip to the gate and drain leads, respectively. Drains, in our case, are connected through three bond wires of 50 μm pitch, since the DC current is too high for a single wire. DC current is not an issue on the gate side, and therefore, only a single bond wire is used there. The distance between the transistor cells is 400 μm. The package would finally be closed by a dielectric lid.

In order to simulate the electromagnetic properties of this structure, it needs to be as close to reality as possible. Therefore, the complete sealed package is considered.[1] The dielectric of the lid has some influence on the package capacitances.

Second, the package is embedded in a standard microstrip environment, as shown in Figure 6.4. At a certain distance from the gate and drain, it can be assumed that the waves are of pure microstrip mode. Any field perturbation in the vicinity of the package will decay as the distance from the package increases. These microstrip lines therefore enable EM simulation of the package, but their impact on the result needs to be removed in the end. This is relatively easy to achieve by de-embedding. The performance of the microstrip lines is well known, and it is straightforward to subtract these from the simulated S-parameters.

The transistor chip inside the package is replaced by a dielectric block with the same dielectric constant as the chip. The structure of this block is seen in Figures 6.4 and 6.5. From the active device, only the metallization is considered in the form of the bonding pads for gate and drain, or as a larger gold area for the source. The active transistors are modeled separately and once the package model is known, the nonlinear model is embedded. What is required are ports also at the gate, drain, and source bond pads that are the interface to the active device. The location of these ports is shown by the 3D arrows in Figure 6.9. It is important to note that the source of the transistor is connected to ground. In the case of the package, ground is the metallic flange. That is why the internal port on the gate and drain side are defined between the bond pad on the chip and the flange. Regarding the transistor cells, the ground and source are on top of the chip, since they are basically in coplanar design and were characterized on-wafer. In this configuration, it is necessary also to define a third port for each transistor cell: the path from transistor source to package ground. The structure in Figure 6.9 shows two of these ports, on the gate and on drain side of the chip. When it comes to package parameter extraction, these two ports are joined without introducing significant errors.

---

[1] Even though the lid is commonly not shown in the figures in order to show the internal structure of the packaged transistor.

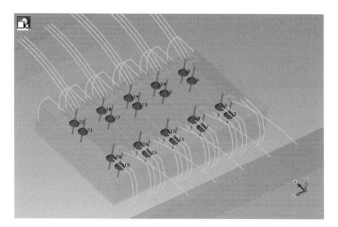

Figure 6.9 Location of the internal ports for a five-cell transistor.

6.3.2   The internal ports

For the microstrip ports at drain and gate, the EM simulation is performed in a quite straight/forward way. Boundary conditions are defined that assume the propagation of microstrip mode waves along the lines, and the EM simulator calculates the resulting fields and the ratios of transmitted and reflected waves. The internal ports, in contrast, cannot be defined in this way. These ports are inside the structure to be calculated, and no known field distribution can be assumed. Instead, thin wires are assumed that also have a certain port impedance, e.g., 50 Ohm.

A wire of a certain length will introduce a magnetic field. It therefore has, due to Maxwell's theory, its magnetic field and an inductance. In our case, the length of the port equals the height of the chip, which cannot be ignored. There also will be electric fields along the port, resulting in a certain capacitance. These reactive components might be unexpected if a 50 Ohm port was defined. In fact, we force current and voltage at the ports, which in turn impacts magnetic and electric fields.

As there is no way to get around the reactive component introduced by the internal ports, it is necessary to determine the port inductance in order to correct for it. Compared to the port inductance, the port capacitance is of only minor relevance.

In order to separate the port inductance from other inductive effects, a test structure is simulated. It is basically a dielectric block, reflecting the properties of the chip and its geometry concerning height, bond pad size, and spacing of the neighboring internal ports. Figure 6.10(a) shows the structure that was used, and in 6.10(b), the dielectric is removed to show the location of the two internal ports. The equivalent circuit for this test structure is shown in Figure 6.11. It is seen that the port inductance is in series with the chosen port impedance. In this configuration a port inductance of $0.63\,\text{pH}/\mu\text{m}$ is determined; see Figure 6.12. Hence, as the chip's height is $365\,\mu\text{m}$, each port shows an inductance of $230\,\text{pH}$.

As is evident from other investigations of internal ports [11, 12], the inductance is also inherently related to the discretization schemes. In this case, this means that for

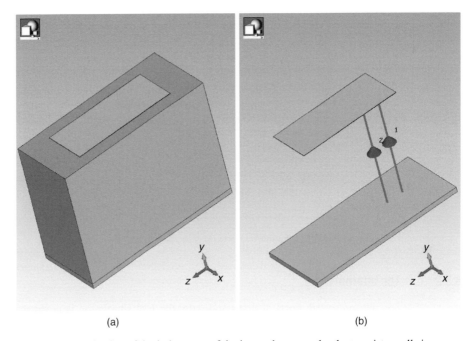

(a)   (b)

Figure 6.10 Investigation of the inductance of the internal ports under the transistor cells in a simplified environment.

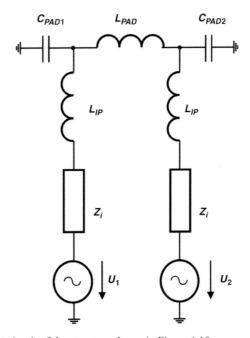

Figure 6.11 Equivalent circuit of the structure shown in Figure 6.10.

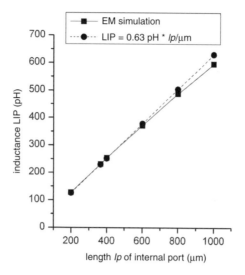

Figure 6.12 Solid line: inductance of the internal port configuration shown in Figure 6.10 as a function of length lp of the internal port. Broken line: inductance values obtained assuming 0.63 pH/μm.

each simulation of the package, the port inductance needs to be resimulated unless the discretization is kept identical in the area of the internal ports.

Once the port inductance is known, it must be subtracted at the respective ports from the simulated S-parameter matrix; this is easily done through a circuit simulation or common mathematical tools. Only with this correction, the result of the EM simulation reflects the electromagnetic properties of the package alone and can it be used to determine the package model.

The backside metallization of the package is supposed to be the common ground for all the internal ports. Of course, this is also an approximation, since the metal of the flange is of high, but not infinite, conductivity and there are some distances between the reference points on the flange. While it is good to keep this in mind during the investigation, it turned out in our case that it was safe to neglect the uncertainties that are introduced in the geometry shown here.

6.3.3 The bondwires

An important part of the package are the bondwires. A simplified test structure as shown in Figure 6.13 is used to investigate how the length of the bondwires impacts the inductance, and in order to determine the appropriate equivalent circuit. EM simulations were carried out for length between 1100 and 2100 μm. Simulated S-parameters for a single wire are shown in Figure 6.14.

Of course, the bondwires will have an inductance, but also a capacitance to ground. Besides the impact of wire length, it is necessary to determine the appropriate equivalent

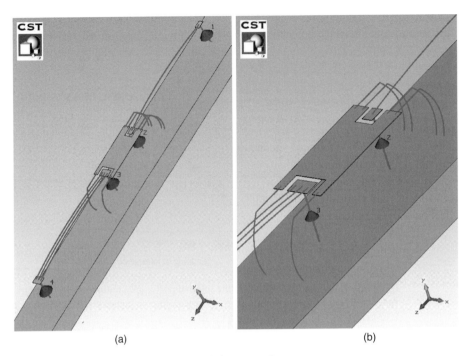

(a)   (b)

Figure 6.13 Test structure to determine bondwire properties.

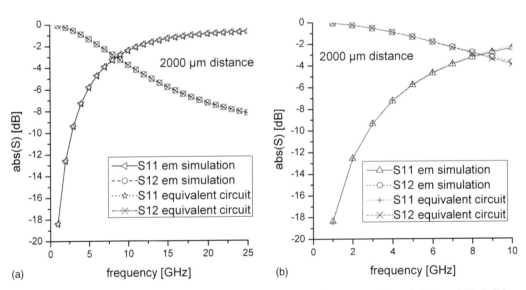

(a)   (b)

Figure 6.14 Typical S-parameters for one bondwire of the structure shown in Figure 6.13. Solid lines: EM simulation, broken lines: equivalent circuit fit.

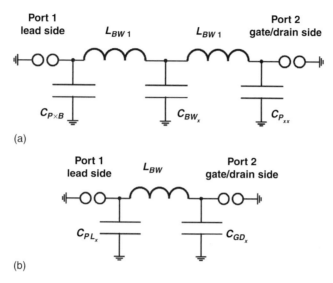

Figure 6.15 Equivalent circuits for one bondwire: (a) broadband circuit; (b) circuit valid up to 10 GHz.

circuit. This means finding the optimum between the simplicity of the circuit and the accuracy of the description. In the case of the bondwires, Figure 6.15 shows two alternatives. In order to achieve broadband accuracy of the bondwire model, it is necessary to account for the capacitance to ground, as shown in Figure 6.15a. The element values for the equivalent circuit elements are shown in Figure 6.16. Obviously the capacitances are rather small, and therefore, relevant only at higher frequencies.

At this point, it is worthwhile to consider that the packaged transistor is designed for a target frequency of 2 GHz. Also, the package is specified to work only below 10 GHz. It is simply too large to perform well at higher frequencies. It is therefore well justified to restrict the model to the frequency range below 10 GHz, and to neglect the capacitance as shown in Figure 6.15b. The resulting prediction of the S-parameters is shown in Figure 6.14b.

In conclusion, there are capacitances between the bondwires and ground and between the parallel bondwires at gate and drain, but their impact on the electrical performance is negligible. Therefore, they will be omitted in the package model. Since the bondwires are in parallel, electrical fields between them will be close to zero and the capacitance has no effect. The capacitance to ground will be lumped together and added to the overall capacitance of the feed metal.

6.3.4   Discretization of the package

The discretization significantly impacts the accuracy of the EM simulation. The mesh needs to be fine enough to resolve all structures to be simulated, but on the other hand

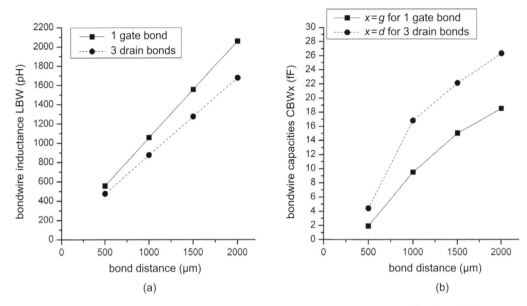

Figure 6.16 Lumped elements values describing bondwires up to 25 GHz, for the equivalent circuit shown in Figure 6.15: (a) bondwire inductance; (b) bondwire capacitance to ground.

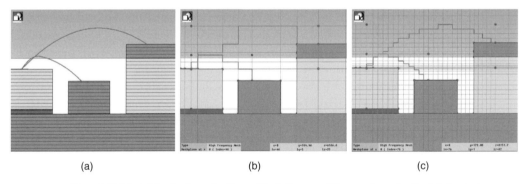

Figure 6.17 Model for the investigation of the influence of the number of cells: (a) physical course; (b, c) approximated course fit to the discretization grid using (b) 0.5 mio. cells, (c) 1.0 mio. cells.

as coarse as possible to reduce the numerical effort. In our case, a rectangular grid is used, which fits well to most of the structure, but the bondwires are of a certain curvature that is not simple to translate into rectangles. In fact, the path approximated by staircase function will be longer than the original shape, and the inductance will be overestimated. Figure 6.17 shows the impact of different discretization schemes on the bondwire geometry. The ideal shape of two bondwires is shown. The shorter one connects the source of the chip with a metallic pedestal, the longer connects gate or drain with the gate or drain lead, respectively. The length of the discretized bondwire depends on the shape and on the resolution. The worst case would be if a wire should cross a

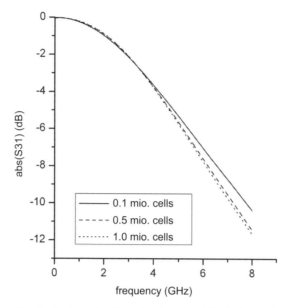

Figure 6.18 S-parameters for the investigation of various densities in the mesh as in Figure 6.17. S_{31} is the relation between input waveguide port and internal port of the investigated gate cell.

rectangular area diagonally, but is approximated by the sides of the rectangle. The error in the length in this case can be directly given by applying the law of Pythagoras.

In order to determine the optimum grid, the package was discretized in 100 000, 500 000, and one million cells. Figure 6.18 shows the transmission factor S_{31} from the waveguide gate port to the internal gate port, over the bondwire. It is observed that the result changes significantly as the number of cells is increased from 100 000 to 500 000. But doubling the number of cells again does not impact the result much. The final discretization was determined through a number of similar preliminary EM simulations. Once the result is no longer improved by a refinement of the grid, it can be assumed that the optimum between the number of cells and the accuracy of the simulation is found.

6.4 Equivalent-circuit package model

The EM simulation as discussed in the previous sections, provides us with an S-parameter matrix of the package. This matrix has a larger number of ports. Besides the external gate and drain ports of the packaged transistor, it provides internal gate and drain ports for each of the power transistor chips. In principle, this S-parameter matrix can directly be used to describe the package in a circuit simulator. It has, however, many advantages for proceeding in the characterization process and determining an equivalent-circuit-based model:

1. The S-parameter file is quite large. For an 11-cell transistor, the matrix has at least 36 ports. This results in 1296 complex S-parameter values for each frequency point that

was simulated. Depending on the number of frequency points, the S-parameter file can easily be about 50 MB large. The equivalent circuit, on the other hand, requires only a few parameter values.

2. The equivalent circuit can extrapolate the package performance beyond the frequency range considered in EM simulation. This holds, especially concerning low frequencies and DC, which is commonly not covered in S-parameter files. Extrapolation towards higher frequencies will also be well behaved. Extrapolation of S-parameters is usually performed in circuit simulators based on a generic algorithm that might show the general trend but cannot be expected to provide much accuracy. Extrapolation, however, is limited to the frequency range below the first package resonance. The resonance itself has not been characterized, and will therefore not be described quantitatively too accurately. At frequencies beyond, the equivalent circuit might be too simple.

3. The equivalent circuit is described in terms of inductances, resistances, and capacitances that can be handled in any kind of circuit simulator. A circuit defined by frequency-domain data such as an S-parameter file will work in Harmonic Balance simulation, but not generally in time-domain simulators.

4. An equivalent circuit model of physically meaningful topology and element values permits better understanding of the package and, for example, opens the possibility to identify show-stoppers such as problematic bondwire inductance values.

In order to take full advantage of the package model, the equivalent circuit needs to be as simple as possible, of course without degrading model accuracy. The main issue is to determine the equivalent-circuit parameters. A generic algorithm will be presented in the following.

Figure 6.19 shows the equivalent-circuit model of gate and drain part of the package. It is well suited for industry-standard packages, such as the A191 package from Kyocera shown in Figure 6.1.

The package basically consists of three parts. The main part is a gold-plated copper-tungsten (CuW) flange. The two holes in the flange are used for mounting purposes. The flange is also required for heat sinking of the power transistor. A dielectric alumina frame is mounted on the flange. Finally, a gold-plated lead is soldered onto the alumina frame. It is of a T-shape with broad lines for connecting the package to the microstrip circuit environment. At the inner side, the leads are connected to the chip by bondwires.

After the preliminary investigations during EM simulation as described above, a quite basic equivalent circuit is expected to provide good accuracy. At gate and drain sides, inductances L_{inter} account for line inductances. The capacitance to ground of the whole T-lead is lumped into a single capacitance C_p. On the lead, the current spreads and is distributed from the microstrip connection to the different bond pads. This current spreading is described by a number of inductances, i.e., L_{const} and the various L_{ij}. This configuration has been found to be appropriate for the package under investigation. However, other configurations might be suited as well. Nevertheless, it is required that the package geometry provides low phase differences between the different transistor chips, which implies low inductance values on the lead.

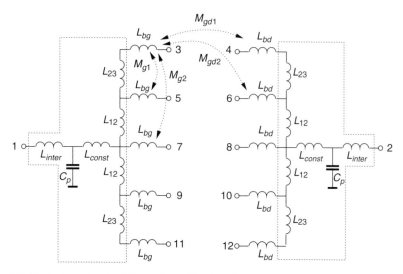

Figure 6.19 Equivalent circuit of the package. For simplicity, only a five-cell design is shown, and the source inductances are omitted. All bondwires L_{bg} and L_{bd} are inductively coupled, which is indicated by the arrows for a few cases.

The inductances L_{bg} at the gate and L_{bd} at the drain side describe the bondwires, respectively. Even though the gates are connected through one bondwire and the drain is connected by three parallel wires, only one parameter is determined for each of the connections. As a result, the values of L_{bg} and L_{bd} will differ.

Finally, all bondwires are inductively coupled, which is accounted for in the model by a high number of mutual inductances:

- M_{gi} describing the mutual inductance between bond wires connecting different transistor cells on the gate side;
- M_{di} the same, but on the drain side;
- M_{gdi} describing the mutual inductance between bondwires connecting the transistor cells at the gate side to bondwires at the drain side.

The indices i denote the distance between the bondwires under consideration, starting with $i = 1$ for neighboring cells on the gate and drain side, and for bondwires directly facing each other in the case of M_{gd1}, respectively. For the sake of simplicity, the mutual coupling is shown only for one bondwire and its nearest neighbors in Figure 6.19. It is assumed that the bondwire model, including its interaction with its environment through mutual inductive coupling, is almost identical no matter which of the cells is regarded. This assumption will be verified during parameter extraction.

The equivalent circuit Figure 6.19 does not show the source connections. Of course, any source inductance will impact the performance of the packaged transistor and must be considered, depending on how the connection of transistor source to package flange is realized. If the transistors provide via connections to the backside, the resulting inductance is not part of the package in a strict sense. If the source is also connected through bondwires, the same strategy as for gate and drain bondwires applies, including inductances and mutual inductances. In this case, the number of parameters increases significantly if it is considered that all bondwires are mutually coupled, but the modeling and extraction strategy is analog for gate, drain, and source ports. Therefore, only the extraction of gate and drain bondwires will be considered in detail and it is assumed that the source bondwires are determined in the same manner.

6.4.1   Analytic parameter extraction strategy

The equivalent circuit has quite a number of elements, and due to the mutual coupling between the bondwires, even more parameters to determine. Therefore, global optimization cannot be considered to be the best way of parameter extraction, nor the fastest or easiest. An analytic extraction routine needs to be determined that yields an unambiguous solution for each of the parameters. An additional benefit of the analytic approach is that it intrinsically provides the possibility to verify whether the equivalent circuit topology is suited [13]. The assumption underlying an equivalent circuit approach is that all elements such as inductors and capacitors are free of dispersion, i.e., characterized by one value for L or C independently of frequency. This assumption should be verified during parameter extraction. Additionally, if it turns out that the equivalent circuit topology relies on simplifications that are appropriate, by applying this condition, one is able to identify where it needs to be refined. How it needs to be refined, however, is a different question, and also, how the parameters of the refined equivalent circuit will be determined.

The package is symmetrical, and many parts are meant to be identical, for example, the many parallel bondwires at the gate. It is to be expected that this is reflected in the parameter values. It is definitely advisable to take these geometrical similarities into account when defining the model, and to use, e.g., only one value for gate bondwire inductance. Reducing the number of model parameters is always desirable, since it makes the model easier to handle and to understand. In this case, it is implied and justified by the geometry.

However, during parameter extraction, it is recommended to get as much information from the EM simulation data as possible. From theory, we will see that many of the parameters can be extracted in various ways. It is also possible to determine, e.g., inductances of the different bondwires independently. Extracting all of these values also gives an insight into whether or not the equivalent circuit topology is appropriate.

Despite the fact that the equivalent circuit consists of many elements and even more parameters are to be determined, a straightforward extraction routine can be derived. It is based on the observation that the equivalent-circuit topology is quite close to a T-type topology of two-ports that is shown in Figure 6.20.

The Z-parameter matrix of the T-topology circuit is given by:

$$\begin{pmatrix} V_1 \\ V_2 \end{pmatrix} = \begin{pmatrix} Z_{11} & Z_{12} \\ Z_{21} & Z_{22} \end{pmatrix} \cdot \begin{pmatrix} I_1 \\ I_2 \end{pmatrix}. \tag{6.2}$$

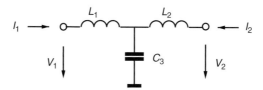

Figure 6.20 A T-topology two-port circuit.

The Z-parameters are defined as follows:

$$Z_{11} = \left.\frac{V_1}{I_1}\right|_{I_2=0} = j\omega L_1 - j\frac{1}{\omega C_3} \tag{6.3}$$

$$Z_{21} = \left.\frac{V_2}{I_1}\right|_{I_2=0} = -j\frac{1}{\omega C_3}, \tag{6.4}$$

and analogously for Z_{21} and Z_{22}.

The parameters can be given right away:

$$C_3 = \text{Im}\left(\frac{1}{\omega Z_{12}}\right) = \text{Im}\left(\frac{1}{\omega Z_{21}}\right) \tag{6.5}$$

$$L_1 = \text{Im}\left(\frac{Z_{11} - Z_{12}}{\omega}\right) = \text{Im}\left(\frac{Z_{11} - Z_{21}}{\omega}\right) \tag{6.6}$$

$$L_2 = \text{Im}\left(\frac{Z_{22} - Z_{12}}{\omega}\right) = \text{Im}\left(\frac{Z_{22} - Z_{21}}{\omega}\right). \tag{6.7}$$

The Z-parameter Z_{ij} is defined as the voltage at port i when the current at port j is forced, and all other ports are open. With respect to a T-topology circuit, it leads to a significant simplification of the equivalent circuit, if a port is open. In the simple example shown here, no current flows through L_2, if $I_2 = 0$. Therefore, L_2 has no impact on Z_{11} and Z_{12}, since no voltage drop occurs here without current flowing.

It is the benefit of a Z-parameter-based approach, and it also works for circuits with many ports. Figure 6.21 shows how C_p and L_{inter} of our package model can be extracted from the EM simulated parameters $Z_{1,1}$ and $Z_{1,11}$. In analogy to the previously discussed twoport, the parameters are given by:

$$C_p = \text{Im}\left(\frac{1}{\omega Z_{1,11}}\right) = \text{Im}\left(\frac{1}{\omega Z_{11,1}}\right) \tag{6.8}$$

$$L_{inter} = \text{Im}\left(\frac{Z_{1,1} - Z_{1,11}}{\omega}\right) = \text{Im}\left(\frac{Z_{1,1} - Z_{11,1}}{\omega}\right). \tag{6.9}$$

All other inductances are not relevant, since the current flows only through L_{inter} and C_p.

The extraction of the package parameters therefore follows a similar approach to the extraction of the extrinsic inductances as discussed in Chapter 3.

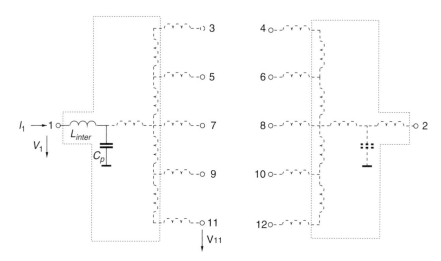

Figure 6.21 Extraction of C_p and L_{inter} from Z-parameters.

6.4.1.1   Lead inductance and capacitance

The first parameters that can be directly extracted as shown in the previous section are C_p and L_{inter}. These describe the inductive behavior of the microstrip feedline, some of the inductive behavior of the lead, and the capacitance from lead to ground.

These parameters are strongly dependent on the location of the reference planes in the EM simulation. As discussed previously, the package will be embedded in a microstrip environment in the EM simulation. Inductance and capacitance of any microstrip line at the gate will add to L_{inter} and C_p on the gate side. The same holds for the drain side. Therefore, it is necessary to perform the de-embedding of the lines from the EM-simulated file carefully. If the reference planes are not where they are expected to be, it will result only in wrong element values, but the algorithm in general will work well.

A second remark on the capacitance value: investigating the bondwire performance through EM simulation, it was observed that the bondwire capacitance can be neglected with respect to bondwire inductance; see Section 6.3.3. In the real package, however, a high number of wires will be placed in parallel. In our case, there are 11 cells, which might slightly increase the value C_p. A second issue are the bond pads on the chip. Also, these rather small squares add to the value of C_p. It needs to be considered how the chip periphery is described through the parasitic elements of the active transistor model, since the bond pads rather belong to the chip. The value of C_p would be overestimated, if the pad capacitances are considered twice, on-chip and again here through package extraction.

Regarding Figure 6.21 again, it is clear that the capacitance C_p can basically be determined from any parameter $Z_{1,i}$ or $Z_{i,1}$, where i stands for the ports 3, 5, 7, 9, and 11. It is useful to determine both parameters from all possible combinations. This issue will be highlighted in the following example regarding bondwire inductances and the spreading inductances of the lead.

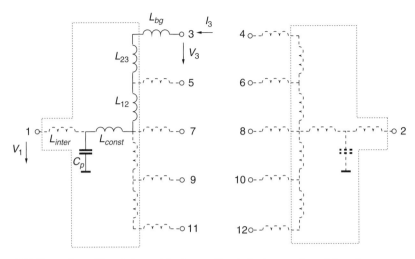

Figure 6.22 Extraction of bondwire and lead spreading inductances, L_b and $L_{12}, L_{23}, \ldots$, respectively, from Z-parameters.

6.4.1.2 Bondwire and spreading inductances

All bondwires are mutually inductively coupled, which makes it difficult to determine the self-inductance of a single gate or drain connection from the performance of the entire package. But the Z-parameter approach is also beneficial in this case, as it allows the different inductive effects to be separated.

Figure 6.22 highlights how it works. Forcing a current i_3 into port no. 3, and leaving the other ports open yields a current flowing only through those elements that are drawn in solid lines. Any voltage that is induced in any of the other bondwires is not leading to any additional current, since all other bondwires are not connected at one of their terminals. Again, the equivalent circuit reduces to a number of T-topology subcircuits.

The inductances of the bondwires and of the spreading inductance, however, are always in series. In the example equivalent circuit, Figure 6.22, the following equations can directly be given:

$$Z_{3,3} - Z_{3,1} = j\omega(L_{const} + L_{12} + L_{23} + L_{bg}) \tag{6.10}$$

$$Z_{5,5} - Z_{5,1} = j\omega(L_{const} + L_{12} + L_{bg}) \tag{6.11}$$

$$Z_{7,7} - Z_{7,1} = j\omega(L_{const} + L_{bg}) \tag{6.12}$$

$$Z_{9,9} - Z_{9,1} = j\omega(L_{const} + L_{12} + L_{bg}) \tag{6.13}$$

$$Z_{11,11} - Z_{11,1} = j\omega(L_{const} + L_{12} + L_{23} + L_{bg}). \tag{6.14}$$

This set of equations is easily solved for the spreading inductances $L_{12} \cdots L_{23}$.

Figure 6.23 shows the extracted spreading inductances for the gate lead of an 11-cell transistor package. Due to symmetry, only five inductance values are obtained, since identical values are calculated for the elements on the left and right sides. It is also

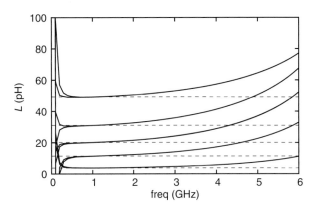

Figure 6.23 Lead spreading inductance $L_{12} \cdots L_{56}$. Solid lines: extracted from EM simulation, dashed lines: frequency-independent value used in the equivalent circuit. The inductance values increase towards the lead edge.

found that the same inductance values are obtained when the lead on the drain side is investigated.

First, it is observed that the inductance increases towards the outer edge of the lead. The values shown in the figure are the absolute inductance values for one of the inductances, not the sum of the inductances from the center of the package to the point where the bondwire is placed.

The shape of the extracted curves is also interesting to look at. Frequency-independent inductance values are a condition for the equivalent circuit to be valid. The constant values that are used as parameters are indicated by the dashed lines, but the extracted lines are not really constant. The extraction of the parameters is, however, justified for the following reasons:

1. Towards very low frequencies, the equation $L = \mathrm{Im}(Z)/\omega$ approaches $0/0$ as ω decreases. The numerical errors prevent calculation of any decent value. Furthermore, there might be increasing numerical uncertainty in the EM simulated data.
2. The increasing inductance values towards higher frequencies are a clear indication that the equivalent circuit is too simple for frequencies $f \gg 2\,\mathrm{GHz}$. This is no real surprise as the manufacturer specified the package only for frequencies below 6–8 GHz. Around 8–10 GHz, the package will show a resonance, rendering it impossible to use it at fundamental frequencies higher than, say, 2 GHz. The dispersion of the inductance value happens as we approach the package resonance frequency, but beyond the target frequency of 2 GHz. Extracting one value of these curves, therefore, is not "picking the value that looks nice," but determining the value for the fundamental frequency; taking into consideration that accuracy will degrade towards higher harmonics. Only if significant dispersion is already observed close to the fundamental frequency, improvements of the circuit topology should be considered.

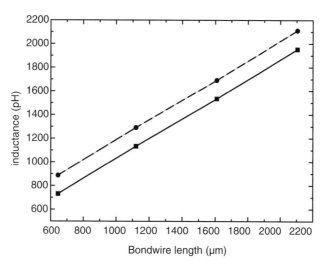

Figure 6.24 Total inductance according to equation (6.10) (dashed curve) and (6.12) (solid curve), respectively, extracted from EM simulations varying bondwire length.

3. These graphs are special in a way, as the lead will show a line resonance close to the resonance frequency of the whole package. When this happens, it is to be expected that an approximation by just a few lumped inductances in series will no longer be sufficient. The lead as a distributed structure shows stronger dispersion than other elements, such as the bondwires, but its inductances are much lower compared to the bondwires, as we will see in the following.

From equations (6.10)–(6.14), all but one spreading inductance can be determined, which is called L_{const}. It is placed in the equivalent circuit to add some freedom, as there might be a common inductance on the lead in addition to the different lateral spreading inductances. Unfortunately, it is in series with the bondwire inductances. It is not just added to the bondwire inductance, which would basically yield the same inductance value in all of the branches like that shown in Figure 6.23. However, this spreading inductance will not be inductively coupled to any of the bondwires. Therefore, it needs to be determined.

Separation of L_{const} from L_{bg} or L_{bd} is not possible from just one EM simulation. Therefore, a number of EM simulations were performed varying the bondwire length. This is relatively easily achieved, since the package dimensions can be varied without any problem in the virtual world. In our case, the package has been stretched. Figure 6.24 shows the resulting total inductance values for the center and for the edge of a package which would refer to (6.12) (solid curve) and (6.10) (dashed curve) in our example equivalent circuit, respectively. The impact of the spreading inductances is clearly seen as the offset between the curves. Both curves show the same constant slope, which is caused by the bondwire inductance that is proportional to length, but there is a small offset when the bondwire length becomes zero. The intersection of the solid curve with the y-axis would give the value of L_{const}.

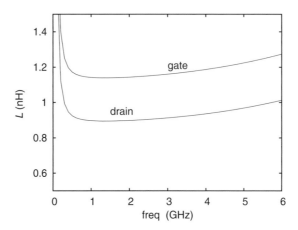

Figure 6.25 Extracted inductance of gate bondwire L_{bg} and total inductance of parallel drain bondwires L_{bd}, respectively.

Once L_{const} is known, the bondwire inductance can be determined. The respective extracted inductance values are shown in Figure 6.25 as a function of frequency. The extracted values are fairly constant over frequency, which proves that the extraction routine and the equivalent circuit are physically significant. The single gate bondwire inductance is around 1.2 nH, or 915 pH/mm, which is fairly close to what would be expected from the well-known rule of thumb that assumes about 1 nH/mm. On the other hand, the three parallel wires at the drain side show a total inductance around 0.9 nH, or 710 pH/mm.

The different values for gate and drain connections result from the fact that the gates are connected by only one bondwire per cell, while three bondwires are used in parallel at the drain side. As the three bondwires are connected in parallel, only one resulting inductance is determined. The three wires on the drain side are obviously so close that the inductance value is lowered by only one third, due to mutual inductive coupling. The three drain bondwires can therefore be interpreted as an approximation of a ribbon, rather than as three individual wires.

6.4.1.3 Mutual inductances of gate and drain bondwires

At the current stage, all self-inductances of the bondwires are known, as well as the parameters describing the lead. What is still missing is quantification of the mutual coupling of the bondwires.

Even if the bondwires are defined as being left open except for at one port in the Z-parameter notation, a voltage is still induced. This property is exploited to determine the mutual inductances. How the mutual coupling of the bondwires is determined at one side, gate or drain, will be explained at the example of the mutual inductance M_{g1}, between the bondwires connecting gate port no. 3 and no. 5 in Figure 6.22. It is necessary to sense voltage v_5 at port 5 while a current is forced to port 3, i.e., to investigate $Z_{3,5}$.

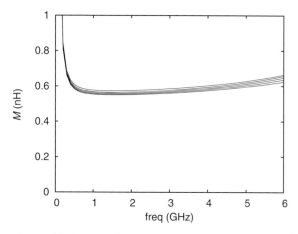

Figure 6.26 Extracted mutual inductances between neighboring bondwires at the gate side as a function of frequency.

Where the Z-parameters read, in terms of equivalent-circuit parameters:

$$Z_{3,5} = j\omega(M_{g1} + L_{12} + L_{const}) + 1/(j\omega C_p) \tag{6.15}$$

$$Z_{5,5} = j\omega(L_{bg} + L_{12} + L_{const}) + 1/(j\omega C_p) \tag{6.16}$$

$$Z_{7,7} = j\omega(L_{bg} + L_{const}) + 1/(j\omega C_p) \tag{6.17}$$

$$Z_{1,7} = 1/(j\omega C_p).$$

Solving for M_{g1} yields:

$$M_{g1} = \text{Im}(Z_{3,5} - Z_{1,7} - Z_{5,5} + Z_{7,7})/\omega - L_{const}. \tag{6.18}$$

In an analog way, the mutual inductance between other bondwires is also determined. It is further necessary to consider the coupling of bondwires at any distance.

In contrast to the parallel drain bondwires connecting one active cell, the bondwires connecting different cells can therefore be investigated individually. Figure 6.26 shows the extracted mutual inductances of all neighboring bondwires at the gate side. Again, the values are fairly constant over frequency. In addition, there is almost no variation of the values depending on the location of the bondwires under investigation. The spread is better observed in Figure 6.27 that shows the extracted mutual inductance values between all bondwires at the gate side. The index given on the x-axes denotes the number of the gate finger, beginning with no. 1 at one side to no. 11 at the other side. Each of the curves corresponds to one fixed distance between the bondwires. As the distance increases, the mutual inductance decreases. The number of extracted values decreases also, according to the number of possible bondwire combinations.

Figure 6.28 shows how the mutual inductance decreases with the distance, for gate and drain side. The symbols denote the extracted values, while the straight line shows the fitting as used later on in the compact package model.

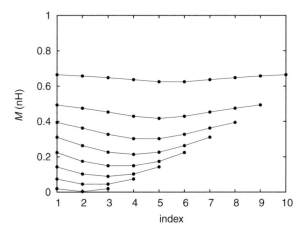

Figure 6.27 Extracted mutual inductance values at the gate side, as a function of the reference bondwire. Numbering starts at one side of the package. The parameter is the distance between the respective bondwires. Neighboring wires are strongly coupled, and the values decrease as the distance increases.

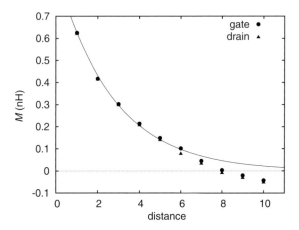

Figure 6.28 Extracted mutual inductance values at the gate and drain side as a function of the distance between the bondwires (symbols). The straight line shows a fitting through a power series.

The extraction reveals quite a strong mutual inductive coupling between the bondwires of neighboring cells. At the gate side, the mutual inductance is slightly below half of the self-inductance, while at the drain side, the coupling is around two-thirds of the self-inductance. Although the coupling decay can be modeled by a power series, as indicated by the straight line, it can still be detected after a distance of three cells.

Figure 6.28 shows the effect that the value of the mutual inductance is not decreasing asymptotically towards zero, but shows a zero-crossing, with increasing negative values

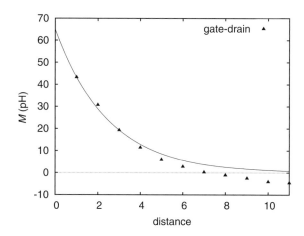

Figure 6.29 Extracted mutual inductance values between gate and drain side as a function of the distance between the bondwires (symbols). The straight line shows a fitting through a power series.

beyond a distance of eight cells. Again, it can be assumed that either limitations inherent to the EM simulation, or the simplicity of the postulated equivalent circuit are the reason for this strange behavior. Therefore, it is again advisable to keep an eye on these results that are in contradiction to the model assumptions. But rather than spending much effort in further investigation, the issue can be neglected as it concerns bondwires that are quite far away. The approximation shown by the straight line has been found to be a good model approximation.

6.4.1.4 Feedback mutual inductances

The magnetical feedback coupling between gate and drain bondwires is determined through much simpler math. For example, the coupling between drain port no. 4 and gate port no. 3 in Figure 6.22 is given by:

$$M_{\text{gd1}} = \text{Im}(Z_{34})/\omega \tag{6.19}$$

since no common current path exists in this setting between the two ports. Hence, the only voltage observable at port no. 4 is the one induced into the bondwire at this port.

Inductive coupling between gate and drain bondwires is highly undesirable, since it leads to a feedback from output to input. Besides crosstalk, it can lead to instability of the transistor, degrade performance, and, in the worst case, to device destruction.

Figure 6.29 shows the extracted mutual inductance values. The x-axis denotes the lateral distance between the respective bondwires, with distance $= 1$ defined as two bondwires directly facing, and distance $= 11$ as one bondwire on one edge of the package, the other one at the opposite edge.

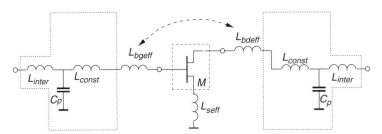

Figure 6.30 Lumped model of the package. The various bondwire inductances are condensed into effective single inductances, the active chip is described by only one large-signal transistor model.

Again, the inductive coupling is not negligible. It is about one order of magnitude below the coupling between the bondwires at the gate or drain side, and it shows a similar decay with increasing distance. However, the drain-to-gate feedback it introduces is highly undesirable.

6.4.2   Deriving a lumped package model

A distributed model as discussed before is of high accuracy and still not too complex. It can be used in circuit design. However, for the practical case, it would be desirable to have a lumped model of the packaged transistor, as shown in Figure 6.30.

The lumped model has the big advantage that it only needs one nonlinear transistor model, in contrast to the one derived so far which needs a nonlinear transistor model for each individual transistor cell. Therefore, simulation speed is significantly higher with the lumped model, and also numerical convergence properties are improved – in fact, the packaged transistor in the lumped form is not really any more complex than a small modeling device.

Of course, a straightforward, formal way to derive the lumped model as discussed in the following requires that an important condition is met. The packaged transistor needs to behave basically like a single big device. All transistor cells must be at the same temperature during operation, and all gates and drains must be connected with the same delay. If one of these preconditions is not met, it means that, e.g., a hot spot is formed, and the performance of some cells is compromised. It could also happen that the individual transistors do not work fully in-phase, and the output power is not added constructively. In any case, as a result, the power transistor will not provide the gain or output power one would expect. It will be lower than the ideal multiple of the small cells connected in parallel. Obviously, if such an effect is observed, the transistor, or the package, or both, need further attention and improvement to provide good performance. A well-engineered, industrial packaged transistor will not show these effects that also compromise the stability and reliability of the packaged device.

The lumped equivalent circuit shown in Figure 6.30 consists of a single nonlinear transistor model embedded into the package model. The nonlinear model is derived by upscaling the model parameters of a single transistor cell by the number of parallel cells

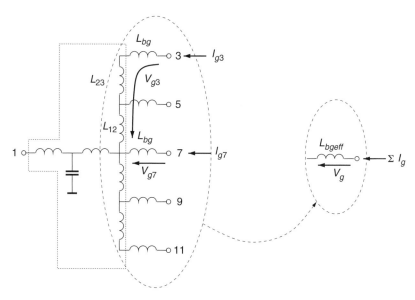

Figure 6.31 Derivation of the lumped package model: effective gate bondwire inductance L_{bgeff}.

of the power transistor. Transistor models usually provide a scaling factor, typically M for this purpose.

For the package, derivation of the lumped model requires the effective gate and drain bondwire inductances, L_{bgeff} and L_{bdeff}, respectively, and their effective mutual inductance M_{gdeff} to be determined.

Figure 6.31 shows how L_{bgeff} is derived. All inductances from the node where the current spreads to the individual bondwires to the internal gate ports will be connected in parallel. In matrix notation, this yields, in our example:

$$j\omega \mathbf{L_g} \cdot \vec{i}_g = \vec{v}_g \tag{6.20}$$

with

$$\mathbf{L_g} = \begin{pmatrix} L_{bg} + L_{12} + L_{23} & M_{g1} & M_{g2} & M_{g3} & M_{g4} \\ M_{g1} & L_{bg} + L_{12} & M_{g1} & M_{g2} & M_{g3} \\ M_{g2} & M_{g1} & L_{bg} & M_{g1} & M_{g2} \\ M_{g3} & M_{g2} & M_{g1} & L_{bg} + L_{12} & M_{g1} \\ M_{g4} & M_{g3} & M_{g2} & M_{g1} & L_{bg} + L_{12} + L_{23} \end{pmatrix}$$

$$\vec{i}_g = \begin{pmatrix} i_{g3} & i_{g5} & i_{g7} & i_{g9} & i_{g11} \end{pmatrix}^{\mathrm{T}}$$

$$\vec{v}_g = \begin{pmatrix} v_{g3} & v_{g5} & v_{g7} & v_{g9} & v_{g11} \end{pmatrix}^{\mathrm{T}},$$

where M_{g1} denotes the mutual inductance between neighboring bondwires, M_{g2} the mutual inductance between a wire and the overnext, and so on.

Deriving L_{bgeff} from equation (6.20) requires first to solve it for $\vec{v}_g$:

$$\vec{i_g} = \frac{1}{j\omega}\mathbf{L_g}^{-1} \cdot \vec{v}_g. \tag{6.21}$$

Next, in the lumped case, all voltages $v_{g3}, \ldots, v_{g11}$ equal v_g. Thus, the matrix elements in each row of $\mathbf{L_g}^{-1}$ can be added. In the parallel connection, the current is $i_g = i_{g3} + i_{g5} + i_{g7} + i_{g9} + i_{g11}$. Therefore, also adding the rows of the matrix, we end up with:

$$i_g = \frac{1}{j\omega}\sum_{\text{all elements}}\mathbf{L_g}^{-1} \cdot v_g. \tag{6.22}$$

The effective inductance is therefore given by:

$$L_{bgeff} = \frac{1}{j\omega}\frac{1}{\sum\mathbf{L_g}^{-1}}. \tag{6.23}$$

The same formalism is applied to obtain the other effective inductances and mutual inductances.

6.4.3 Testing the model

Testing the model before using it in circuit design is an essential step in the modeling procedure. In the case of the package model, it means to measure and simulate a real device. So far, the parts were investigated only separately. The transistor cells, for example, need to be measured and modeled beforehand, just as explained elsewhere in this book. Second, thermal coupling needs to be determined; third, the package equivalent circuit is to be determined. The last two steps that are specific to the package are mainly performed through numeric simulation. This is, in this chapter, the first time that measurement comes into play. Verification by measurement is therefore even more important than for any measurement-extracted model. If the model is extracted from measurement, the aim of verification is to show that it is also working under different conditions. In our case, it is necessary to show that the assumptions and simplifications made in simulation hold, and that the model works at all.

As an example, an 11-cell, 60 W, 2 GHz GaN HEMT is chosen. It was fabricated and packaged completely at the Ferdinand-Braun-Institut, Leibniz-Institut für Höchstfrequenztechnik, Berlin, Germany [14]. It is the same transistor that was investigated throughout this chapter. The 11 active cells have a gate width of $8 \times 250\,\mu\text{m}$ each and are modeled by an Angelov GaN-HEMT model [15, 16].

The comparison of measured and simulated S-parameters looks promising, as seen in Figure 6.32. The graph compares measurements (symbols) in phase and magnitude to the distributed (solid lines) and lumped model (dashed lines) results. The power transistor is operated at $V_{ds} = 24$ V, $I_d = 0.8$ A. The target frequency of the device is at 2 GHz, therefore, S-parameters are shown up to the fifth harmonic. Only towards a multiple of the target 2 GHz, is some deviation between measurement and the two models observed.

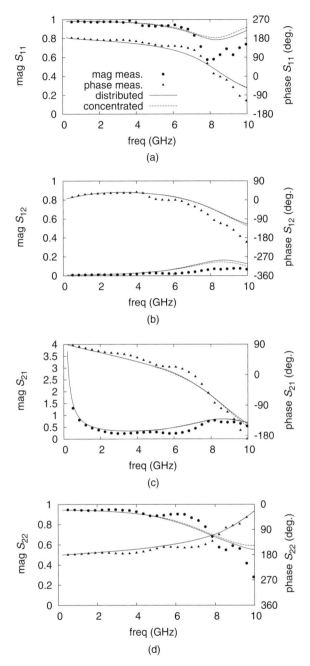

Figure 6.32 *S*-parameters of the packaged HEMT, at $V_{ds} = 24$ V, $I_d = 0.8$ A. Measurement in magnitude (bullet) and phase (triangles), and simulated with distributed (solid lines) and lumped (dashed lines) model.

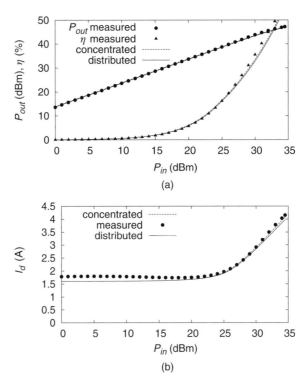

Figure 6.33 Load-pull behavior of the packaged transistor, measured (symbols) and simulated with lumped (dashed lines) and distributed (solid lines) model.

Load-pull results are shown in Figure 6.33, for 2 GHz, $V_{ds} = 24$ V, and $I_d = 1.6$ A. The simulations (lines) agree very well with the measurements (symbols) regarding output power, self-biasing, and drain efficiency.

References

[1] D. Brody and G. R. Branner, "A modeling technique for internally matched bipolar microwave transistor networks," *Proc. 37th Midwest Circuits Syst. Symp.*, Lafayette, LA, Aug. 1994, pp. 1224–1226.

[2] T. Johansson and T. Arnborg, "A novel approach to 3-D modeling of packaged RF power transistors," *IEEE Trans. Microw. Theory Tech.*, vol. 47, pp. 760–68, June 1999.

[3] T. Liang, J. A. Plá, P. H. Aaen, and M. Mahalingam, "Equivalent-circuit modeling and verification of metal-ceramic packages for RF and microwave power transistors," *IEEE Trans. Microw. Theory Tech.*, vol. 47, pp. 709–714, June 1999.

[4] K. Mouthaan, "Modeling of RF high power bipolar transistors," Ph.D. dissertation, Dept. Microelectron. Comput. Eng., Delft Univ. Technol., Delft, The Netherlands, 2001.

[5] P. H. Aaen, J. A. Plá, and C. A. Balanis, "On the development of CAD techniques suitable for the design of high-power RF transistors," *IEEE Trans. Microw. Theory Tech.*, vol. 53, pp. 3067–3074, Oct. 2005.

[6] P. H. Aaen, J. A. Plá, and C. A. Balanis, "Modeling techniques suitable for CAD-based design of internal matching networks of high-power RF/microwave transistors," *IEEE Trans. Microw. Theory Tech.*, vol. 54, pp. 3052–3059, July 2006.

[7] P. H. Aaen, J. A. Plá, and J. Wood, *Modeling and Characterization of RF and Microwave Power FETs*. Cambridge Univ. Press, 2007.

[8] J. Flucke, F.-J. Schmückle, W. Heinrich, and M. Rudolph, "An accurate package model for 60W GaN power transistors," *Proc. 4th Eurp. Microw. Integr. Circ. Conf. (EuMIC)*, 2009, pp. 152–155.

[9] M. Rudolph, F. Schnieder, and W. Heinrich, "Investigation of thermal crunching effects in fishbone-type layout power GaAs-HBTs," *Dig. 12th GAAS symp.*, 2004, pp. 435–438.

[10] K. S. Yee, "Numerical solution of initial boundary value problems involving Maxwell's equations in isotropic media," *IEEE Trans. Antennas Propag.*, vol. 14, pp. 302–307, 1966.

[11] W. Thiel and W. Menzel, "Full-wave design and optimization of mm-wave diode-based circuits in finline technique," *IEEE Trans. Microw. Theory Tech.*, vol. 47, no. 12, pp. 2460–2466, Dec. 1999.

[12] P. K. Talukder, "Finite-difference-frequency-domain simulation of electrically large microwave structures using PML and internal ports," Ph.D. dissertation, Fakultät IV, Berlin Institute of Technology, Jan. 30, 2009.

[13] J. Flucke, F.-J. Schmückle, W. Heinrich, and M. Rudolph, "On the magnetic coupling between bondwires in power-transistor packages," *Proc. 5th German Microw. Conf. (GeMiC)* 2010.

[14] J. Würfl, R. Behtash, R. Lossy, A. Liero, W. Heinrich, G. Tränkle, K. Hirche, and G. Fischer, "Advances in GaN-based discrete power devices for L- and X-band applications," *Proc. 36th Eur. Microw. Conf. (EuMC)*, 2006, pp. 1716–1718.

[15] I. Angelov, L. Bengtsson, and M. Garcia, "Extension of the Chalmers nonlinear HEMT and MESFET model," *IEEE Trans. Microw. Theory Tech.*, vol. 44, pp. 1664–1674, Oct. 1996.

[16] I. Angelov, V. Desmaris, K. Dynefors, P. A. Nilsson, N. Rorsman, and H. Zirath, "On the large-signal modeling of Al-GaN/GaN HEMTs and SiC MESFETs," *Proc. 13th GAAS Symp.*, Paris, 2005, pp. 309–312.

7 Nonlinear characterization and modeling of dispersive effects in high-frequency power transistors

Olivier Jardel,[1,2] Raphael Sommet,[1] Jean-Pierre Teyssier,[1] and Raymond Quéré[1]

[1] University of Limoges, France
[2] Now with 3–5 Lab

7.1 Introduction

Power amplifiers (PAs) are key elements of telecommunications and radar front ends at radio frequencies. Recent evolutions are driving the use of those power amplifiers in more and more complex conditions. These result from an increased complexity of signals that are fed into the PA on the one hand, and from the increased density of power that solid state PAs are required to support on the other hand. In both cases – modulated signals used in telecommunications systems and complex pulsed signals used in radar systems – there exist low-frequency components which excite the dispersive phenomena that are present in electronic devices. From the point of view of systems designers, the dispersion phenomena appear as memory effects in PAs. Those memory effects can be classified as short-term memory (STM) effects and long-term memory (LTM) effects [1]. A typical simplified schematic of a radio frequency (RF) PA is given in Figure 7.1, where the bias networks, the matching networks (Q_e and Q_s) as well as the embedding thermal network ($Z_{th}(\omega)$) are shown. Typically, STM effects result from input and output matching networks as well as microwave electrical time constants involved in the device itself. They lie in the picosecond (ps) or nanosecond (ns) range. On the other hand, LTM effects are due to biasing networks, self-heating and trapping effects as well as networks which are dedicated to the overall management of the PA. They are in the microsecond (µs) to the millisecond (ms) range and appear as feedback terms in the working of the RF PA. Thermal and trapping effects are inherent in the nature of the device and its packaging, while bias effects are inherent in the design of the PA.

For system simulation, different approaches have been proposed to take into account those LTM effects in a unified way. However, those approaches do not distinguish between the various origins of LTM and are of no use in circuit simulation during the design phase of the PA. During this phase the designers require circuit level models that are able to deal with thermal and trapping effects. Specifically, for thermal behavior the model must include a fourth terminal, loaded by the thermal impedance, as shown in Figure 7.1. This implies that nonlinear models must take into account the dispersive effects that are responsible of the LTM effects. Thus, those models consist of classical nonlinear models driven by the two control voltages at the terminals of the device

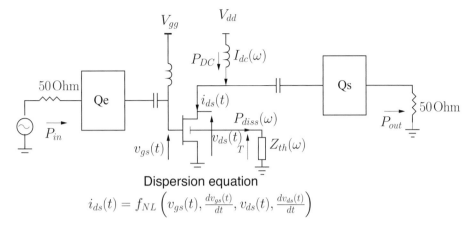

Figure 7.1 Simplified schematic of a RF-PA showing the various frequency-dependent elements which are responsible for memory effects. In particular, LTM effects are due to thermal effects through the thermal impedance, the trapping effects taken into account through the dispersion equation and bias frequency-dependent effects.

augmented at one terminal for temperature and specific circuits dealing with thermal and trapping effects.

This chapter aims to propose such models that are extracted through specific characterization or simulation techniques. It is organized as follows. In Section 7.2, the nonlinear electrothermal modeling of field effect transistors (FETs) is presented with a specific emphasis on gallium nitride high electron mobility (GaN HEMT). Some details are given on the determination of the thermal effects and it is shown how 3D simulation can be used to generate the embedding thermal circuit for use in circuit simulators. Section 7.3 will describe how trapping effects can be taken into account in a circuit simulator approach and how they affect the large signals characteristics of the devices. The description of characterization techniques in Section 7.4 will allow a better understanding of the requirements of the approach proposed in terms of specific measurements to be performed on dispersive devices.

7.2 Nonlinear electrothermal modeling

In the electrothermal model extraction, the first step is to derive the stationary nonlinear characteristics at various temperatures as well as the equivalent thermal circuit. Provided that the equivalent thermal circuit is not frequency dependent, the model obtained is referred to as stationary, as it does not present any dispersive effects.

7.2.1 Electrothermal model extraction

The equivalent circuit of a FET is shown in Figure 7.2. The different elements are extracted following the steps listed below.

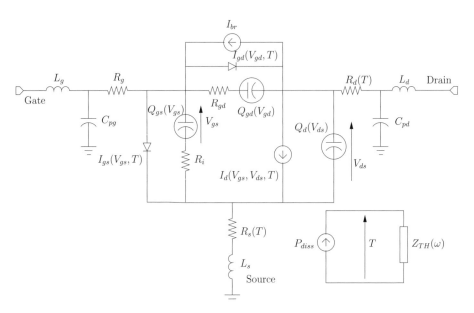

Figure 7.2 Nonlinear electrothermal equivalent circuit of a microwave FET.

1. Extraction of IV characteristics. Those current-voltage characteristics are extracted from pulsed measurements IV (PIV) which allow the currents and voltages in the whole working domain of the transistor, including breakdown region and high power regions, to be measured. The main current source, diodes currents as well as break-down currents are obtained through those pulsed measurements. Moreover, the PIV technique allows quasi-isothermal conditions to be maintained and the trapping effects to be investigated, as will be shown later.

2. Extraction of parasitic access elements. The parasitic elements correspond to the access lines and contact resistances. L_g, C_{pg}, L_d, C_{pd}, L_s, R_g, R_d, R_s are extracted from S-parameter measurements either at the quiescent bias point or in the cutoff or diode conduction modes.

3. Extraction of the nonlinear charges. The nonlinear charges are then extracted from S-parameter multibias measurements in pulsed mode. Those pulsed S-parameter measurements allow the whole I–V domain to be investigated while maintaining small temperature varaiations in the device.

4. Extraction of the thermal impedance. In order to complete the nonlinear electrother-mal behavior, the thermal impedance must be extracted either from 3D thermal simulation or from thermal measurements.

Although the different steps for the extraction of nonlinear electrothermal models will be described for a FET device, they remain valid in the case of an HBT. For FETs and, in particular, for AlGaN/GaN HEMT, the modeling process of stationary characteristics follows steps 1–3 listed above. The first step allows the current voltage characteristics of the drain current source to be obtained versus the three independent variables which are the gate voltage, the drain voltage, and the temperature $I_{ds} = f(V_{gs}, V_{ds}, T)$. A number of expressions of this fundamental characteristic of the transistor have been proposed [2–4] which allow the behavior of the current to be accurately described. More recently [5], a new model has been proposed which allows the negative drain source voltages to be take into account and which derivatives are continuous up to infinity. The I–V equations used in reference [6] are recalled here.

$$
\begin{cases}
I_d = I_{dt} \cdot \gamma_{md} \\[4pt]
\gamma_{md} = 1 + \beta_{gm} \cdot (V_{ds} + V_{dm}) \cdot \left(1 + \tanh\left(\alpha_{gmx}(V_{gs} - V_{gm})\right)\right) \\[4pt]
Id_t = \dfrac{I_{DSS}}{1 - \frac{1}{m}(1 - e^{-m})} \left[V_{GSN} - \dfrac{1}{m}\left(1 - e^{-m \cdot V_{GSN}}\right) \right] \\[4pt]
\qquad \cdot \left[1 - e^{-V_{DSN}\left(1 - a \cdot V_{DSN} - b \cdot V_{DSN}^2\right)} \right] \\[4pt]
V_{GSN} = 1 + \dfrac{V_{gs}(t - \tau) - V_\phi}{V_P} \\[4pt]
V_{DSN} = \dfrac{V_{ds}}{V_{DSP}\left(1 + w \cdot \dfrac{V_{gs}(t - \tau)}{V_P}\right)} \\[4pt]
V_P = V_{P0} + p \cdot V_{DSP} + V_\phi
\end{cases}
\tag{7.1}
$$

and the diodes equations are given by:

$$
\begin{cases}
I_{dgs} = I_{sgs} \cdot \exp\left(\dfrac{q \cdot V_{gs}}{k \cdot T_{sgs} \cdot N_{gs}}\right) \\[6pt]
I_{dgd} = I_{sgd} \cdot \exp\left(\dfrac{q \cdot V_{gd}}{k \cdot T_{sgd} \cdot N_{gd}}\right).
\end{cases}
\tag{7.2}
$$

The temperature dependence of the characteristics is obtained through IV measurements at different temperatures and the parameters which are temperature dependent are modeled under the form $X(T) = X(T_0)(1 + \alpha_T(T - T_0))$ for parameters I_{DSS}, P, R_S, R_D, N_{gs}, and N_{gd}. For the saturation currents of the diodes, the temperature dependence is $Y = Y_0 + \beta_T \cdot e^{T/T_a}$. Equations (7.1) and (7.2) are the fundamental stationary equations which will be modified to take into account trapping phenomena as seen in Section 1.5. Dispersion due to the thermal time constants is due to the thermal impedance which is determined either by measurements or by simulation.

In the second step of the extraction procedure, intrinsic reactive components are extracted from multibias pulsed S-parameter [7] measurements after the de-embedding of parasitics elements. The use of pulsed S-parameters allows a better consistency between the first derivatives of IV characteristics obtained from equations (7.1) and the transconductance and output conductance obtained from S-parameters. All the elements of the intrinsic part of the transistor model are obtained through direct calculations using the method proposed in reference [8] with the constraint that those intrinsic elements remain constants over the whole frequency range to ensure the physical meaning of equivalent circuit topology. Once the values of the gate source and gate drain capacitances are obtained from measurements, it still remains to represent the nonlinear variations of those capacitances in the harmonic balance (HB) circuit simulator. This poses the problem of charge conservation as raised in reference [9]. Gate source and gate drain capacitances represent the first derivatives of the nonlinear function representing the gate charge $Q_g(V_{gs}, V_{ds})$. This function of the two control voltages must be a state function to satisfy the charge conservation principle [9] and thus meet the Cauchy conditions, i.e.:

$$C_{gs}(v_{gs}, v_{gd}) = \frac{\partial Q_g}{\partial v_{gs}}$$

$$C_{gd}(v_{gs}, v_{gd}) = \frac{\partial Q_g}{\partial v_{gd}}$$

$$\frac{\partial C_{gs}(v_{gs}, v_{gd})}{\partial v_{gd}} = \frac{\partial C_{gd}(v_{gs}, v_{gd})}{\partial v_{gs}}. \tag{7.3}$$

In practice these conditions are difficult to satisfy due to measurement and extraction errors which are inherent in the extraction process. The solution is then to fit the derivatives of a well-chosen state function $Q_g(V_{gs}, V_{ds})$. However, the choice of such a function is difficult and the dependency of the gate charge on two variables often leads to convergence problems in the simulator. This is why another modeling method has been chosen.

Considering the fact that in working conditions the transistor is loaded by the optimum impedance Z_{opt}, this loading condition imposes a relation between the two commands V_{gs} and V_{ds} in order to travel the loadline. In reference [10] the charge was dynamically evaluated from table-based models of the capacitances. This implies a complicated coding task to implement the model in the simulator. A simpler version of the same idea is to keep the values of the capacitances only along this loadline; they become functions of their controlling voltage, that is to say: $C_{gs}(V_{gs})$ and $C_{gd}(V_{gd})$. Then it becomes easy to find nonlinear functions that fit the variations of the capacitances. Very simple expressions have been proposed in reference [11] which are recalled here for convenience.

$$C_{gx} = C_{gx0} + \frac{C_{gx1} - C_{gx0}}{2}$$
$$\cdot [1 + \tan h(a_{gx} \cdot (Vm_{gx} + Vgx))]$$
$$- \frac{C_{gx2}}{2} \cdot [1 + \tan h(b_{gx} \cdot (Vp_{gx} + Vgx))], \tag{7.4}$$

where "x" stands for "d" or "s". These expressions lead to very good convergence properties of the harmonic balance algorithm.

A careful examination of the errors induced by this simplification has been proposed in reference [6] that shows that the simplified model remains valid in a large domain of the IV characteristics.

To conclude on this HEMT stationary model, notice that this model has to be made as simple as possible in order to preserve good convergence properties. Pulsed IV and pulsed S-parameter measurements allow quasi-isothermal characteristics to be extracted and accurate yet simple capacitance models to be derived. However, as the next section will show, trapping effects are to be taken into account through specific pulsed measurements.

7.2.2   Thermal impedance determination

Thermal phenomena, such as self-heating and thermal coupling, are known to impact and affect the semiconductor device efficiency, especially in the high-power applications domain such as those addressed by GaN devices. This electrothermal feedback is a low-frequency mechanism which can be taken into account through the additional thermal circuit represented in Figure 7.2. In that case, thermal modeling becomes crucial. The approach can be performed either by measurement or by simulation. Several models have been investigated [12–16] and many characterization methods have been developed and proposed [17–22]. The choice of characterization method is very important, especially for power devices often covered by a rather thick Au thermal drain. The main idea of these methods consists in making a specific choice for the temperature-sensitive electrical parameter. In parallel to the measurement approach, the simulation approach allows complex structures to be dealt with. This approach can rely on analytic expressions of thermal impedance in the most simple cases or on 3D methods such as finite element method (3D-FEM), which provides very accurate results on complex structures at the cost of a long simulation time that is not compatible with circuit-aided design requirements. Therefore, the coupling of the thermal behavior with the electrical one must be done through the well-known analogy between thermal flux and electrical current as well as between temperature and voltage, thus defining the thermal impedance concept. Classically, it may be a simple thermal resistance or a more complicated circuit using several resistance-capacitance (RC) cells. The main difficulty is to reduce the structural complexity of the device into an equivalent circuit which remains compatible with CAD requirements in computer time. The solution relies on a physics-based approach and a model-order reduction (MOR) technique which allows the thermal simulation to be directly linked to the CAD equivalent circuit.

7.2.2.1   Definition of the thermal admittance

In order to generate an accurate thermal model that will be able to give a correct estimation of the temperature distribution in the device, a 3D thermal simulation tool is required.

The thermal system is governed by the heat equation:

$$\nabla \cdot (\kappa(T)\nabla T) + g = \rho C_p \frac{\partial T}{\partial t}, \tag{7.5}$$

where κ is the thermal conductivity, T the temperature, ρ the density, C_p the specific heat, and g the volumetric heat generation.

Assuming that κ is constant and equal to the conductivity at 300 K, the FE formulation of equation (7.5) leads to the semidiscrete heat equation:

$$\mathbf{M}\dot{\mathbf{T}} + \mathbf{K} \cdot \mathbf{T} = \mathbf{G}, \tag{7.6}$$

where the conductivity matrix $\mathbf{K}$ and the specific heat matrix $\mathbf{M}$ are n-by-n symmetric and positive-definite matrix, $\mathbf{T}$ the n-by-1 temperature vector at mesh nodes, $\mathbf{G}$ the n-by-1 load vector which takes into account the power generation and the boundary conditions.

In the frequency domain, equation (7.6) reads:

$$(j\omega\mathbf{M} + \mathbf{K}) \cdot \mathbf{T} = \mathbf{Yth} \cdot \mathbf{T} = \mathbf{G}, \tag{7.7}$$

where $\mathbf{Y_{th}} = \mathbf{Z_{th}}^{-1}$ represents the n-by-n thermal admittance matrix.

Unfortunately, the dimension of this admittance matrix is very large (n of the order 10^4 to 10^5), which makes its direct integration into a circuit simulator prohibitive. Moreover, we do not need to know all the temperature nodes in order to perform an electrothermal simulation; only those that are connected to the electrical circuit. This is why a MOR technique has been used to extract automatically fundamental information for selected ports from the large numerical system.

7.2.2.2   Model-order reduction

Most reduction techniques (Padé, Schur technique, and so on) [23–26] are based on extracting a small set of dominant poles (eigenvalues and their corresponding eigenvectors) to represent the original system, which may contain thousands of poles. In general, these methods are computational intensive and require special care in order to avoid unstable poles. It has been shown [27–29] that the Ritz vector approach will ensure that important response modes are not neglected and yields improved accuracy with fewer vectors as compared to the use of eigenvectors. Moreover, this method ensures that the steady-state temperature is exact. The MOR consists of projecting the vector of temperature $\mathbf{T} \in \mathbb{R}^{n \times 1}$ into a transformed vector $\mathbf{P} \in \mathbb{R}^{m \times 1}$ with $m \ll n$.

$$\mathbf{T} = \Phi_m \mathbf{P}, \tag{7.8}$$

where the projection matrix, constituted by m Ritz vectors, satisfies the M-orthogonalization relationship:

$$\Phi_m^t \mathbf{M} \Phi_m = \mathbf{I}_m, \tag{7.9}$$

where $\mathbf{I}_m$ is the $m \times m$ identity matrix. The algorithm for the Ritz vectors generation may be found in reference [28].

Applying transformation (7.8) to equation (7.6) and multiplying by Φ_m^T, a reduced system is obtained:

$$\begin{aligned} \Phi_m^T \mathbf{M} \Phi_m \dot{\mathbf{p}} + \Phi_m^T \mathbf{K} \Phi_m \mathbf{p} &= \Phi_m^T \mathbf{F} \\ \mathbf{I}_m \dot{\mathbf{p}} + \mathbf{K}^* \mathbf{p} &= \Phi_m^T \mathbf{F}. \end{aligned} \tag{7.10}$$

The first Ritz vector represents the static response to the load vector $\mathbf{G}$, and the others concern the dynamic response of the structure; m is directly linked to the precision of the transient response and is often several orders lower than n.

The next step consists of finding the eigenvalues λ_i in order to obtain a set of m independent differential equations.

Then let Ψ be an m-by-m projection matrix such as $\Psi^T \cdot \Psi = I_m$, which satisfies $p = \Psi t$, where t is an m-by-1 temperature vector in the diagonalized system.

The diagonalized reduced system governing equation is then:

$$I_m \dot{t} + \Lambda_m t = \Psi^T \Phi_m^T F, \tag{7.11}$$

where $\Lambda_m = diag\,[\lambda_1 \cdots \lambda_m]$.

Equation (7.8) can be written in the Fourier domain as:

$$\begin{bmatrix} j\omega + \lambda_1 & 0 & 0 \\ 0 & \ddots & 0 \\ 0 & 0 & j\omega + \lambda_m \end{bmatrix} t(\omega) = \Psi^T \Phi_m^T \cdot F. \tag{7.12}$$

In the electrothermal simulation, only r nodes corresponding to the boundary conditions and to the power injection will be retained, so we introduce an r-by-n selection matrix S, which satisfies $d_r = S \cdot d$ and $F = S^T F_r$.

The dissipated power profile in a multifinger transistor may change during operation, so Ritz vectors evaluated for a specified load vector direction will not always lead to an accurate temperature response. This is why, in order to account for the load vector variation, we express the load vector F_r as the superposition of unitary spatial load vectors $Fu_i = [0 \quad \cdots \quad 1 \quad \cdots \quad 0]^T$, where Pi, the power dissipated at node i, is able to change during operation.

$$F_r = \sum_{i=1}^{r} Pi(\omega) \cdot Fu_i. \tag{7.13}$$

The global response corresponds to the linear combination of each individual response with the Pi coefficients.

7.2.2.3 Implementation and equivalent circuit

The reduction software has been written in C-ANSI and uses the BLAS and LAPACK libraries. These libraries are well suited because the matrices are symmetric band and are positive-definite. The inputs are F, M, K, the nodes corresponding to the fixed baseplate temperature as well as the retained nodes for the outputs of the reduced model.

This data proceeds from the FE tool MODULEF [30] or ANSYS [31]. The outputs are the eigenvalues of K^*, which are directly linked to the thermal time constants and the r matrices $A_i = S\Phi\Psi$ obtained for each unitary power injection. These outputs are used in order to generate automatically a SPICE format file for the thermal subcircuit. This subcircuit is composed of resistors $(1/\lambda_{ij})$ in parallel with unitary capacitors, current controlled current sources to inject power and voltage controlled voltage sources to collect temperatures at retained nodes. It translates equation (7.10) and the basic

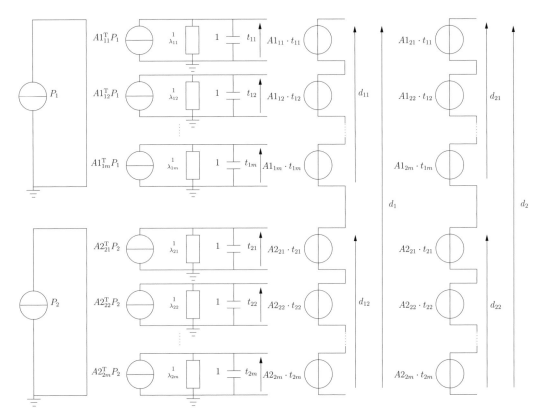

Figure 7.3 SPICE subcircuit description: the variables and the symbols are respectively: P_i the input power at node i ($1 \leq i \leq r$), $\mathbf{A_i} = \mathbf{S} \cdot \Phi \cdot \Psi$ where i is the index of the node of injected power ($1 \leq i \leq r$), Ai_{kl} is the $Ai[k, l]$ term of the matrix $\mathbf{A_i}$, t_{ij} the contribution of input power i for the reduced variable $\mathbf{t}$ ($1 \leq i \leq r$, $1 \leq j \leq m$), d_{ij} the contribution of input power i for the temperature node j ($1 \leq i \leq r$, $1 \leq j \leq m$), d_i the complete contribution of all input powers (superposition theorem) for node i ($1 \leq i \leq r$).

transformation to return to the initial coordinate system. Figure 7.3 shows the schematic of the SPICE circuit for $r = 2$. This schematic can be easily extended to a greater value of r.

We can observe in Figure 7.3 several blocks of m RC cells. m gives typically the number of time constants that are considered for the reduction. It corresponds also to the number of Ritz values. For each input power i, we can observe r blocks, which correspond to the contribution of the considered power to each of the r output temperatures, it meaning the coupling terms. d_{ij} represents the contribution of the i input power for the temperature at the j node. The complete contribution of all input powers is obtained owing to the superposition theorem.

In this case, the total number of passive elements used for the SPICE circuit is $r.r.m$. For example, if we consider an eight-finger transistor with one selected temperature per finger and 20 Ritz vectors, 1280 passive elements will be used to generate the thermal subcircuit.

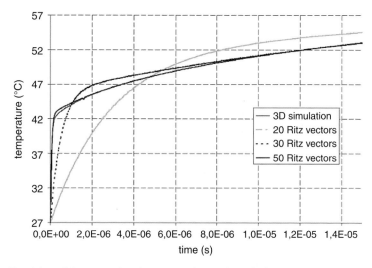

Figure 7.4 Precision of the approximation versus the number of Ritz vectors. Increasing the number of vectors improves the accuracy at short timescale. Whatever the number of vectors, the final solution is the one given by the first Ritz vector.

7.2.2.4   Model validation for an AlGaN/GaN HEMT

In the case of an AlGaN/GaN HEMT, owing to the simultaneous presence of traps and self-heating, it is still very difficult to extract the thermal impedance from measurements. However, it is possible to calibrate the simulation using a measurement of the DC value of the thermal impedance. To do that, the method proposed in reference [32] can be applied. In this method, the thermometer is the "ON" resistance of the device usually called R_{ON}. It is shown that the thermal resistance of the device can be accurately determined provided that some assumptions on the trapping behaviour of the device are verified [33]. This method makes use of the linear dependency of the "ON" resistance of the device with temperature. By measuring it in pulsed conditions at different temperatures and then at different DC power levels, the thermal resistance can be extracted with a reasonable accuracy.

FE simulations of the measured structure shown in Figure 7.5 have been performed with linear materials. The thermal impedance extracted from the MOR method is plotted in Figure 7.6. It shows that in GaN HEMTs, the increase of temperature is faster than in GaAs devices. The calibration process, using measured value, requires the nonlinear thermal conductivities of GaN and SiC to be taken into account. Moreover, a thermal boundary resistance (TBR) of $2.2\,10^{-8}$ has been taken into account to fit the measurement results. This in consistent with the TBR obtained in reference [34].

7.3   Trapping effects

This section is devoted to the modeling of the trapping effects in FETs. Since the trapping effects are harmful to the achievable power and to the linearity of FET-based ICs, many

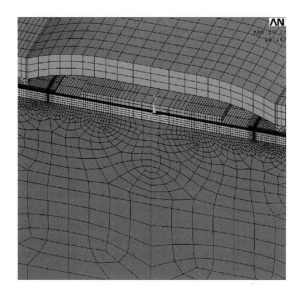

Figure 7.5 Mesh ofAlGaN/GaN HEMT.

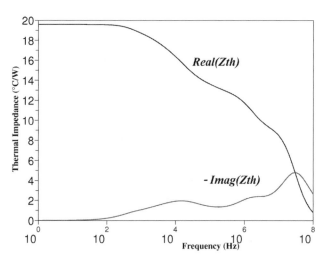

Figure 7.6 Thermal impedance of a 8 × 75 μm AlGaN/GaN HEMT obtained after the MOR process.

efforts have been made since the late 1980s to develop and include those effects in transistors models. The first trapping models were developed for GaAs MESFETs, in which the physical mechanisms of trapping were sufficiently well understood to provide the necessary physical basis for the model designers. Today, GaN HEMTs replace GaAs MESFETs in most applications, even if strong trapping effects commonly hamper them from reaching their expected outstanding power performances. This renews the interest for such models, all the more so as some trapping issues are intrinsically due to GaN HEMTs' peculiar features, and are not likely to be solved rapidly. With the GaN HEMT technology, the model designers are facing a new challenge, because the trapping mechanisms are more complex than in MESFETs, and consequently are not so well understood. On the other hand, and hopefully, they induce globally the same effects on devices' electrical characteristics, and the trapping models' architectures that were developed for GaAs-based devices remain valid for this new technology.

In the first part of this chapter, the physical aspects of trapping effects will be examined (focusing on GaN HEMTs), as an introduction to the development of models and to present what important features they need to take into account. Then, pulsed-IV measurements will be presented and discussed, as a means to highlight the dispersion effects and to get the data for the extraction of current sources and trapping circuit parameters in nonlinear FET models. Finally, an overview of the most common trapping models will be given, showing their characteristics, their interests, and their limitations. Special attention will be drawn to the trapping model developed by the authors, that will serve as a basis for the presentation of the parameter extraction methods.

7.3.1   Physical mechanisms of trapping effects in power FETs

From an electrical point of view, traps can be defined as energy levels included in the semiconductor bandgaps. In FETs, they induce electrical anomalies, known as current collapse, or transconductance and output conductance frequency dispersion [35–37]. Indeed, due to their ability to capture and release free charges with slow time constants in the orders of microseconds to seconds or even more, they induce a delayed response of the channel's current to fast electrical command signals, as is the case under RF drive. The presence and the identification of traps in GaAs-based devices have been extensively studied in recent decades [38]. Specific characterization methods have been developed to provide their cartographies, in both activation energy and physical location in the devices structures [39–41]. These works helped to deduce their physical origins, a necessary step to remove them or at least reduce their densities by improving the fabrication processes. In GaN HEMTs, the identification of traps proveds more difficult, due to the nonreproducibility of measurements, or even to their unvalidity in some cases [42]. Three main factors explain this:

- the varying material qualities and the non-stabilized growth processes;
- the larger trap densities: in the device structures, high dislocation densities are induced by the use of lattice-mismatched Si or SiC substrates, owing to the unavailability of bulk GaN; at the free surfaces, the presence of large ionized donor densities is at the origin of the positive charge needed for the creation of the 2D electron gas in the channel (2DEG) [43];
- the presence of high electric fields in wide bandgap devices that enable specific mechanisms, such as Poole–Frenkel effects, to occur [44].

Knowing the characteristics of the traps present in devices also leads to a better understanding of the physical mechanisms involved in the observed electrical anomalies, often thanks to the help of physical simulations. The latter were used [45–47] to understand how traps, with given densities, given activation energies, and at a given location in the transistors structures, were able to modify their electrical characteristics and to be at the origins of the electrical anomalies. In other words, they helped to make the link between the physics of an isolated trap and the macroscopic so-called trapping effects in devices that is clearly nonobvious, especially due to the NL electric field distribution in FET structures, a function of the command voltages.

However, the rate equation of the Shockley–Read–Hall (SRH) statistics, presented in equation (7.14) for an electron trap, emphasizes the important features of traps: their occupancy factor is determined by a balance between two processes, the capture and emission of free electrons.

$$\frac{df_T}{dt} = n \cdot C_n(1 - f_T) - e_n \cdot f_T, \tag{7.14}$$

with

$$C_n = \sigma_n \cdot vth_n, \tag{7.15}$$

where f_T is the electron occupancy ratio for deep traps, n is the electron concentration, σ_n is the electron capture cross-section, and vth_n is the electron thermal velocity. The capture rate is given by $n \cdot C_n$, whereas the emission rate e_n is determined by the Arrhenius law presented in equation (7.16).

$$e_n = \frac{1}{\tau_{emission}} = A \cdot T^2 \cdot e^{-E_A/k \cdot T}, \tag{7.16}$$

where A is a constant, and T the temperature. $\tau_{emission}$ and E_A are, respectively, the trapped charge emission time constant and the activation energy.

Thus, the capture rate is proportional to the electron concentration, unlike the emission rate, which has a strong temperature dependence. This set of equations shows that the emission time constants are generally several orders of magnitude longer than the capture time constants. This peculiar behavior of the traps is very important in order to understand the transistors' electrical characteristics under RF drive, in transient but also at steady state.

From an electrical point of view, the electrical anomalies that occur in FETs can be separated into two effects: gate-lag and drain-lag. They correspond respectively to the delayed response of the current to a modulation of the gate and of the drain potentials. This separation is practical, each effect being attached to only one command voltage (V_{gs} or V_{ds}). Consequently, in the frequency domain, the impact of gate-lag is noticeable on the transconductance (the partial derivative of the output current with respect to V_{gs}), and the impact of drain-lag on the output conductance (the partial derivative of the current with respect to V_{ds}).

Those effects can be seen in the current measurements of a GaN HEMT presented in Figure 7.7. In Figure 7.7a, V_{gs} is pulsed from $V_p = -6$ V up to $V_{gs} = -5$ V, V_{ds} being kept constant at 5 V. The transient is thus related to gate-lag effects. In Figure 7.7b, V_{ds} is pulsed from 22 down to 18 V, and V_{ds} is kept constant at -5 V. Here, the transient is due to drain-lag effects. Separating these two effects has also a physical meaning, each of them being ruled by different mechanisms: gate-lag effects are mainly due to the surface

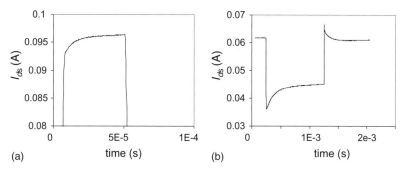

Figure 7.7 Observation of the drain current when the gate-source voltage and the drain-source voltage are pulsed in a GaN HEMT: (a) the current transient is referred as gate-lag, (b) as drain-lag. In both cases, the thermal variations are negligible and cannot explain the current transients.

traps, whereas drain-lag effects are mainly due to buffer or substrate traps [48, 49]. The mechanisms will be presented briefly in the following paragraphs.

7.3.1.1 Drain-lag (DL) effects

In GaAs MESFETs, the presence of deep traps underneath the active layer has been identified as inducing trapping effects. The output current in MESFETs is determined by the channel's effective thickness, modulated by the gate potential. However, in the presence of ionized traps in the substrate [50], or at the substrate–channel interface [51], it is also modulated by the extension of the depletion layer into the active layer. As the injection of free electrons in the substrate is determined by the drain potential, this mechanism leads to drain-lag effects. It has also been called "self-backgating", the traps acting as a pseudo backgate terminal [52].

The self-backgating has been the main dispersive effect in GaAs MESFETs and has been widely studied. The difficulty of reducing it and its large impact on the electrical performances has resulted in the elaboration of the first trapping models, which will be detailed further.

In GaN HEMTs, the phenomenon of drain-lag is quite similar. However, the current in such devices is modulated by the density of free electrons in the channel, not by its effective thickness. The presence of the 2DEG is necessary in order to compensate the positive (piezoelectric) charge σ_{pos} at the AlGaN–GaN interface. However, if the resultant charge of ionized traps in the vicinity of the AlGaN–GaN interface is not null and becomes negative, the equilibrium is kept by changing the 2DEG density (n_s). This is shown in the relation (7.17), in the presence of donors and acceptors:

$$+\sigma_{pos} = q \cdot n_s - q \cdot Nd^+ + q \cdot Na^-, \qquad (7.17)$$

where Nd^+ is the density of ionized donors, and Na^- the density of ionized acceptors.

The schematic given at Figure 7.8 illustrates the impact of the traps on the 2DEG density (n_s), when a drain voltage pulse is applied. The conduction band diagrams (under the gate) are shown at three different situations: (a) at the initial state, (b) after a positive

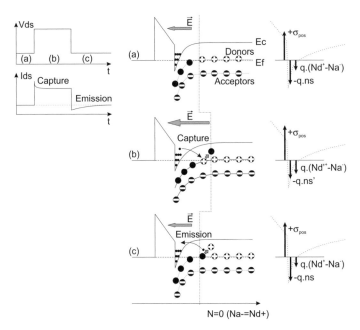

Figure 7.8 Description of the trapping and detrapping mechanisms by traps located in a GaN HEMT buffer and the consequences on the 2DEG density, which evolves to maintain the charge equilibrium. Note that another trap configuration in the buffer would lead to the same reasoning and to the same results.

drain-voltage variation, and (c) back at the initial voltage. It is assumed here that deep acceptor and donor traps are present in the buffer layer – a probable case in compensated semi-insulating buffers – with respective densities Na and Nd; and also that $Nd > Na$ (imperfect compensation). Hence, to maintain the flat-band condition, the Fermi level is clipped to the donors' energy level in the absence of an electric field:

$$Nd^+ - Na^- = 0 \text{ when } E = 0. \tag{7.18}$$

On the other hand, the electric field at the AlGaN–GaN interface is not null, and donors are filled with electrons. The equilibrium is reached when the total amount of charges is null, leading to (case (a)):

$$n_s = Nd^+ - Na^- + \sigma_{pos}/q. \tag{7.19}$$

When the drain voltage is pulsed high (case (b)), the increase of the electric field vertical component induces a deeper scattering of the electrons from the 2DEG, which can be captured by donor traps. Hence, the density of ionized donors decreases (a filled donor is neutral) to a value noted Nd'^+, resulting in a new charge equilibrium equation and a reduced 2DEG density, noted n_s':

$$n_s' = Nd'^+ - Na^- + \sigma_{pos}/q. \tag{7.20}$$

When the drain voltage is pulsed down (case (c)), the inverse phenomenon happens: some donors are no longer submitted to any electric field, and then emit their trapped

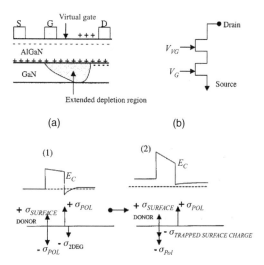

Figure 7.9 Concept of the virtual gate, from reference [54], (a) due to the partial neutralization of the sheet positive charge at the AlGaN surface. In (b), the equivalent HEMT circuit, with two gate potentials: V_G for the metal gate, V_{VG} for the virtual gate. Underneath, the band diagrams of a GaN HEMT in the presence of ionized surface donors, inducing the presence of the 2DEG (1), and when charges are trapped at the surface, inducing a subsequent decrease of this 2DEG (2).

electrons. The 2DEG density can then be expressed again by the formulation (7.19), as soon as the slow emission process is ended. However, the schematic presented here is too simplistic to make the drain-lag effects totally well understood in HEMTs: the scattering of electrons into the buffer not only occurs under the gate, but also in a large part of the gate-drain region, owing to the electric field orientation. Therefore, the 2DEG density may also be reduced in the drain access region, leading to the increase of the access resistance, expressed as:

$$R_{access}(\mathrm{Ohm}/square) = 1/(q \cdot n_s \cdot \mu_n), \tag{7.21}$$

where μ_n is the electron mobility in the channel. This explains the knee voltage increase often observed in the presence of drain-lag effects during pulsed-IV measurements.

7.3.1.2 Gate-lag (GL) effects

This spatial extension does not concern so much the gate-lag effects, which are mainly due to the presence of traps at the semiconductor free surfaces, and particularly at the drain edge of the gate, where the electric field is the strongest, and where the gate can provide free electrons to fill these surface states.

The involvement of the surface states in the current dispersion has been studied by Ladbrooke and Blight [53] in GaAs MESFETs. In GaN HEMTs, the concept of "virtual gate" was introduced by Vetury *et al.* [54], and explains how the probable presence of deep donors at the drain edge of the gate and the neutralization of their charge by trapped electrons is able to reduce the 2DEG density, i.e., to be at the origin of current collapse. Figure 7.9 shows an illustration of Vetury's virtual gate concept (a), and the equivalent electrical schematic of the device, in which the surface states act as a second gate electrode (b). The charge equilibrium is also shown in two cases: when the surface donors are ionized, leading to the presence of the 2DEG (1), and when a partial neutralization of the surface charge induces a reduction of the 2DEG density (2). The free electrons at the surface are provided by the tunnel current controlled by the gate potential [42], and hence, the current dispersion is clearly linked to gate-source voltage variations, inducing so-called "gate-lag" effects.

The physical aspects of the trapping mechanisms that occur in GaAs- and GaN-based devices have been presented here. The rate equation of the SRH statistics rules the trap behavior, and shows that capture and emission are very asymmetrical mechanisms.

However, this single-trap physics is not able to explain alone the current dispersion. The trap locations in the structures as well as their electrical environment have to be considered. Then, two different mechanisms can be identified: trapping in the buffer states and trapping in the surface states, which lead respectively to drain-lag and gate-lag phenomena. However, the reality is much more complicated. For example, the physical separation between these two effects is not totally clear in some cases: GL can be due to bulk effects at high reverse voltages, as encountered under RF overdrive [55]. The surface traps are also likely to be sensitive to drain potential variations that influence the electric field, thus enhancing the tunneling of electrons from the gate [56]. More generally, electrons dynamics through the trapping centers is strongly affected by electric fields. Thus, capture and emission rates in GaN HEMTs cannot be well described in all the cases by the simple SRH statistics, as presented in equation (7.14). For example, the filling of surface states can be slowed down by the tunelling mechanism that supplies the electrons, and their emission rate can be enhanced by Poole–Frenkel effects that reduce their apparent activation energy [44].

However, this short introduction to trapping effects physics is sufficient to approach their characterization and their modeling, which will be presented in the following sections.

7.3.2   Pulsed-IV characterizations for the trapping effects quantification and FETs modeling

One of the most difficult tasks in the extraction of a nonlinear model of a transistor is to get a proper description of the current source. The output current is not only dependent on the RF voltages, but also on nonlinear effects such as self-heating and trapping that induce low-frequency (LF) dispersion. Pulsed-IV measurements are the most common means used to extract the transistor models' current sources parameters. An explanation of their interest can be found in the schematic presented in Figure 7.10. The time evolution of the dispersive effects that occur under RF drive are shown here. In GaN HEMTs, for example, the thermal effects can last several tenths of milliseconds and the detrapping process several seconds. The trapping process is considered here to be a high-frequency effect, and is assumed to be fast enough to be able to follow the RF signal variations. The most common measurement setups are also presented in this schematic, according to their characteristic measurement time. This schematic shows

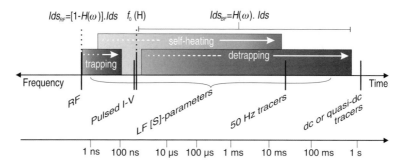

Figure 7.10 Position of the most common measurement tools in transistor's typical frequency range. The pulsed-IV measurements are set in order to characterize only the RF part of the current, with reduced self-heating effects and fixed trapping effects. The low-frequency S-parameters are presented in this schematic, as a powerful means to characterize the DC to RF transconductance and/or the output conductance dispersion.

two points: first, if the pulsed-IV characteristic time is well chosen to be comprised between the processes of capture and emission, the two effects can be separated, thanks to methods that will be presented in the first part. In this part, the measurement issues that can appear will also be addressed with a particular emphasis. Second, the pulsed-IV measurement's characteristic frequency is, ideally, close to the cutoff frequency of the lowpass function that can be used to separate the current in a high frequency and a low frequency part [57], as presented in equation (7.22).

$$Ids = Ids_{LF} + Ids_{RF} = [(1 - H(\omega))] \cdot Ids + H(\omega) \cdot Ids. \tag{7.22}$$

Therefore, such means are able to separate the LF and RF characteristics, and then to provide the RF characteristics at a fixed (and chosen) LF state, unlike those of the DC (a part of self-heating is, however, unavoidable). Nevertheless, the choice of the good LF state, i.e., the one that gives the best representation of the device current RF characteristics, can prove to be difficult. All the more so if the low-frequency state dynamics cannot be easily reproduced, thus leading to measurement uncertainties. This will be explained in detail in the second part.

7.3.2.1 Trapping effects quantification

Pulsed-IV measurement principles will be presented in Section 7.4. Due to the very short pulse lengths used, they constitute a powerful means to rapidly evaluate the trapping effects, and even to discern gate-lag and drain-lag effects [58]. Indeed, assuming that the emission process is longer than the pulse duration and that capture is almost instantaneous leads to the possibility of choosing whether the trap state is determined by the instantaneous voltages (v_{gsi}, v_{dsi}) or remains frozen at a level determined by the quiescent voltages (V_{gsq}, V_{dsq}) during the pulses. Then, three quiescent bias configurations are particularly interesting:

1. ($V_{gsq} = 0$, $V_{dsq} = 0$): in this case, the description of the IV characteristics ensures that the gate-source voltage pulses are negative and the drain-source voltage ones positive.

The capture process predominates during pulses, and then the gate-lag and drain-lag-related traps rapidly reach their steady state, determined by the instantaneous voltages v_{gsi} and v_{dsi}. In fact, the characteristics obtained are similar to DC ones, except, of course, from the thermal aspect.

2. ($V_{gsq} = V_p$, $V_{dsq} = 0$): here, the drain-source voltage pulses are positive, as in the first case. However, the gate-source voltage pulses are positive, and emission predominates. As this is long if compared to the pulse duration, the charge of gate-lag-related traps will have no time to evolve during the pulses. Thus, these traps will remain overcharged at a level determined by the quiescent level V_{gsq}, and the current dispersion will be enhanced in the IV measurements.

3. ($V_{gsq} = V_p$, $V_{dsq} = V_{ds_0}$ = voltage of operation): here, emission predominates for both gate-lag and drain-lag-related traps, when $V_{dsi} < V_{ds_0}$ (when $v_{dsi} > V_{ds_0}$, this case is similar to the second one). The traps remain overcharged to levels determined by V_{gsq} and V_{dsq}, and then the current dispersion is enhanced for both gate-lag and drain-lag-related trapping effects. The dispersion related to gate-lag effects is evaluated by comparing cases (1) and (2), the one related to drain-lag effects by comparing (2) and (3).

It has to be noted that the thermal state is the same in the three cases, the quiescent dissipated power being always null. Therefore, the measured differences are only due to traps-related dispersion.

7.3.2.2   Measurement issues

This section addresses the potential problems related to the measurements of pulsed-IV characteristics of transistors subject to trapping effects. As explained previously, the possibility of highlighting the dispersion effects thanks to pulsed-IV measurements is based on the fact that the capture process duration is very short if compared to the pulse one, and that the emission process has not begun before the end of the pulses. However, these assumptions appear as an ideal case that may be not always verified, due to the complex behavior of the trapping effects in transistors and to the complex variations of the time constants.

Four transient simulations performed with a transitor model that includes GL effects (with only one trap considered) present hereafter what the impact on the measurements could be if these conditions were not respected. The motivation of such study is that measurements performed on GaN HEMTs show that capture-like transients are sometimes observed in the microsecond range and that very long emission processes (several seconds or even minutes) commonly occur. However, no consideration to the physical possibility of these configurations will be given here. It will be shown that measurement results can be strongly dependent on the traps-related time constants and on the pulse settings.

The instantaneous current i_{dsi} ($v_{gsi} = -4$ V, $v_{dsi} = 20$ V, $t = 350$ ns after the beginning of the pulses) is evaluated for different trapping time constants values, and for two different gate-source quiescent voltages, in order to quantify the gate-lag-related dispersion effects. The pulses' duration is fixed at 400 ns, and the pulse repetition period at 10 μs. The simulations are presented in Figure 7.11: on the left graphs, the pulses when $V_{gsq} = -8$ V and $V_{dsq} = 20$ V are presented; on the right graphs, the pulses when $V_{gsq} = 0$ V and $V_{dsq} = 20$ V.

– *Capture is fast ($\tau_{capture} = 10$ ns) – Emission process is long if compared to the pulse duration, but shorter than the pulse repetition period ($\tau_{emission} = 2\,\mu s$):* (case (a) in Figure 7.11).

This case may appear as the ideal one, the capture process being finished at the measurement time and the emission process before the next pulse, when $V_{gsq} = 0$. However, when $V_{gsq} = -8$ V, the traps are not totally frozen during the pulses and the current increases, leading to a lesser difference than expected between the instantaneous currents measured at $V_{gsq} = -8$ V and $V_{gsq} = 0$ V, and hence an underestimated gate-lag-related dispersion.

– *Capture is fast ($\tau_{capture} = 10$ ns) – Emission process is long if compared to the pulse duration and to the pulse repetition period ($\tau_{emission} = 100\,\mu s$):* (case (b) in Figure 7.11).

The capture process is finished at the measurement time, but the emission process is not finished before the coming pulse, when $V_{gsq} = 0$. Thus, the charge state of the traps has not retained to its steady-state value (determined by V_{gsq}) at the beginning of the pulses. However, that has no impact on the charge at the measurement time, as the fast capture in the pulses is only dependent on the instantaneous voltages, and is not affected by the traps' state before the pulses.

Moreover, unlike in the previous case, traps are well frozen in the measurements at $V_{gsq} = -8$ V. The dispersion effects are then correctly evaluated. This result, showing that the presence of traps with long emission time constants is not harmful to the measurement results, is a very important report for the measurements of GaN HEMTs.

– *Capture is fast ($\tau_{capture} = 10$ ns) – Emission process is short if compared to the pulse duration ($\tau_{emission} = 300$ ns):* (not presented in Figure 7.11).

In this case, capture and the emission occur during the pulses, and no difference can be observed between the two configurations. Everything happens as in the case of DC measurements, the pulses being longer than both capture and emission characteristic times.

– *Capture process is long ($\tau_{capture} = 1\,\mu s$) if compared to the pulse duration – Emission process is long if compared to the pulse duration and to the pulse repetition time ($\tau_{emission} = 100\,\mu s$):* (case (c) in Figure 7.11).

This case is more complex. As observed previously (in case (b)), the traps are well frozen when $V_{gs} = -8$ V. However, as the capture process is slow, the traps state (i.e., the current) does not reach its final value during a pulse, and has either no time to go back to its steady state during the time between two pulses, when the emission process predominates. This induces a slow drift of the current value during the pulses, and hence its measured value depends on the time, since the first pulse (see graph at the bottom

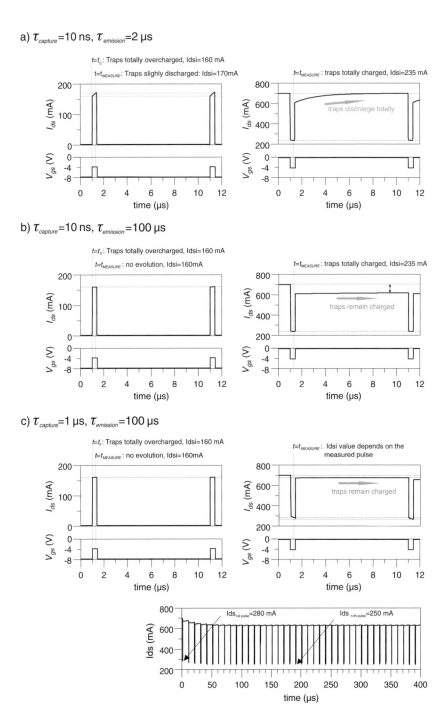

Figure 7.11 Transient simulations with a generic GaN HEMT model including the gate-lag effects, considering one trap level, with a different capture time constants $\tau_{capture} = 10$ ns, $\tau_{emission} = 2$ μs (case a); $\tau_{capture} = 10$ ns, $\tau_{emission} = 100$ μs (case b); $\tau_{capture} = 1$ μs, $\tau_{emission} = 100$ μs (case c). The pulse duration is 400 ns, the pulse repetition period is 10 μs. For the graphs on the left, V_{gs} is pulsed from $v_{gsq} = v_p = -8$ V to $v_{gsi} = -4$ V. For the graphs on the right, $V_{gsq} = 0$ V, $v_{gsi} = -4$ V. In all cases, $V_{dsq} = v_{dsi} = 20$ V. Self-heating effects are not modeled.

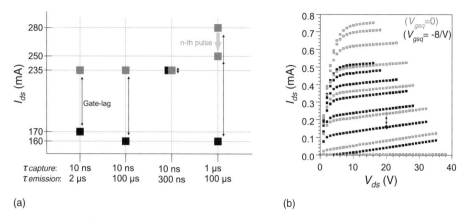

Figure 7.12 Summary of the current measured for each case at $V_{gsi} = -4$ V, $V_{dsi} = 20$ V. The gray points show the current for $V_{gsq} = 0$ V, the black points for $V_{gsq} = -8$ V. For each configuration, the level of gate-lag is different. On the right, two simulated IV networks showing the gate-lag effects, in the configuration of the second case ($\tau_{capture} = 10$ ns, $\tau_{emission} = 100$ μs).

of Figure 7.11, which corresponds to the graph on the right with a larger number of pulses). Hence, a pulsed-IV bench with a fast acquisition capability will give different results from a bench with a longer one, or from a bench with measurements averaged over several pulses.

Figure 7.12 summarizes the measurement differences in the four cases studied here. The gray points correspond to the instantaneous current for the quiescent bias point $V_{gsq} = 0$ V; the black points for $V_{gsq} = -8$ V. The dispersion effects (i.e., the difference between these currents) are differently evaluated in all the cases (the best one being the second case). In Figure 7.12b, two simulated IV networks at the two different quiescent bias points ($V_{gsq} = 0$ and $V_{gsq} = -8$ V) illustrate the second case and show IV characteristics, as typically measured.

This study has shown the possible measurement issues that may appear with nonideal transistors, or that can be created due to inaccurate pulses and period durations settings. The real case is far more complicated than this single trap example, several trapping levels (with different time constants) being commonly active in real devices. Moreover, V_{gs} and V_{ds} are simultaneously pulsed, whereas V_{ds} has been maintained constant here.

Other evaluation errors of the dispersion effects are likely to be made when the pulses are not well square-shaped, and particularly when voltage overshoots appear at the beginning or at the end of the pulses, as the fast capture process can occur during the short overshoots duration. Baylis *et al.* [59] also remarked that the pulse generators' impedance could have a significant influence on the measurement results, when the emission time constants are longer than the pulse periods. Then, to avoid most of the measurements issues, it is generally preferable to choose as long as possible pulse repetition periods. Moreover, when the choice is possible, little devices generating low current levels are preferable in order to limit the creation of overshoots.

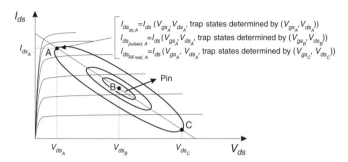

Figure 7.13 Illustration of the approximation made by using low-frequency measurements to determine RF characteristics. Pulsed measurements are more accurate than DC ones, but are not able to reproduce the real RF functioning and the dynamic evolution of the trapped charge.

7.3.2.3 Characterizations for nonlinear modeling

The other use of pulsed-IV measurements is dedicated to transistor modeling. Indeed, the possibility to set a quiescent thermal point and then to obtain quasi-isothermal measurements is of great interest to improve the accuracy of the current source parameters extraction in electrical models. The same reasoning has been applied to the trapping effects: they can be fixed and determined by the quiescent bias point, leading to a more accurate modeling of the current source. However, under large-signal RF excitation, the traps' charge level is not determined by the quiescent bias point, but more likely by the peak voltages $V_{gs\,peak}$ and $V_{ds\,peak}$, owing to the large dissymmetry between the emission and capture time constants. Indeed, the traps charge fast to the peak values, and have no time to discharge during the RF period.

To illustrate the problems encountered in the correct modeling of the current source, large signal load cycles are superimposed on to a IV network in Figure 7.13 (the thermal effects are not considered here). Here, the current at point A under RF conditions is determined by the peak voltages $V_{gs\,peak}$ and $V_{ds\,peak}$ that are reached at point C. These voltages depend on the input power, but also on the input and load impedances. If the current source is modeled from DC measurements or pulsed-IV measurements at $V_{gsq} = 0$ and $V_{dsq} = 0$, the trap states at point A are determined by the voltages V_{gs} and V_{ds} at point A, which are very different from the ones at point C. If the current source is modeled from pulsed-IV measurements at a quiescent bias point B (the nominal operating point of the application), then the traps at point A are determined by the quiescent voltages $V_{gs\,B}$ and $V_{ds\,B}$. Pulsed measurements are therefore a better means to obtain realistic IV characteristics. However, the dynamical aspect of the trapping effects (i.e., the peak voltages dependence on P_{in} and Z_{in}, Z_{out}) cannot be reasonably reproduced.

This example highlights the advantage of pulsed-IV over DC measurements, when the quiescent bias is chosen at the nominal operating point. From a thermal point of view, such a quiescent biasing is also interesting, as the devices temperature in RF amplification is determined by the nominal operating point at a first level of approximation, especially in class A or AB. This explains why such measurements have been successfully used to define the current source parameters in models in which trapping and thermal effects are not modeled [4].

However, pulsed-IV measurements do not enable the real characteristics "seen" by the RF signal under large-signal drive to be reached. The differences between the real and the measured LF dispersive effect states explain the difficulties sometimes encountered in modeling accurately the large-signal measured characteristics, even with very precise IV modeling. Therefore, to include a nonlinear description of the trapping effects in models improves their accuracy, and particularly under large-signal excitation, where the real trap charge deviates the most from the one determined by the quiescent bias point.

The methods for quantifying the dispersion effects and a brief overview of the measurement issues have been presented here, showing that the pulsed-IV characterizations have to be performed with a good understanding of the trapping effects and of the measurement setups, and that the data obtained has to be exploited with great care.

Moreover, it has been shown that pulsed measurements are the most accurate means to obtain RF IV characteristics, as they allow the LF effects to be fixed at chosen states. However, the trap state is not constant under RF drive, but depends on the signal excursions in a complex manner. To improve the model accuracy, trapping circuits have been developed. They are presented in Chapter 8.

7.3.3   Trapping effect models

As presented in the previous sections, the trapping effects may considerably impact the FETs' electrical performances. Therefore, several trapping models have been developed since the early days of nonlinear device modeling. Following the general improvement in technique and the need for more accuracy, these models have evolved from small-signal to large-signal capability.

In this chapter, the functioning of the most common trapping models will be described, from the simplest to the most accurate ones, allowing a progressive development in the explanation of their functioning. A nonlinear model optimized for harmonic balance simulation has been proposed by the authors. It will be described here with a special emphasis on the parameter extraction methods. Finally, the impact and the interests of the trapping models in terms of model accuracy improvement will be presented.

7.3.3.1   Overview of the published models

The next paragraph focuses on drain-lag models, for the reason that the first models were designed to take into account the self-backgating, i.e., drain-lag effects. However, some of the circuits presented hereafter are also used to model the gate-lag effects.

"RC branch"-type models

The first trapping models were developed to take into account the difference (due to self-backgating effects) between the output conductance extracted from small-signal RF characteristics and that obtained from DC or low-frequency IV measurements. They

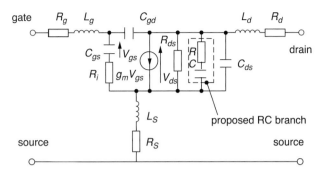

Figure 7.14 Small-signal model proposed in reference [60] to model the LF to HF dispersion of the output conductance in a MESFET's small-signal model.

consisted in adding, at high frequency, a correction term to the DC output conductance (given by the main current source of the model).

At low frequency,

$$G_{ds\,LF} = G_{ds\,DC} \qquad (7.23)$$

At high frequency,

$$G_{ds\,HF} = G_{ds\,DC} + \Delta G_{ds}. \qquad (7.24)$$

The transition from LF to HF is determined by the traps' emission time constants, the emission being limiting the process to the ability of the traps to "follow" or not the signal variations. The first small-signal model of this type was developed by Camacho-Penalosa and Aitchison [60]. It consists of an additional RC circuit added in parallel to the current source, as presented in Figure 7.14. The resistor and capacitor values are chosen in order that $1/R = \Delta G_{ds}$, and $RC = \tau_{emission}$.

This simple approach to modeling the self-backgating effects is efficient and suitable for small-signal models, but is no more valid under large-signal conditions, because the correction term ΔG_{ds} is constant, whatever the bias conditions. This can induce inconsistencies when such models are used under large-signal conditions: the RC circuit impedance being low to provide a noticeable correction of the output conductance, a non-negligible current may flow into it. If this current is higher than the current source one, as may occur for V_{gs} values close to the pinch-off voltage, this results in a negative drain current. Indeed, the current in "RC branch" models is expressed as:

$$I_{ds} = I_{ds\,DC} + \Delta G_{ds} \cdot (v_{ds} - v_T), \qquad (7.25)$$

where v_T is the voltage representing the traps' level of charge (and is equal to the capacitor voltage).

Figure 7.15 illustrates this problem. A FET model is simulated in transient to reproduce pulsed-IV networks. In Figure 7.15a, without the RC circuit to modify the output conductance, the model is able to fit the measurements done at $V_{dsq} = 0\,\mathrm{V}$ (which is equivalent to a DC measurement without thermal effects). In Figure 7.15b, with a RC

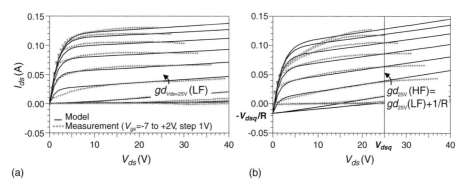

Figure 7.15 Problems encountered by using low-impedance RC branches in large-signal simulations: (a) measurement at $V_{dsq} = 0$ V compared to a transistor model without RC branch; (b) measurement at $V_{dsq} = 25$ V, compared to a model with a RC branch: the RF output conductance in the vicinity of $V_{ds} = 25$ V is better reproduced, but the current I_{ds} becomes negative near pinch-off.

circuit added (R = 1500 Ohm, RC = 100 μs), the model is able to fit the output conductance of the network measured at $V_{dsq} = 25$ V (which gives the RF output conductance in the vicinity of $V_{ds} = 25$ V). However, the current I_{ds} becomes negative in a part of the simulated IV network, which is, of course, inconsistent. To overcome this problem, Matsunaga *et al.* [61] propose another approach: the correction is done at the main current source level through an additional term v_T in its gate command voltage, which depends on the low-frequency value of the drain-source voltage. The latter is provided by a high impedance RC branch.

$$I_{ds} = f(v_{gs} + v_T, v_{ds}), \tag{7.26}$$

with

$$v_T = Ed \cdot v_{ds\,LF}, \tag{7.27}$$

where Ed is a fitting parameter.

Nonlinear "RC branch" models

To avoid such inconsistencies, but also to improve the relatively poor accuracy of the simple "RC branch" models, other models have been proposed. They consist in putting a nonlinear correction term, depending on the instantaneous command voltages v_{gs} and v_{ds}, in place of the single value correction term ΔG_{ds}, thus allowing the output conductance RF value to be yielded over the whole characteristics. The first model of this type was introduced by Filicori *et al.* [62].

The measurement presented in Figure 7.15b illustrates their interest: the IV curves are not accurately fitted with a single RC branch model (especially in the area delimited by $(v_{gsi} = [0, +2\,\text{V}], v_{dsi} = [0, 10\,\text{V}])$, that could be improved with the use of a nonlinear correction term. Moreover, the negative current problem highlighted previously can be avoided if the correction term is null in the range of $(v_{gsi} = V_p, v_{dsi} = [0, V_{dsq}])$.

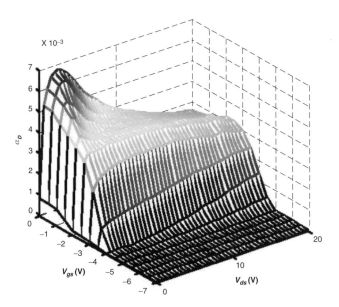

Figure 7.16 Output conductance correction term (ΔG_{ds}, named here α_D) extracted from pulsed measurements of a $8 \times 125\,\mu$m GaN HEMT [66]. Note that the correction term is null at $V_{gs} = V_p$, hence avoiding negative currents, and is the highest in the knee region, as needed in the example presented in Figure 7.15.

In references [62] and [63], the correction term is directly introduced into the main current source. Then the (high-Z) RC branch is kept to probe the LF value vd_{LF} of the drain-voltage). In other cases, [64,65], the resistor is directly replaced by a nonlinear current source $I_{ds\,TRAP}(v_{ds}, v_{gs})$ in a RC branch, resulting in the same expanded validity domain of the correction term.

Be that as it may, the current source formulation is then expressed as:

$$I_{ds} = I_{ds\,DC}(v_{gs}, v_{ds}) + f_d(v_{gs}, v_{ds}) \cdot (v_{ds} - v_{d_{LF}}), \qquad (7.28)$$

where $f_d(v_g, v_d)$ represents the nonlinear correction term that replaces the fixed ΔG_{ds} in simple RC branch models (in equation (7.25)). An example of the correction term is presented in Figure 7.16, extracted from pulsed-IV measurements of a GaN HEMT device [66]. They can be modeled thanks to nonlinear formulations [65] or lookup tables [67].

However, they do not have a fully large-signal behavior. This is because the process of capture is not considered, whereas it determines, as well as the emission process, the response of the device in time domain and under large-signal excitation. Indeed, taking into account only the emission process amounts to considering that the density of trapped charges is determined by the average value of the traps' command voltage. However, as explained in the previous section, the density of trapped charges tends to be determined by the peak command voltages more than by their mean values. One can retort that the pulsed-IV measurements are, however, large-signal, and that they are well reproduced by these "emission-only" models. However, such characterizations correspond to a very

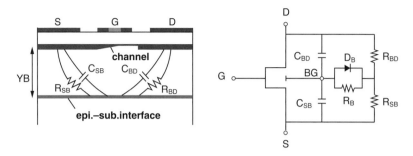

Figure 7.17 Schematic diagram of the parasitic elements in an HJFET, and equivalent circuit model with a pseudo-backgate terminal in reference [47]. Like in the previously presented models, it is assumed that changes in the backgate voltage are proportional to changes in the drain voltage, which has been experimentally observed (e.g., see reference [73]).

specific and unique case, in which the traps' state is determined by the quiescent bias V_{dsq}, which is also equal to the LF value of the drain voltage, owing to the short duration of the pulses.

Despite all, these models constitute a clear improvement in that the correction term is described over a large domain of definition. Moreover, some improvements were also done on the transient part: for example, the emission time constant is temperature-dependent in reference [68]. Brady *et al.* [69] also proposed a three-pole filter, in order to take into account the presence of three trap levels with three different emission time constants.

Due to the relative simplicity of the correction terms extraction procedure, from a reduced set of pulsed-IV measurements [62], and to their good accuracy on both small-signal and pulsed-IV measurements, such models are widely used [70–72]. Moreover, most of them also take into account the dispersion effects due to the gate-lag effects, in a very similar manner.

Large-signal models
Other trapping models capable of distinguishing the capture and emission processes in the function of the drain voltage variations have been developed. The first fully large-signal model was developed by Kunihiro and Ohno [47], who designed a self-backgating subcircuit from the calculation of the charge variation at the backgate terminal (located underneath the channel at the buffer–substrate interface). The time-dependent deviation of the traps charge was derived from the rate equation of the SRH statistics (see equation (7.14)), leading to:

$$\frac{d\Delta Qb}{dt} = -JsB \cdot (e^{e \cdot \Delta Qb / Cb \cdot kb \cdot T} - 1) - \frac{\Delta Qb}{Rb \cdot Cb}, \tag{7.29}$$

where ΔQb is the deviation of the traps' occupancy ratio.

This equation shows that the electron capture (the first term) and emission (the second term) can be expressed by currents through a diode and a resistor, respectively, which charge and discharge the backgate capacitance C_b (see Figure 7.17, where

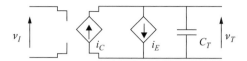

Figure 7.18 Schematic diagram of the traps' model presented in reference [74]. Two current sources are competing to charge or discharge the capacitor C_T, which voltage represents the traps' state.

$C_b = C_{sb} + C_{bd}$). The charge at the backgate (Q_b) terminal corresponds to the traps' charge. Then, according to the self-backgating mechanism in MESFETs, the backgate voltage is added to the threshold voltage of the main current source, in order to emulate the constriction of the channel.

Another large-signal model has been developed by Rathnell and Parker [74]. Here, the emission and capture currents are created by nonlinear equation-based sources, which are in the opposite direction and then are competing to charge and discharge a capacitor (see the schematic presented at Figure 7.18). The contribution of the trapping effects is implemented by adding the trap charge equivalent voltage (i.e., the capacitor voltage) to the gate command voltage of the main current source:

$$I_{ds} = f(v_{gs} + v_t, v_{ds}). \tag{7.30}$$

The two current sources modeling capture and emission are directly derived from the SRH statistics, and are commanded by a potential v_i, that corresponds to the difference between the traps' activation energy and the Fermi level.

These two models have a clear physical behavior, as has been derived from the SRH statistics. However, they have to model the effects of the traps on the device's electrical characteristics in the function of the command voltages, which are nonobviously linked to single trap levels and their own physics, as detailed in the first paragraph of this section. In the first model, this link is made by physical and geometrical considerations that are appropriate to the MESFETs. However, the parameter extraction is better achieved thanks to transient measurements. In the second one, this link is purely phenomenological and extracted from pulsed-IV measurements, all the more so, since the model is used for GaN HEMTs with complex trapping effects [75].

Therefore, if using the physical definition ensures a physical behavior of the models, purely phenomenological models can be preferred: the methods to extract the parameters are finally the same, but the models are more flexible and can be adapted according to the measured characteristics. Leoni *et al.* [73] proposed a model which can be seen as a "2 path RC branch": a diode is used to switch from one path to the other one, whether the voltage variation is positive or negative, i.e., whether capture or emission predominates. Therefore, two resistances placed in each path have very different values and allow a fast charge of the capacitor (capture), and a slow discharge (emission), to take into account the dissymmetry between the capture and the emission time constants. Extensive pulsed-IV characterizations with different pulse duration and repetition rates, at different gate-drain and gate-source voltages, provide at the same time the nonlinear correction terms, the nonlinear variations of the emission time constants, and the number

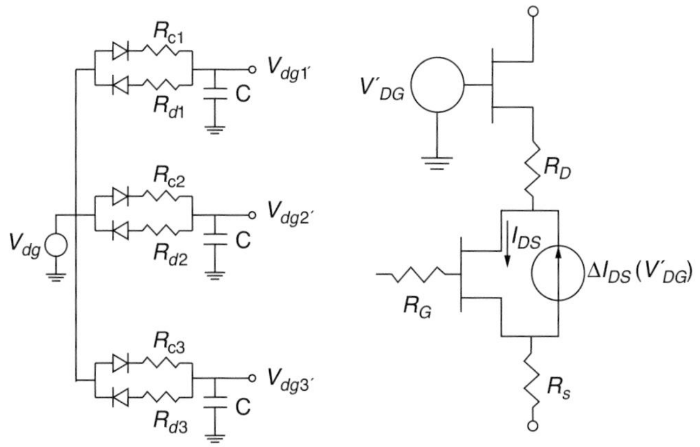

Figure 7.19 Architecture of the Leoni model for drain-lag effects in MESFETs. Three trapping levels are identified, leading to three two-path RC branch circuits. The parasitic MESFET and the current source are used to increase the drain resistance and decrease the saturated current, as measured in reference [73].

of trapping levels. After some adequate simplifications, trapping circuits are designed: they reproduce the nonlinear behavior of the trapping effects, model the dependence on the command voltages of the current dispersion (i.e., have a nonlinear correction term), and of the emission-related time constants.

Figure 7.19 shows Leoni's model for drain-lag, in a case where three trapping levels are identified. As measured (and expected from drain-lag effects, see Section 7.3.1), the saturated current decreases and the drain access resistance increases, leading to the implementation of the traps' charge-dependent circuits to reproduce these effects (a current source to decrease the current, a parasitic MESFET to increase the drain resistance).

Fully large-signal model for harmonic balance simulation
The previous model is the most accurate, as it is totally nonlinear in both current dispersion and time domains. The price to pay for such accuracy is the need for large sets of measurements and long data exploitation. However, most power device ICs are designed in HB simulators, in which only the steady-state value of the trap's charge matters (unlike in transient or envelope transient simulations). Then, the trap's asymmetrical behavior is fundamental, as it determines the trap charge steady state; the absolute values of the trapping time constants are of less importance, so much of the trapping process is faster than the emission and the system is unbalanced.

Therefore, the authors proposed some simplifications with respect to the transient part of the models. They have designed a large-signal trapping model that has a realistic large-signal behavior in HB simulations with a reduced set of parameters, and consequently, with a minimum of additional measurements needed to extract them [6]. Its architecture, its functioning, and the methods to extract its parameters will be detailed here.

In this model, the trapping effects contribution on the current is taken into account by adding a parasitic voltage to the gate command of the main current source (as

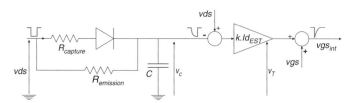

Figure 7.20 Complete drain-lag subcircuit presented in reference [6]. The left part of the circuit synthesizes the equivalent charge of the traps, the second part treats the latter voltage to modify the current source command voltage v_{gs}. The same architecture can be used for modeling the gate-lag effects with the diode inverted and $k < 0$.

in references [61] and [74]). This allows the trap contribution to be added without modifying the main current source, hence simplifying its implementation. Indeed,

$$I_{ds} = f(v_{gs}, v_{ds}, v_T(t)) \tag{7.31}$$

can be rewritten as:

$$I_{ds} = f(v_{gs\,T}, v_{ds}), \tag{7.32}$$

with

$$v_{gs\,T} = f(v_{gs}, v_T(t)). \tag{7.33}$$

The drain-lag subcircuit schematic is presented in Figure 7.20. It can be separated into two parts: an envelope detector that reproduces the asymmetrical behavior of the traps and the two processes of capture and emission. It is repeated as many times as trap levels are considered. It is similar to the "2-path RC branch" model presented previously. However, there is only one diode, as $R_{emission} \gg R_{capture}$. Then, when the diode is open (i.e., the drain voltage variation is positive), the current flows almost only through $R_{capture}$ and charges the capacitor, in order to model the capture process. The emission is modeled by the discharge of the capacitor through $R_{emission}$, the diode being blocked during a drain voltage negative variation. The capacitor voltage V_C is then related, as is the previous models, to the density of trapped charges, and the two time constants are then defined as:

$$\begin{cases} \tau_{capture} \approx R_{capture} \cdot C \\ \tau_{emission} = R_{emission} \cdot C \text{ as } R_{capture} \ll R_{emission}. \end{cases} \tag{7.34}$$

The second part of the circuit achieves the treatment of the capacitor voltage, in order to synthesize the trap's contribution. Then:

$$V_T = k \cdot Id_{EST} \cdot (v_{ds} - v_C) \tag{7.35}$$

and:

$$V_{gs\,int} = v_{gs} + k \cdot Id_{EST} \cdot (v_{ds} - v_C), \tag{7.36}$$

where Id_{EST} represents the estimated output current, introduced to take into account the fact that the current dispersion is a ratio of the total output current (at first-order

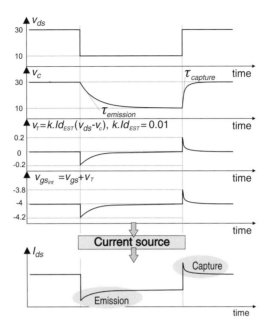

Figure 7.21 Chronogram of the drain-lag subcircuits internal voltages and the output current, for a drain pulse from 30 V to 10 V, and $k.Id_{EST} = 0.01$.

approximation), and then to make V_T linearly dependent on it. For simplicity reasons, Id_{EST} is expressed as:

$$Id_{EST} = Gm_{EST} \cdot (v_{gs} - Vp) \ if \ v_{gs} > V_p, \ else \ 0, \tag{7.37}$$

with a smooth transition at $v_{gs} = V_p$. This leads, at the main current source level, to:

$$I_{ds} = f(v_{gs} + k \cdot Id_{EST} \cdot (v_{ds} - v_C), v_{ds}). \tag{7.38}$$

The chronogram presented in Figure 7.21 shows the subcircuit internal voltages and the output current when a drain voltage pulse is applied from 30 to 10 V, and $k \cdot Id_{EST} = 0.01$. The transient capacitor voltage exhibits the slow emission and fast capture processes. It is then treated to give v_T, which represents the trapping effects contribution to the command v_{gs}.

Except during the transients, the command voltage $v_{gs\,int}$ is equal to the external command voltage v_{gs}, and then the current is determined by the main current source, which models the pulsed-IV network at the quiescent point ($V_{gsq} = 0$, $V_{dsq} = 0$), for which the current measurement is done at the trap's steady state.

Note that the expression (7.38) can be derived in a form that is equivalent to that ruling the models previously presented in equation (7.25), but in an indirect manner. To show it, a simple calculation can be done considering the main current source to be linearly dependent on V_{gs}, as:

$$I_{ds} = GM \cdot (v_{gs} - V_p). \tag{7.39}$$

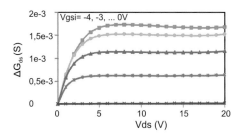

Figure 7.22 Correction of the output conductance brought by the drain-lag subcircuit, which shows a dependence on both v_{gs} and v_{ds}, even if Id_{EST} is simply expressed as a function of v_{gs} only. In this example, the device is a 2×50 μm GaN HEMT with substantial lag effects. The parameters values are $k = 1$, $Gm_{EST} = 0,02$.

Then, equations (7.38) and (7.39) give:

$$I_{ds} = GM \cdot \left[v_{gs} + k \cdot Id_{EST} \cdot (v_{ds} - v_C) - V_p \right] \tag{7.40}$$

$$I_{ds} = GM \cdot (v_{gs} - V_p) + GM \cdot k \cdot Id_{EST} \cdot (v_{ds} - v_C) \tag{7.41}$$

$$I_{ds} = I_{ds\,DC} + \Delta G_{ds} \cdot (v_{ds} - v_C). \tag{7.42}$$

Hence, this formulation is equivalent to (7.25), with:

$$\Delta G_{ds} = GM \cdot k \cdot Id_{EST}(v_{gs}). \tag{7.43}$$

However, in the real cases, the main current source also depends on v_{ds}, implying that the expression of ΔG_{ds} also depends on v_{ds} (then the calculation can no more be simply done as in equation (7.43)). Figure 7.22 shows an example of the equivalent correction term ΔG_{ds} of this drain-lag model, in a transistor model including a "modified Tajima" current source [4]. ΔG_{ds} is then obtained from two pulsed-IV network simulations at different drain-source quiescent voltages, and the following calculation:

$$\Delta G_{ds} = \frac{idsi_{V_{dsq}=25\,\mathrm{V}} - idsi_{V_{dsq}=0}}{v_{dsi} - 25}, \tag{7.44}$$

This figure shows that despite the simple expression of the estimated current $Id_{EST}(V_{gs})$, the model synthesizes a nonlinear correction term dependent on v_{gs}, but also v_{ds}. Its typical transistor IV characteristics shape has been observed in reference [73] or in [66] (see Figure 7.16).

7.3.4 Parameter extraction

As indicated previously, there is no point in modeling accurately the trap's time constants for HB simulations, as long as the capture is much faster than the emission. Another reason for little interest being taken in the time constants is the difficulty in extracting them correctly and simply, due to the combination of emission-related exponential terms and self-heating related ones. The last reason is the complexity of the trap's behavior in GaN HEMTs (see Section 7.3.1).

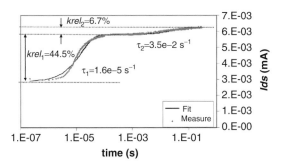

Figure 7.23 Current transient measured for a drain voltage pulse around the nominal operating point, in emission mode. Two traps are identified [6].

Then, the time constants are considered as fixed values. The capture time constants are not measured, as they are generally so fast that it would require specific equipment. The emission time constants are extracted from current transient measurements for voltage pulses in the vicinity of the nominal operating point. Thus, this model has a small-signal transient description, validated only in the vicinity of the nominal bias point.

Figure 7.23 shows a drain current transient measured for a drain voltage pulse of a GaN HEMT [6] in order to activate the emission process close to the operating point $V_{ds\,0} = 25\,V$ (V_{ds} is pulsed from 30 to 20 V). Two time constants are isolated, and a model with two traps is then implemented. The gate-source bias voltage V_{gs} is very close to the pinch-off, in order to avoid self-heating and its related transients.

The gate-lag-related emission time constants are extracted similarly, by pulsing V_{gs} close to $V_{gs\,0}$. In this case, the thermal variations are often harder to avoid, but in emission mode, the thermal-related transients tend to decrease the current, whereas the emission-related transients tend to increase it, allowing the two processes to be discerned more easily. These transient measurements provide other information: the relative current dispersion ratio k_{rel1} and k_{rel2} of each trap. As two traps are considered, the transient voltage v_T is now expressed as:

$$v_T = A_{DL} \cdot Id_{EST}(vgs) \cdot [k_{rel1} \cdot (v_{ds} - v_{C1}) + k_{rel2} \cdot (v_{ds} - vc_{C2})], \qquad (7.45)$$

where A_{DL} is a fitting parameter.

With practice and for more convenience, A_{DL} and the parameter Gm_{EST} (introduced in equation (7.37)) are merged in one single fitting parameter, which is set in order to fit the pulsed-IV characteristics at different bias voltages. At the highest level of simplification – if the model designer is sure that his model will not be used for transient simulations – only one trap can even be considered (then $k_{rel1} = 100\%$), the time constants can be set to arbitrary values (keeping $\tau_{capture} \ll \tau_{emission}$), and A_{DL} and Gm_{EST} can be merged in a single parameter that sets the current dispersion, without inducing any loss of precision in HB simulations. Then, the trapping effects are modeled thanks to a single parameter circuit (the pinch-off voltage of Id_{EST} having the same value as the main current source one).

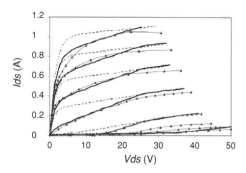

Figure 7.24 Comparison between simulated (transient) and measured pulsed-IV curves of a transistor with strong drain-lag effects, at $V_{gsq} = V_p$ and $V_{dsq} = 25$ V. The trapping model is able to switch from capture to emission, and hence reproduces the brutal change in the output conductance (i.e., the current slope) at $v_{dsi} = V_{dsq}$. The slight difference between the measurements and the model for $V_{ds} > 25$ V are mainly due to gate-lag effects, which are not modeled here [6].

7.3.5   Improvements in transistor model accuracy

It has been shown here that the "nonlinear RC branch" models are often the most accurate to reproduce the pulsed-IV characteristics, especially as they are directly extracted from them. However, they reproduce the characteristics when emission predominates and when the trap charges depend on quiescent voltages, but are not able to fit correctly the measurements if the trapping processes change. Conversely, the fully large-signal models presented here are able to fit the measurements, whatever the predominating process. Figure 7.24 shows the comparison between the measured and simulated pulsed-IV characteristics of a GaN HEMT device having strong drain-lag effects. The transition between the two processes of emission and capture, which is materialized by change of the curves' slope at $v_{dsi} = V_{dsq}$ is clear.

The fully large-signal models presented in this section have been validated by comparison with transient measurements [47,73,74]. It has also been shown that FET power characteristics obtained from load-pull and time domain load-pull (LSNA) setups can be better reproduced when the trapping effects are taken into account, and particularly on a larger range of load impedances [76]. Figure 7.25 shows a comparison between measured and modeled characteristics, with and without trapping effects (gate- and drain-lag), for different load impedances at a voltage standing wave ratio (VSWR) of 2.5 and 1.6 around the optimum power load impedance of a $8 \times 75\ \mu m$ GaN HEMT measured in class AB at 10 GHz [76]. The dramatic improvement of the simulated power characteristics is demonstrated here. It is particularly sensitive on the mean drain current, which impacts directly on the power consumption and then on the PAE. The shape of the mean drain current is explained in Figure 7.27, and the contribution of each trapping model in Figure 7.26.

At low-input power level, where the voltage swings are very small, the trapped charge densities (by gate- and drain-lag-related traps) are rapidly determined by the

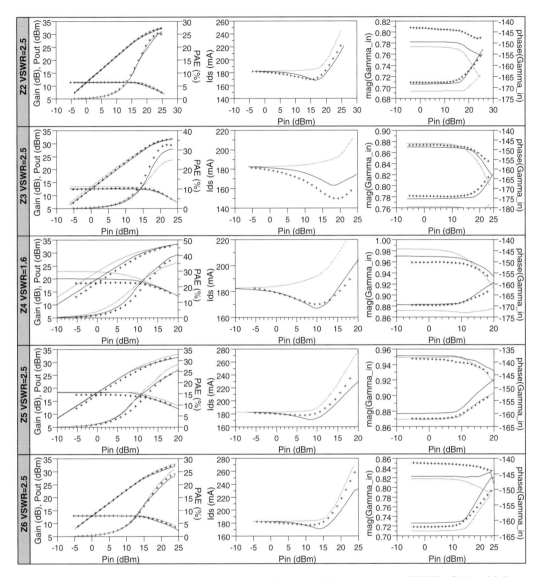

Figure 7.25 Measurements and simulations for the load impedances at a VSWR of 1.6 and 2.5 around the optimum power load impedance, with the model including trapping effects and the model not including them (crosses: measurements, black lines: model with trapping effects, gray lines: model without trapping effects) [76].

peak voltages $v_{gs\,PEAK}$ and $v_{ds\,PEAK}$, which in this case are almost the bias voltages (case (a)). If the input power increases (case (b)), the peak RF voltages reach higher values, and so do the trapped charges' density. Therefore, the trapping effects are more pronounced. The whole IV characteristics are degrading, as well as the current at the bias point decreasing.

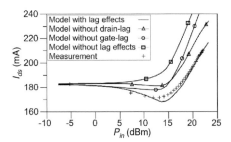

Figure 7.26 Mean output current of an 8×75 µm GaN HEMT measured at 10 GHz in class AB on its optimum load impedance. This particular shape can be reproduced only by adding trapping subcircuits in the nonlinear model [76].

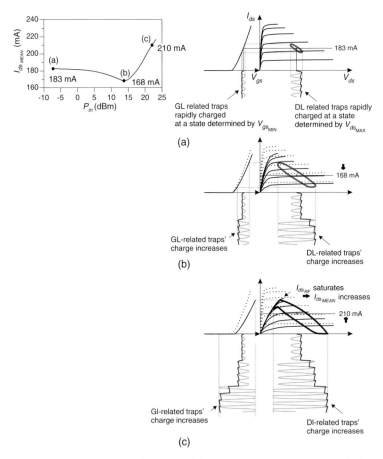

Figure 7.27 Explanation of the particular shape of the mean output current versus the input power level for a FET device subject to trapping effects. The more the input power increases, the more the IV characteristics "seen" by the RF signal are degraded, leading first to a mean current decrease (b), then to a slower increasing of the current that predicted without traps, when the self-biasing effect competes with the current dispersion (c).

This case corresponds to the weakest mean current. However, as the input power continues to increase (case (c)), the traps are more and more charged, the dispersion effects are increasingly important, and the IV characteristics degrade more. However, the current saturates, self-biasing appears, and tends to increase the mean current. The two effects (dispersion and self-biasing) compete, resulting, however, in a general increase of the mean current, but with a lower slope than if there were no traps (Figure 7.26 shows that this slope is mainly due to the gate-lag effects). Note that the mean voltage V_{ds} remains constant (the voltage supply maintains a fixed value voltage). Therefore, the "emission only" models, in which the trapping level is determined by the mean DC values of the command voltages, are not able to reproduce this dynamic trapping effects' behavior, and are of limited interest for improving the large-signal simulation results.

7.3.6 Conclusions

Some of the most common trapping model architectures have been presented here. The simplest models have been designed to improve the small-signal models' accuracy, by adding correction terms to the output conductance (drain-lag models) and the transconductance (gate-lag models) values.

The interest of models that take both capture and emission processes into account has been demonstrated and explained. Those models are the only ones to be able to reproduce the trapping effects under large-signal RF drive, although some "emission-only" trapping models have sometimes been called "large-signal", due to the fact that they provide nonlinear correction terms defined over the whole command voltage range. The authors' model has been presented, as well as the process of extracting its parameters. Its main interest is that it is able to provide a realistic nonlinear description of the trapping effects, with a reduced set of parameters that can even be reduced to a single one if it is designed for nontransient large-signal simulations like HB. Despite these simplifications, the gain of accuracy has been demonstrated on IV and load-pull measurements.

7.4 Characterization tools

The modeling process presented in the previous sections requires some specific measurement tools and methods in order accurately to characterize the transistors. On the one hand, the stationary model, which constitutes the basis of the dispersive model, must be extracted in conditions that ensure the "quasi isothermal" behavior of the transistor and a known status of traps. Moreover, the overall operating area must be investigated, including high-current zones as well as breakdown limits. As the transistor behavior is determined by the three independent variables, the input and output voltages and the temperature, characterization must be performed at different base-plate temperatures. All these conditions are met by pulsed I–V and pulsed S-parameter measurements. In addition, changing the parameters of the pulse such as duration, recurrence, or starting values of the pulses allows a thorough investigation of the trapping effects as shown in Section 7.2. In order to perform such characterization, one has to implement a versatile

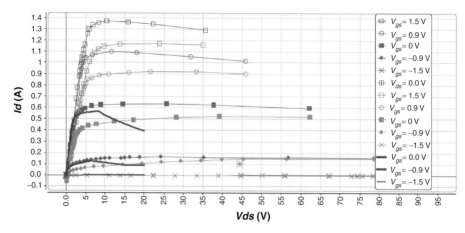

Figure 7.28 PIV measurement of a 10 W GaN device: comparison of DC (lines up to 20 V) and pulsed characteristics, starting from $(V_{ds}, I_d)=(0\ V, 0\ mA)$ for highest I_d currents and from (45 V, 100 mA) for smaller Id currents.

tool that is capable of performing on-wafer measurements. This kind of tool will be described below.

On the other hand, model verification and refinement require dynamic RF measurements to be performed. Indeed, the pulsed measurements mentioned previously allow both the stationary model as well as the equivalent circuit that takes into account the dynamics of the traps to be extracted, but it remains crucial to use the overall nonlinear dispersive model in the large-signal conditions that are close to those encountered by the transistor in real applications. These conditions imply a strong interaction between nonlinear dynamic phenomena. Load-pull measurements provide a convenient means to characterize all these effects. Indeed, recent developments of load-pull techniques concerning multiharmonic tuning, pulsed measurements, and time domain large-signal measurements provide a wide range of characterization methods which allow all the parasitic dispersive effects to be excited. Note, however, that the signals used for model extraction are usually quite simple and very different from the signals that the transistor will see in a final application.

7.4.1   Pulsed measurements

7.4.1.1   I–V measurements

The technique of PIV characterization was pioneered in the late 1980s and further developed during the 1990s [7,77–79]. Figure 7.28 shows the result of a practical measurement performed on a 10 W GaN FET, with a comparison between DC and pulsed measurements. Two pulsed measurements are proposed, where each one has a different initial bias condition. One can clearly see that the IV relationship is a strong function of the initial conditions, and that the characteristics are very different from the DC measurements. Note that the DC measurements have been performed over a much smaller bias region than the pulsed measurements. This is typically the case, since DC measurements stress the device much more than pulsed measurements. PIV measurements can measure in some regions were DC measurements would cause permanent damage to the device because of excessive heating or electric field or breakdown.

Although the basic principle of PIV measurements is relatively simple, it is still a real challenge to carry them out every day. The self-heating phenomenon inside transistors can be very fast, for example with a high-power device driven close to its limits, an increase of $20\,^\circ$C has been reported in about 50 ns, and an increase of $50\,^\circ$C in about 400 ns. It means that isothermal PIV data acquisitions must be done in a very short time. Practically speaking, PIV measurements are not easy to do with a pulse duration smaller than 200 ns; therefore, those measurements must be referred to as "quasi-isothermal" measurements. There are two main challenges when performing PIV measurements: to generate and measure fast bias pulses; and to keep the characterization process safe from the device point of view.

The first challenge consists in generating and measuring the fast high-power bias excitation pulses, which need transition times that are only a fraction of the pulse width, say 30 ns. The very new high-power microwave transistors are now handling very high currents and voltages. Only very recent MOSFET or GaN-based pulsers can deliver pulses up to 300 V and 10 A. When performing such measurements, one has to be very cautious about the presence of parasitic inductive and capacitive elements. Because of the hard slopes of the voltage and the current pulses, parasitic inductive and capacitive elements can easily cause ringing effects in the applied pulses. In fact, one needs a really careful design of the cabling between the pulses and the transistors; good practice is to locate the pulsed bias generators as close as possible to the transistor terminals. A complete characterization of a transistor requires the application of extreme pulsed bias conditions, for example near the transistor breakdown area. Under such conditions, especially when using a low-impedance generator, sudden breakdown effects may generate a spectacularly big current spike through the transistor, with an immediate destruction of the device and, sometimes, of the pulser. To prevent this from happening, one can introduce a resistive network between the transistor and the pulser. The resistances will then provide a robust protection for breakdown current spikes. Moreover, this network provides access for the measurements and allows the bias impedance to be changed, depending on the device limits. Such networks can be adjusted to reduce/avoid parametric oscillations of a particular device.

7.4.1.2 Pulsed S-parameter measurements

In order to extract the equivalent circuit of the transistor for the overall operating zone and at known temperatures, the S-parameters must be measured during the short biasing pulses applied to the transistor. Such measurements – isothermal pulsed-bias S-parameter measurements – were pioneered in the early 1990s [78–80], and are still an interesting literature topic today [81, 82].

The pulsed S-parameter measurements are very challenging because they need to be made under fast pulse conditions. In 1990, the VNA Anritsu Wiltron 360-PS20 was able to measure RF pulses as short as 100 ns, but not without severe desensitization. Indeed, the reduction of the measurement's dynamic range is given by $-20 \cdot \log \tau$ where

τ is the duty cycle. It was based on a high Q filtering at intermediate frequency. This VNA was the only one on the shelf with the capability to do very fast RF measurements. Today, improved ways of measuring pulsed S-parameters have become available with the advent of a new generation of VNAs. In addition to having better hardware, the dynamic range of pulsed measurements is further improved, but there is still a desensitization function of the duty cycle. Nevertheless, it is now possible to make 0.001% duty cycle S-parameter measurements in the X-band with a dynamic range better than 50 dB.

Another tricky point for synchronized PIV and pulsed S-parameter setups is the electric separation of LF and RF frequency domains: the bias tee. The requirements for these passive components are extreme: wideband RF on the one hand, very large voltage and current for the bias path on the other hand, in addition to a wide low-frequency bandwidth in order to keep the PIV short pulse capability. The bias tee is frequently one of the most restraining components of pulsed setups.

PIV combined with pulsed S-parameter measurements are frequently made under extrinsic temperature control, with a thermochuck or a thermal chamber. Such setups have provided data for parameter extraction of a large number of good-quality microwave high-power transistor models all over the world.

7.4.2   Load-pull measurements

7.4.2.1   Frequency domain load pull measurements

As described in reference [83], load-pull systems are typically classified into two main categories: active and passive. In a passive load-pull system, the load impedance is controlled by passive tuners. The passive tuner is usually mechanical, whereby a metal part is moving in a waveguide in order to create controllable reflections. A good example is described in reference [84]. The major drawback of using a passive structure is that one cannot compensate for any power that is dissipated between the device under test and the passive structure that generates the controllable reflection. This power dissipation inevitably occurs in all components that are placed between the transistor terminal and the tuning elements such as probes, cables, couplers, diplexers, and so on. As a result, the maximum amplitude of the reflection coefficient, as seen by the transistor, will always be smaller than one. Depending on the amount of inevitable losses in the measurement setup, the maximum amplitude of the reflection coefficient may become too small for properly loading the transistor.

This problem may be solved by using so-called active load-pull setups. These are setups in which one introduces one or more amplifiers to generate wave signals that are sent towards the DUT output terminals. The amplifier can potentially compensate for any losses and generate reflection coefficients with amplitudes equal to and even bigger than one. A good example, illustrating the problems and benefits of active load-pull is described in reference [85]. The power-handling capability of active load-pull setups is limited by the amplifier that is used.

In contrast to the active load-pull setups, the maximum power-handling capability and frequency range of passive load-pull setups is only limited by passive structures like cables, couplers, etc. Passive tuners typically operate across multiple octaves and can handle power levels larger than 100 W. Today, the vast majority of the load-pull setups use passive mechanical tuners. Note that the typical mismatch of a GaN transistor is low and suitable for most simple mechanical tuners, in contrast with silicon LDMOS transistors, which exhibit strong mismatches.

Most classical load-pull systems control the load impedance and perform power measurements only at the input signal fundamental frequency. Any large-signal excitation of the power transistor will not only generate output power at the fundamental frequency, but also at multiples of the fundamental frequency, the harmonics. The overall behavior of the transistor will not only depend on the load impedance at the fundamental frequency but will also depend on the load impedances at the harmonic frequencies. Some load-pull systems control the load impedance at the second and even the third harmonic frequencies [86, 87]. Such systems are called harmonic load-pull systems.

A frequent extension of load-pull systems consists in applying modulated signals. Signals with two tones are of great interest as one can measure some intermodulation characteristics or memory effects. Pulsed bias and/or pulsed RF signals might be required if the target application of the transistor under test is pulsed.

7.4.2.2 Time-domain load-pull (TDLP) waveform measurements

An indepth model validation for highly nonlinear effects can be achieved by measuring the time-domain voltage and current waveforms under realistic large-signal operating conditions, and by comparing them with the time-domain waveforms resulting from a simulation.

Load-pull systems having the capability to provide such time-domain voltage and current waveforms were first developed in the late 1990s [88, 89] by adding tuning technology to large-signal network analyzers [90]. Nonlinear or large-signal network analyzer (NVNA or LSNA) receiver technology itself was developed during the late 1980s and the 1990s [91–93]. Note that all of the power information that is measured by a classical load-pull system can easily be derived from the time-domain voltage and current waveforms. In a typical modern time-domain load-pull (TDLP) system, the incident and reflected waves are sensed between the tuner and the transistor terminals by two dual directional coupling structures and are measured in the time domain by a broadband receiver. The wave signals are then converted into voltage and current waveforms V1, I1, V2, and I2. The oldest approach for a broadband receiver is to use a mixer based receiver [92]. Such a receiver is leveraged from existing VNAs and measures the fundamental and the harmonics one by one, aligning the phases of all harmonics by means of a reference channel that is excited by a multiharmonic reference signal. A modern version of the mixer-based time domain receiver is described in reference [94]. An efficient alternative solution is based on the use of a four-channel sampling frequency convertor [95].

Note that in a TDLP system, one always puts the directional coupling structure between the transistor and the tuner. The advantage of this approach is that the measured voltage waves, just by themselves, completely determine the voltage waves at the transistor terminals and it is not necessary to know the *S*-parameters of the tuner. As such, the TDLP setup no longer requires any tuner calibration. A second advantage is that the

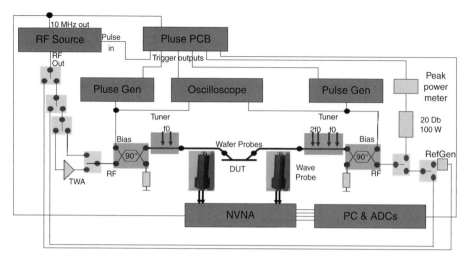

Figure 7.29 Schematic of a pulsed TDLP setup with harmonic tuning capabilities.

setup can always sense all significant harmonics signals. This is not true with the classical load-pull configuration where the harmonic information is often blocked by the tuner and, as such, cannot be sensed after it. But putting the directional coupling structure between the transistor and the tuner may cause losses that result in a degradation of the best available reflection coefficient. This issue is addressed by using extremely low-loss directional coupling structures. Currently, there are basically two solutions: one can use a specialized distributed coupler design [96] or one can use a wave probe. The wave probe is a loop coupler with a very tiny loop, the loop being significantly smaller than a quarter wavelength of the highest frequency to be measured. The principle was published more than 60 years ago [97, 98]. The loop introduces virtually no insertion loss, yet has a directivity that is sufficient for all load-pull applications. The coupling factors range from about -20 dB to 40 dB. The increase of the coupling factor for higher frequencies is beneficial for harmonic calibrated measurements. One can control the distance between the wave probe and the center conductor of the transmission line, which allows the coupling factor to be optimized for a specific power level. Figure 7.29 shows a picture of the pulsed TDLP system with wave probes at XLIM/University of Limoges (France).

Figure 7.30 shows a complete example of TDLP measurements that have been performed on such setups. An input power sweep has been applied at 4 GHz to a AlGaN/GaN transistor for a given set of input and output impedances. Harmonic frequencies up to 16 GHz are taken into account, the measurements planes are extrinsic (wafer probes calibration). All the four electrical parameters V_{gs}, I_g, V_{ds}, I_d are shown versus time, in addition to input and output load lines. The power level increase leads the transistor progressively into nonlinear regime. Note that the highly nonlinear effects in the waveforms cannot be characterized by classic load-pull techniques. Prior to the existence of TDLP systems, there was no way to measure these dynamic load lines and they were only useful in simulators.

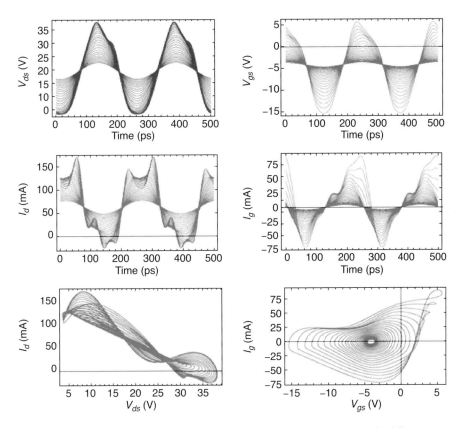

Figure 7.30 Time-domain waveforms of V_{ds} and I_d, V_{gs} and I_g, and the relevant load lines.

The TDLP systems are now extended to pulsed, burst pulsed, or modulated signals measurement in several advanced labs [99–102]. These new and still difficult measurement technologies are working with test signals that are even closer to real demanding application signals such as radar applications or wideband modulated signals communications.

7.5 Conclusions

As explained in this chapter, the modeling of dispersive effects exhibits three phases which are: the extraction of an isothermal stationary nonlinear model, the determination of the thermal embedding circuit, and the extraction of an equivalent circuit which brings the low-frequency dynamical properties of the trap's dispersive effects. Each of the three phases requires specific measurements such as I–V and S-parameter pulsed measurements as well as TDLP measurements. Such dispersive models can deal with dispersive effects that are observed either in continuous wave (CW) measurements or in more complex signals like pulsed RF signals.

However, there is still work to do in order to better understand all those dispersive effects. One of the biggest challenges consists of being able to electrically characterize the dynamical thermal effects. While it is now possible to extract the thermal impedance of bipolar transistors through low-frequency specific measurements [103]; for GaN devices the experimental separation of trapping and thermal effects through low-frequency measurements remains an open problem, as the time constants associated with those two effects lie in the same range. Moreover, the impact of the dispersive effects on the behavior of the transistor fed by complex modulated signals remains to be assessed through simulations and measurements.

An interesting way to overcome the complexity of model extraction techniques is to directly extract models from large-signal measurements, which can actually be considered as a superset of S-parameter measurements [104–106]. Although this is an appealing idea, it turns out to be difficult to apply in practice because it is hard to explore the whole operating region of the transistor by means of large-signal waveforms, in contrast to using a pulsed bias technique, and the mathematical representation of weak nonlinearities is still problematic.

From the characterization point of view, the techniques used are rapidly evolving in order to keep track of new power transistor technology. The main issue with pulsed-bias pulsed S-parameter characterization is the need to apply pulses with ever-increasing amplitude (up to 200 V and 10 A) and ever-decreasing pulse width (smaller than 400 ns). The load-pull measurements can be done with a variety of setups, with active or passive approaches, and with or without the handling of harmonic frequencies. The challenge of load-pull system developments is to offer time domain voltage and current waveforms at the transistor terminals, in addition to the capability to present low input impedances (down to 1 Ohm) and to handle high power levels (up to hundreds of watts) with multitones or modulated signals.

Acknowledgment

The authors acknowledge Jean Claude Jacquet from 3–5 Lab for physical simulations as well as fruitful discussions.

References

[1] E. Ngoya, C. Quindroit, and J. Nebus, "On the continuous-time model for nonlinear-memory modeling of RF power amplifiers," *IEEE Trans. Microw. Theory Tech.*, vol. 57, no. 12, pp. 3278–3292, Dec. 2009.

[2] W. Curtice, "A MESFET model for use in the design of gaas integrated circuits," *IEEE Trans. Microw. Theory Tech.*, vol. 28, no. 5, pp. 448–456, May 1980.

[3] I. Angelov, H. Zirath, and N. Rosman, "A new empirical nonlinear model for HEMT and MESFET devices," *IEEE Trans. Microw. Theory Tech.*, vol. 40, no. 12, pp. 2258–2266, Dec. 1992.

[4] J. Teyssier, J. Viaud, and R. Quéré, "A new nonlinear I(V) model for FET devices including breakdown effects," *IEEE Trans. Microw. Guid. Wave Lett. (see also IEEE Microw. and Wireless Components Lett.)*, vol. 4, no. 4, pp. 104–106, Apr. 1994.

[5] O. Jardel, G. Callet, C. Charbonniaud, J. Jacquet, N. Sarazin, E. Morvan, R. Aubry, M.-A. Di Forte Poisson, J.-P. Teyssier, S. Piotrowicz, and R. Quéré, "A new nonlinear HEMT model for AlGaN/GaN switch applications," *Microw. Integrated Circuits Conf., 2009, EuMIC 2009, Eur.*, Sep. 2009, pp. 73–76.

[6] O. Jardel, F. De Groote, C. Charbonniaud, T. Reveyrand, J. P. Teyssier, R. Quéré, and D. Floriot, "A drain-lag model for AlGaN/GaN power HEMTs," *Microw. Symp., 2007. IEEE/MTT-S Int.*, June 3–8, 2007, pp. 601–604.

[7] J.-P. Teyssier, P. Bouysse, Z. Ouarch, D. Barataud, T. Peyretaillade, and R. Quéré, "40-GHz/150-ns versatile pulsed measurement system for microwave transistor isothermal characterization," *IEEE Trans. Microw. Theory Tech.*, vol. 46, no. 12, pp. 2043–2052, Dec. 1998.

[8] G. Dambrine, A. Cappy, F. Heliodore, and E. Playez, "A new method for determining the FET small-signal equivalent circuit," *IEEE Trans. Microw. Theory Tech.*, vol. 36, no. 7, pp. 1151–1159, July 1988.

[9] D. Root, "ISCAS2001 tutorial/short course amp; special session on high speed devices amp; modelling," *IEEE Int. Symp. Circuits and Systs., 2001. Tutorial Guide: ISCAS 2001*, 2001, pp. 2.7.1–2.7.8.

[10] J. Collantes, P. Bouysse, J. Portilla, and R. Quéré, "A dynamical load-cycle charge model for RF power FETs," *IEEE Trans. Microw. Wireless Compon. Let. (see also IEEE Microw. and Guided Wave Lett.)*, vol. 11, no. 7, pp. 296–298, July 2001.

[11] S. Forestier, T. Gasseling, P. Bouysse, R. Quéré, and J. Nebus, "A new nonlinear capacitance model of millimeter wave power PHEMT for accurate AM/AM-AM/PM simulations," *IEEE Trans. Microw. Wireless Compon. Lett. (see also IEEE Microw. and Guided Wave Lett.)*, vol. 14, no. 1, pp. 43–45, Jan. 2004.

[12] H. Vinke and C. Lasance, "Compact models for accurate thermal characterization of electronic parts," *IEEE Trans. Compon., Packag. Manuf. Technol. A*, vol. 20, no. 4, pp. 411–419, Dec. 1997.

[13] C. Lasance, D. Den Hertog, and P. Stehouwer, "Creation and evaluation of compact models for thermal characterisation using dedicated optimisation software," *Semiconductor Thermal Measurement and Management Symp., 1999. 15 Annual IEEE*, 1999, pp. 189–200.

[14] M.-N. Sabry, "Compact thermal models for electronic systems," *IEEE Trans. Compon. Packag. Technol.*, vol. 26, no. 1, pp. 179–185, Mar. 2003.

[15] D. Zweidinger, S.-G. Lee, and R. Fox, "Compact modeling of BJT self-heating in SPICE," *IEEE Trans. Comput.-Aided Des. of Integr. Circuits Sys.*, vol. 12, no. 9, pp. 1368–1375, Sep. 1993.

[16] M. Rencz and V. Szekely, "Dynamic thermal multiport modeling of IC packages," *IEEE Trans. Compon. Packag. Technol.*, vol. 24, no. 4, pp. 596–604, Dec. 2001.

[17] O. Mueller, "Internal thermal feedback in four-poles especially in transistors," *Proc. IEEE*, vol. 52, no. 8, pp. 924–930, Aug. 1964.

[18] S. Marsh, "Direct extraction technique to derive the junction temperature of HBT's under high self-heating bias conditions," *IEEE Trans. Electron Devices*, vol. 47, no. 2, pp. 288–291, Feb. 2000.

[19] V. Szekely, "A new evaluation method of thermal transient measurement results," *Microelectronics J.*, thermal investigations of ICs and microstructures, vol. 28, no. 3, pp. 277–292, 1997.

[20] N. Bovolon, P. Baureis, J.-E. Muller, P. Zwicknagl, R. Schultheis, and E. Zanoni, "A simple method for the thermal resistance measurement of AlGaAs/GaAs heterojunction bipolar transistors," *IEEE Trans. Electron Devices*, vol. 45, no. 8, pp. 1846–1848, Aug. 1998.

[21] J. Lonac, A. Santarelli, I. Melczarsky, and F. Filicori, "A simple technique for measuring the thermal impedance and the thermal resistance of HBTs," *Gallium Arsenide and Other Semiconductor Application Symp., 2005. EGAAS 2005. European*, Oct. 3–4, 2005, pp. 197–200.

[22] A. De Souza, J.-C. Nallatamby, M. Prigent, and R. Quéré, "Dynamic impact of self-heating on input impedance of bipolar transistors," *Electron. Lett.*, vol. 42, no. 13, pp. 777–778, June 22, 2006.

[23] B. Moore, "Principal component analysis in linear systems: controllability, observability, and model reduction," *IEEE Trans. Autom. Control*, vol. 26, no. 1, pp. 17–32, 1981.

[24] C. D. Villemagne and R. E. Skelton, "Model reduction using a projection formulation," *Int. J. Contr.*, vol. 46, no. 6, pp. 2141–2169, 1987.

[25] M. Safonov and R. Chiang, "A Schur method for balanced-truncation model reduction," *IEEE Trans. Autom. Control*, vol. 34, no. 7, pp. 729–733, 1989.

[26] A. Rachid and G. Hashim, "Model reduction via Schur decomposition," *IEEE Trans. Autom. Control*, vol. 37, no. 5, pp. 666–668, 1992.

[27] J. T. Hsu and L. Vu-Quoc, "A rational formulation of thermal circuit models for electrothermal simulation. i. finite element method [power electronic systems]," *IEEE Trans. Circuits Syst. I: Fundam. Theory Appl.*, vol. 43, no. 9, pp. 721–732, Sep. 1996.

[28] J. T. Hsu and L. Vu-Quoc, "A rational formulation of thermal circuit models for electrothermal simulation. II. Model reduction techniques [power electronic systems]," *IEEE Trans. Circuits Syst. I: Fundam. Theory Appl.*, vol. 43, no. 9, pp. 733–744, 1996.

[29] E. Wilson and M. W. Yuan, "Dynamic analysis by direct superposition of Ritz vectors," *Earthquake Eng. Structural Dynamics*, vol. 10, no. 6, pp. 813–821, 1982.

[30] INRIA, *MODULEF Manual*. INRIA, 1991.

[31] ANSYS, *ANSYS Manual*. ANSYS, 2008.

[32] J. Joh, J. del Alamo, U. Chowdhury, T.-M. Chou, H.-Q. Tserng, and J. Jimenez, "Measurement of channel temperature in GaN high-electron mobility transistors," *IEEE Trans. Electron Devices*, vol. 56, no. 12, pp. 2895–2901, Dec. 2009.

[33] G. Mouginot, R. Sommet, R. Quéré, Z. Ouarch, S. Heckmann, and M. Camiade, "Thermal and trapping phenomena assessment on AlGaN/GaN microwave power transistor," *Proc. 2010 Eur. Microw. Integrated Circuit Conf. (EuMIC)*, pp. 110–113, 2010.

[34] A. Sarua, H. Ji, K. Hilton, D. Wallis, M. Uren, T. Martin, and M. Kuball, "Thermal boundary resistance between GaN and substrate in AlGaN/GaN electronic devices," *IEEE Trans. Electron Devices*, vol. 54, no. 12, pp. 3152–3158, Dec. 2007.

[35] M. Khan, Q. Shur, M. S. nad Chen, and J. Kuznia, "Current/voltage characteristic collapse in AlGaN/GAN heterostructure insulated gate field effect transistors at high drain bias," *Electron. Lett.*, vol. 30, p. 2175, 1994.

[36] S. Binari, P. Klein, and T. Kazior, "Trapping effects in GaN and SiC microwave FETs," in *Invited Paper, Proc. IEEE*, vol. 90, no. 6, June 2002, pp. 1048–1058.

[37] I. Daumiller, D. Theron, C. Gacquière, A. Vescan, R. Dietrich, A. Wieszt, H. Leier, R. Vetury, U. Mishra, I. Smorchkova, S. Keller, C. Nguyen, and E. Kohn, "Current instabilities in GaN-based devices," *IEEE Electron Device Lett.*, vol. 22, p. 62, 2001.

[38] M. Rocchi, "Status of the surface and bulk parasitic effects limiting the performances of GaAs ICs," *Physica*, vol. 129B, pp. 119–138, 1985.

[39] D. Lang, "Deep level transient spectroscopy: a new method to characterize traps in semiconductors," *J. of Applied Physics*, vol. 45, pp. 3023–3032, 1974.

[40] D. Look, *Electrical Characterization of GaAs Materials and Devices*. John Wiley & Sons, 1989.

[41] P. Audren, J. Dumas, M. Favennec, and S. Mottet, "Etude des Pièges dans les transistors haute mobilité électronique sur GaAs l'aide de la méthode dite de "relaxation isotherme". Corrélation avec les anomalies de fonctionnement," *J. Physique III, France*, vol. 3, pp. 185–206, Feb. 1993.

[42] O. Mitrofanov and M. Manfra, "Mechanisms of gate-lag in GaN/AlGaN/GaN high electron mobility transistors," *Superlattices and Microstructs*, vol. 34, pp. 33–53, 2003.

[43] J. Ibbetson, P. Fini, K. Ness, S. DenBaars, J. Speck, and U. Mishra, "Polarization effects, surface states, and the source of electrons in AlGaN/GaN heterostructure field effet transistors," *Applied Phys. Lett.*, vol. 77, no. 2, pp. 250–252, July 2000.

[44] O. Mitrofanov and M. Manfra, "Poole–Frenkel electron emission from the traps in AlGaN/GaN transistors," *J. Applied Physics*, vol. 95, no. 11, pp. 6414–6419, June 2004.

[45] K. Horio and Y. Fuseya, "Two-dimensional simulations of drain current transients in GaAs MESFET's with semi-insulating substrates compensated by deep levels," *IEEE Trans. Electron Devices*, vol. 41, no. 8, pp. 1340–1346, Aug. 1994.

[46] R. E. Anholt, *Electrical and Thermal Characterization of MESFETs, HEMTs, and HBTs*. Artech House Publishers, 1995.

[47] K. Kunihiro and Y. Ohno, "A large-signal equivalent circuit model for substrate induced drain-lag phenomena in HJFET's," *IEEE Trans. Electron Devices*, vol. 43, no. 9, pp. 1336–1342, Sep. 1996.

[48] W. Mickanin, P. Canfield, E. Finchem, and B. Odekirk, "Frequency-dependent transients in GaAs MESFETs: process, geometry and material effects," 11th Annual, *Gallium Arsenide Integrated Circuit (GaAs IC) Symp.*, 1989, pp. 211–214.

[49] R. Yeats, D. D'Avanzo, K. Chan, N. Fernandez, T. Taylor, and C. Vogel, "Gate-slow transients in GaAs MESFETs – causes, cures, and impact on circuits," in *Proc. Int. Electron Device Meeting*, 1988, pp. 842–845.

[50] P. Hower, W. Hooper, D. Tremere, W. Lehrer, and C. Bittman, "The Schottky barrier GaAs field effect transistors," *Gallium Arsenide and Related Compounds, Conf. Ser. no 7*. Institute of Physics, 1968.

[51] T. Itoh and Y. Yanai, "Stability of performance and interfacial problems in GaAs MESFET's," *IEEE Trans. Electron Devices*, vol. ED-27, no. 6, pp. 1037–1045, June 1980.

[52] N. Scheinberg, R. Bayruns, and R. Goyal, "A low frequency GaAs MESFET circuit model," *IEEE J. Solid-State Circuits*, vol. 23, no. 2, pp. 605–608, 1988.

[53] P. Ladbrooke and S. Blight, "Low-frequency dispersion of transconductance in GaAs MESFET's with implications for other rate-dependent anomalies," *IEEE Trans. Electron Devices*, vol. 35, no. 3, pp. 257–267, Mar. 1988.

[54] R. Vetury, N. Zhang, S. Keller, and U. Mishra, "The impact of surface states on the DC and RF characteristics of AlGaN/GaN HFETs," *IEEE Trans. Electron Devices*, vol. 48, no. 3, pp. 560–566, Mar. 2001.

[55] K. Horio, A. Wakabayashi, and T. Yamada, "Two-dimensional analysis of substrate-trap effects on turn-on characteristics in GaAs MESFETs," *IEEE Trans. Electron Devices*, vol. 47, no. 3, pp. 617–624, Mar. 2000.

[56] T. Barton and P. Ladbrooke, "The role of the surface in the high voltage behaviour of the GaAs MESFET," *Solid State Electron.*, vol. 29, no. 8, pp. 807–813, 1986.

[57] D. Root, S. Fan, and J. Meyer, "Technology independent large-signal non quasi-static FET model by direct construction from automatically characterized device data," *Eur. Microw. Conf. Dig.*, 1991, pp. 927–932.

[58] C. Charbonniaud, S. De Meyer, R. Quéré, and J. Teyssier, "Electrothermal and trapping effects characterization of AlGaN/GaN HEMTs," in *11th GaAs Symp.*, Oct. 2003, pp. 201–204.

[59] C. Baylis, L. Dunleavy, P. Ladbrooke, and J. Bridge, "The influence of pulse separation and instrument input impedance on pulsed IV measurement results," in *63rd ARFTG Conf. Dig.*, 2004, pp. 1–6.

[60] C. Camacho-Peñalosa and C. Aitchison, "Modelling frequency dependence of output impedance of a microwave MESFET at low frequencies," *Electron. Lett.*, vol. 21, no. 12, pp. 528–529, June 1985.

[61] N. Matsunaga, M. Yamamoto, Y. Hatta, and H. Masuda, "An improved GaAs device model for the simulation of analog integrated circuit," *IEEE Trans. Electron Devices*, vol. 50, no. 5, pp. 1194–1199, May 2003.

[62] F. Filicori, G. Vannini, A. Santarelli, A. Mediavilla-Sánchez, A. Tazón, and Y. Newport, "Empirical modeling of low-frequency dispersive effects due to traps and thermal phenomena in III-V FET's," *IEEE Trans. Microw. Theory Tech.*, vol. 43, no. 12, pp. 2972–2978, Dec. 1995.

[63] A. Parker and D. Skellern, "A realistic large-signal MESFET model for SPICE," *IEEE Trans. Microw. Theory Tech.*, vol. 45, no. 9, pp. 1563–1571, Sep. 1997.

[64] T. Brazil, "A universal large-signal equivalent circuit model for the GaAs MESFET," *21st Eur. Microw. Conf.*, Stuttgart, Germany, vol. 2, Dec. 1991, pp. 921–926.

[65] T. Fernandez, Y. Newport, J. Zamamillo, A. Tazón, and A. Mediavilla, "Extracting a bias-dependent large-signal MESFET model from pulsed IV measurements," *IEEE Trans. Microw. Theory Tech.*, vol. 44, no. 3, pp. 372–378, Mar. 1996.

[66] A. Jarndal, B. Bunz, and G. Kompa, "Accurate large-signal modeling of AlGaN-GaN HEMT including trapping and self-heating induced dispersion," *Proc. 18th Int. Symp. Power Semiconductor Devices & IC's*, June 2006.

[67] A. Raffo, A. Santarelli, P. Traverso, M. Pagani, F. Palomba, F. Scappaviva, G. Vannini, and F. Filicori, "Improvement of PHEMT intermodulation prediction through the accurate modelling of low-frequency dispersion effects," *IEEE MTT-S Int. Microw. Symp. Dig.*, 2005, pp. 465–468.

[68] P. Canfield, S. Lam, and D. Allstot, "Modeling of frequency and temperature effects in GaAs MESFET's," *IEEE J. Solid-State Circuits*, vol. 25, no. 1, pp. 299–306, Feb. 1990.

[69] R. Brady, G. Rafael-Valdivia, and T. Brazil, "Large-signal FET modeling based on pulsed measurements," *IEEE MTT-S Symp.*, June 2007, pp. 593–596.

[70] A. Santarelli, V. Di Giacomo, A. Raffo, F. Filicori, G. Vannini, R. Aubry, and C. Gacquière, "Nonquasi-static large-signal model of GaN FETs through an equivalent voltage approach," *Int. J. RF and Microwave Computer-Aided Eng.*, pp. 507–516, Nov. 2007.

[71] I. Angelov, V. Desmaris, K. Dynefors, P. Nillson, N. Rorsman, and H. Zirath, "On the large-signal modelling of AlGaN/GaN HEMTs and SiC MESFETs," *13th GAAS Symp.*, Paris, 2005, pp. 309–312.

[72] T. Roh, Y. Kim, Y. Suh, W. Park, and B. Kim, "A simple and accurate MESFET channel-current model including bias dependent dispersion and thermal phenomena," *IEEE Trans. Microw. Theory Tech.*, vol. 45, no. 8, pp. 1252–1255, Aug. 1997.

[73] R. Leoni, III, M. Shirokov, J. Bao, and J. Hwang, "A phenomenologically based transient SPICE model for digitally modulated RF performance characteristics of GaAs MESFETs," *IEEE Trans. Microw. Theory Tech.*, vol. 49, no. 6, pp. 1180–1186, June 2001.

[74] J. Rathnell and A. Parker, "Circuit implementation of a theoretical model of traps centres in GaAs and GaN devices," *Microelectron: Des., Technol., and Packag. III, Proc. SPIE Conf. Microelectronics, MEMS and Nanotechnology*, vol. 6798, 2007, pp. 67 980R (1–11).

[75] S. Albahrani, J. Rathnell, and A. Parker, "Characterizing drain current dispersion in GaN HEMTs with a new trap model," *Proc. 4th Eur. Microw. Integrated Circuits Conf.*, Sep. 2009, pp. 339–342.

[76] O. Jardel, F. De Groote, T. Reveyrand, J. Jacquet, C. Charbonniaud, J. Teyssier, D. Floriot, and R. Quéré, "An electrothermal model for AlGaN/GaN power HEMTs including trapping effects to improve large-signal simulation results on high VSWR," *IEEE Trans. Microw. Theory Tech.*, vol. 55, no. 12, pp. 2660–2669, Dec. 2007.

[77] M. Paggi, P. H. Williams, and J. M. Borrego, "Nonlinear GaAs MESFET modeling using pulsed gate measurements," *IEEE Trans. Microw. Theory Tech.*, vol. 36, no. 12, pp. 1593–1597, 1988.

[78] J. F. Vidalou, F. Grossier, M. Camiade, and J. Obregon, "On-wafer large signal pulsed measurements," in *Proc. IEEE MTT-S Int. Microw. Symp. Dig.*, 1989, pp. 831–834.

[79] B. Taylor, M. Sayed, and K. Kerwin, "A pulse bias/RF environment for device characterization," *Proc. 42nd ARFTG Conf. Dig.*, vol. 24, Fall, 1993, pp. 57–60.

[80] A. Parker, J. Scott, J. Rathmell, and M. Sayed, "Determining timing for isothermal pulsed-bias S-parameter measurements," *Proc. IEEE MTT-S Int. Microw. Symp. Dig.*, vol. 3, 1996, pp. 1707–1710.

[81] P. Aaen, J. A. Pla, and J. Wood, *Characterization of RF and Microwave Power FETs*. Cambridge Univ. Press, 2007.

[82] L. Betts, "Tracking advances in pulsed S-parameter measurements," *Microw. and RF J.*, Sep. 2007.

[83] F. Deshours, E. Bergeault, F. Blache, J.-P. Villotte, and B. Villeforceix, "Experimental comparison of load-pull measurement systems for nonlinear power transistor characterization," *IEEE Trans. Instrum. Meas.*, vol. 46, no. 6, pp. 1251–1255, 1997.

[84] C. Tsironis, "Harmonic rejection load tuner," Oct., 2001. US patent no. 6297649. Field Oct. 2001.

[85] A. Ferrero, "Active load or source impedance synthesis apparatus for measurement test set of microwave components and systems," Jan. 2003. US patent no. 6509 743. Field Jan. 2003.

[86] Maury, *Device Characterization with Harmonic Source and Load Pull*, Maury Microwave Corporation, Dec., 2000, application Note 5C-044.

[87] Focus, *Load Pull Measurements on Transistors with Harmonic Impedance Control*, Focus Microwaves, Aug., 1999.

[88] D. Barataud, F. Blache, A. Mallet, P. P. Bouysse, J.-M. Nebus, J. P. Villotte, J. Obregon, J. Verspecht, and P. Auxemery, "Measurement and control of current/voltage waveforms of microwave transistors using a harmonic load-pull system for the optimum design of high efficiency power amplifiers," *IEEE Trans. Instrum. Meas.*, vol. 48, no. 4, pp. 835–842, 1999.

[89] J. Benedikt, R. Gaddi, P. J. Tasker, and M. Goss, "High-power time-domain measurement system with active harmonic load-pull for high-efficiency base-station amplifier design," *IEEE Trans. Microw. Theory Tech.*, vol. 48, no. 12, pp. 2617–2624, 2000.

[90] J. Verspecht, "Large-signal network analysis," *IEEE Microwave J.*, vol. 6, no. 4, pp. 82–92, 2005.

[91] M. Sipila, K. Lehtinen, and V. Porra, "High-frequency periodic time-domain waveform measurement system," *IEEE Trans. Microw. Theory Tech.*, vol. 36, no. 10, pp. 1397–1405, 1988.

[92] U. Lott, "Measurement of magnitude and phase of harmonics generated in nonlinear microwave two-ports," *IEEE Trans. Microw. Theory Tech.*, vol. 37, no. 10, pp. 1506–1511, 1989.

[93] J. Verspecht and D. Schreurs, "Measuring transistor dynamic loadlines and breakdown currents under large-signal high-frequency operating conditions," *Proc. IEEE MTT-S Int. Microw. Symp. Digest*, vol. 3, 1998, pp. 1495–1498.

[94] P. Blockley, D. Gunyan, and J. B. Scott, "Mixer-based, vector-corrected, vector signal/network analyzer offering 300kHz-20GHz bandwidth and traceable phase response," *Proc. IEEE MTT-S Int. Microw. Symp. Digest*, 2005.

[95] J. Verspecht, "The return of the sampling frequency convertor," *Proc. 62nd ARFTG Microw. Meas. Conf.* Fall, pp. 155–164.

[96] V. Teppati and A. Ferrero, "A new class of nonuniform, broadband, nonsymmetrical rectangular coaxial-to-microstrip directional couplers for high power applications," *IEEE MWC*, vol. 13, no. 4, pp. 152–154, 2003.

[97] H. C. Early, "A wide-band directional coupler for wave guide," *Proc. IRE*, vol. 34, no. 11, pp. 883–886, 1946.

[98] R. F. Schwartz, P. J. Kelly, and P. P. Lombardini, "Criteria for the design of loop-type directional couplers for the L band," *IRE Trans. Microw. Theory Tech.*, vol. 4, no. 4, pp. 234–239, 1956.

[99] J. Teyssier, S. Augaudy, D. Barataud, J. Nebus, and R. Quéré, "Large-signal time domain characterization of microwave transistors under RF pulsed conditions," *57th ARFTG Conf. Dig.*, vol. 39, May 2001, pp. 1–4.

[100] J. Faraj, F. De Groote, J.-P. Teyssier, J. Verspecht, R. Quéré, and R. Aubry, "Pulse profiling for AlGaN/GaN HEMTs large signal characterizations," *38th Eur. Microw. Conf., 2008. EuMC 2008.*, Oct. 27–31, 2008, pp. 757–760.

[101] F. De Groote, P. Roblin, Y.-S. Ko, C.-K. Yang, S. J. Doo, M. V. Bossche, and J.-P. Teyssier, "Pulsed multi-tone measurements for time domain load pull characterizations of power transistors," *Proc. 73rd ARFTG Microw. Meas. Conf*, 2009, pp. 1–4.

[102] M. Abouchahine, A. Saleh, G. Neveux, T. Reveyrand, J.-P. Teyssier, D. Rousset, D. Barataud, and J.-M. Nebus, "Broadband time-domain measurement system for the characterization of nonlinear microwave devices with memory," *IEEE Trans. Meas. Tech.*, vol. 58, no. 4, pp. 1038–1045, 2010.

[103] A. El Rafei, R. Sommet, and R. Quere, "Electrical measurement of the thermal impedance of bipolar transistors," *IEEE. Electron Device Lett.*, vol. 31, no. 9, pp. 939–941, 2010.

[104] J. Verspecht, D. Schreurs, A. Barel, and B. Nauwelaers, "Black box modelling of hard nonlinear behavior in the frequency domain," *Proc. IEEE MTT-S Int. Microw. Symp. Dig.*, vol. 3, 1996, pp. 1735–1738.

[105] M. C. Curras-Francos, P. J. Tasker, M. Fernandez-Barciela, Y. Campos-Roca, and E. Sanchez, "Direct extraction of nonlinear FET Q-V functions from time domain large signal measurements," *IEEE Trans. Microw. Wireless Compon. Lett. (see also IEEE Microw. Guided Wave Lett.)*, vol. 10, no. 12, pp. 531–533, 2000.

[106] J. Verspecht, D. Gunyan, J. Horn, J. Xu, A. Cognata, and D. E. Root, "Multi-tone, multi-port, and dynamic memory enhancements to PHD nonlinear behavioral models from large-signal measurements and simulations," *Proc. IEEE/MTT-S Int. Microw. Symp*, 2007, pp. 969–972.

8 Optimizing microwave measurements for model construction and validation

Dominique Schreurs,[1] Maciej Myslinski,[2] and Giovanni Crupi[3]

[1] K.U. Leuven, Belgium
[2] previously with K.U. Leuven, Belgium
[3] previously with K.U. Leuven, Belgium, now with University of Messina, Italy

8.1 Introduction

The process of modeling microwave devices is based on either technology computer aided design (TCAD) simulations or measurements. The former is the basis for the so-called compact models, which is the preferred modeling technique for devices aiming for digital and low-frequency (i.e., not microwave) analog designs. This book focuses on the second technique, namely the use of measurements in the model construction process. This chapter will demonstrate that acquiring measurements for efficient and accurate model construction requires thorough experiment design. This chapter focuses on the use of linear and nonlinear microwave measurements for model construction and – importantly – model validation. Noise measurements are beyond the scope of this chapter, so readers are referred to the dedicated (Chapter 10) on noise modeling.

The chapter starts with a brief review on linear and nonlinear microwave measurements. Subsequently, as on-wafer devices are embedded in a layout structure to enable on-wafer measurements, the effect of pads and access transmission lines should be de-embedded from the measurements. This is an important step, especially in the case of devices on silicon substrates, as an over- or under de-embedding may result in nonphysical device parameter values. Due to differences in the mathematical procedures, linear and nonlinear measurements are discussed separately.

We proceed by explaining issues in collecting linear measurements aimed at device modeling. If the purpose is to extract the bias-dependency of the model parameters, e.g., to advance to a nonlinear model at a later stage, the number of required measurements can easily become extensive, especially in the case of power devices which have a wide operating range. We explain a technique to collect the measurement data in an efficient way.

Next, we proceed to model validation, both regarding linear and nonlinear measurements. We introduce metrics to quantify the model accuracy. We also explain how

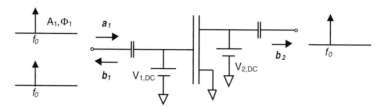

Figure 8.1 Schematic representation of microwave linear vector measurements.

nonlinear measurements can be represented best to deduce the cause of model inaccuracies.

In the final part, we discuss the use of nonlinear measurements in the modeling process. We make the distinction between models based on time-domain data and on frequency-domain data.

Throughout the chapter, examples are included to illustrate the various techniques for experiment design. However, performing measurements for device modeling is not a simple push-button process. The reader should be aware that the experimental boundary conditions (bias range, frequency range, and so on) are device-technology dependent, as a result of which the experiment design has to be tailored for each new device under consideration.

8.2   Microwave measurements and de-embedding

8.2.1   Linear versus nonlinear microwave measurements

The principle of linear microwave measurements is depicted in Figure 8.1. The device is biased under the operation condition of interest and a small incident traveling voltage wave a_1 is applied to the device while port 2 is terminated in 50 Ohm. In response, the device scatters back a scattered traveling voltage wave towards both its port 1 and port 2, which are denoted as b_1 and b_2, respectively. In case the device responds linearly to a given excitation, the measured quantities b_1 and b_2 contain only one spectral component and this at the same frequency f_0 as the frequency of the excitation signal. The excitation signal can also be applied at port 2 of the device, denoted as a_2, and in such case port 1 is terminated in 50 Ohm.

To characterize the linear behavior of a microwave device, S-parameter measurements are used. For reference, the definition of S-parameters is listed in equation (8.1). S-parameters are frequency-dependent complex numbers. In the case of active devices, such as transistors, S-parameters are also bias-dependent. Note that to obtain S-parameters, only ratios of scattered and incident traveling voltage waves are required and not their absolute values. As a result, S-parameters can be accurately obtained from relatively simple microwave measurement and calibration techniques. The instrument

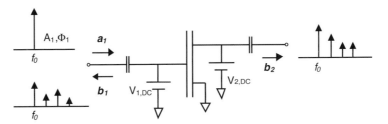

Figure 8.2 Schematic representation of microwave nonlinear vector measurements.

that enables linear measurements is the vector network analyser [1].

$$
\begin{aligned}
S_{11} &= \left.\frac{b_1}{a_1}\right|_{a_2=0} & S_{12} &= \left.\frac{b_1}{a_2}\right|_{a_1=0} \\
S_{21} &= \left.\frac{b_2}{a_1}\right|_{a_2=0} & S_{22} &= \left.\frac{b_2}{a_2}\right|_{a_1=0}.
\end{aligned}
\tag{8.1}
$$

To characterize nonlinear behavior of a microwave device, nonlinear measurements are required. In this case, a larger excitation is applied to the device (Figure 8.2). As a result, the spectra of the scattered traveling voltage waves have spectral components at not only the excitation frequency, but also at its harmonics. The physical cause is that the device's characteristics are nonlinear. An illustration is depicted in Figure 8.10 in Section 8.4. As long as the excitation is small, the local device characteristics around the DC operating point can be approximated as being linear. When the excitation signal gets larger, the actual shape starts to play a role. The response is a distorted time-domain waveform, e.g., the clipped waveform of the drain-source current when the instantaneous gate voltage reaches pinch-off condition (Figure 8.14). A distorted time-domain waveform corresponds to a spectrum with harmonic components in frequency domain.

In terms of characterization, the most complete information is obtained if both the amplitude and phase of each of the spectral components can be measured in an absolute way. The instrument that enables this is the large-signal network analyzer. The original architecture is based on sampling [2], but more recently a mixer-based setup has been also developed more recently [3]. The same measurement can be accomplished in time domain by using an oscilloscope. In this case, a four-channel model is preferred, not only to enable the characterization of the four traveling voltage waves simultaneously, but also for calibration purposes [4]. In Section 8.4.2, we will show that the simultaneous measurement of the four traveling voltage waves can provide useful information about the cause of model inaccuracies.

8.2.2 De-embedding

A typical layout of a transistor is shown in Figure 8.3. The actual device is depicted by the dashed ellipse. In order to enable measurements, the device is embedded in a structure consisting of pads and short sections of transmission line to connect the pads to the

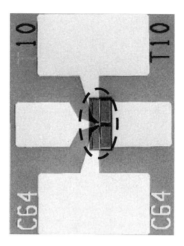

Figure 8.3 Transistor layout by which the actual device is indicated by the dashed ellipse.

device. The pad is in coplanar waveguide configuration, namely ground–signal–ground, by which the ground–signal distance is compatible to the pitch size of the measurement probe. In this particular example, the signal pad is slightly extended beyond the ground section to enable easier wirebonding of the device in its final application. The specific difficulty is now the fact that the measurements are calibrated up to the probe tips whereas we want to characterize and model the device within the area of the dashed ellipse. So the reference plane has to be shifted from the probe tips to the actual device ports, and this process is called de-embedding. A question is why de-embedding cannot be part of the calibration procedure. Actually, techniques exist by which the calibration is performed directly on the substrate on which the DUT is located [1]. This approach requires the presence of a set of calibration standards on the same wafer. As one cannot be sure about the fabrication accuracy of a 50 Ohm resistor (i.e., actually realized as two 100 Ohm resistors in parallel to be compatible with the coplanar waveguide configuration), the methods are based on the use of multiple transmission lines with varying lengths. The range in lengths determines the frequency range over which the calibration is valid. As this results in an area-consuming layout, it is often not possible to have this set of calibration standards integrated on the wafer, and therefore calibration (performed on dedicated calibration substrates) and de-embedding are decoupled.

A second note is that de-embedding is of particular importance for devices on silicon substrates (MOSFETs, GaN HEMTs on Si, and so on). In case of devices on semi-insulating substrates, such as GaAs, InP, etc., a section of transmission line can be approximated by a network consisting of a parallel capacitor and an inductor and resistor in series. Often, these elements get incorporated in the extrinsic core of the device's model, and so-called cold measurements (i.e., DC drain–source voltage is equal to 0 V) suffice to extract their values. In case of silicon substrates, the model of a section of transmission line is more complex due to the lossy substrate. As the latter results in too

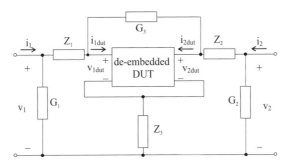

Figure 8.4 Schematic representing the de-embedding network.

many unknowns to be straightforwardly extracted from S-parameter measurements on the device, a dedicated de-embedding step is essential.

The pad and access transmission lines are represented as shown in Figure 8.4. At both port 1 and port 2, the section of transmission line gets approximated by a network consisting of a series impedance and parallel admittance. Note that this scheme may have to become distributed (i.e., the network is repeated multiple times) at higher frequencies. The frequency at which this happens depends on the transmission line dimensions, but usually the schematic with only one network can be used up to at least 50 GHz. In the schematic, there is also a coupling admittance between port 1 and port 2, and an impedance connecting the device's source to the ground. It is debatable whether the latter two components are part of the de-embedding network or part of the device itself. It depends on the actual device type and its layout whether or not these two elements are included in the de-embedding process. The overall goal is to de-embed the measurement pads and access transmission lines up to a reference plane corresponding to how the device will be inserted in an actual circuit design. All components in the de-embedding network are complex and frequency dependent.

In order to determine the values of the de-embedding components, one makes use of so-called dummy structures, which are simple passive components such as open, short, thru, etc. By combining the measurements on these dummy structures, one has the necessary information to determine the values of the de-embedding components. Depending on device layout, target frequency range, and available space on the wafer, the number of dummy structures typically varies between one and four, and the corresponding mathematical procedure is adjusted accordingly [5–7]. Note that the dummy structures consume less area on the wafer as compared to the multiple transmission lines, which one would need for an on-wafer calibration.

Once the values of the de-embedding components are known, the de-embedding schematic can be subtracted from the measurements, as a result of which the reference plane will be moved from the probe tips to the device ports. The mathematical procedure is different for linear and nonlinear measurements.

In the case of linear measurements, the schematic is subtracted by means of transformations between S-parameters, Y-parameters, and Z-parameters. Equation (8.2) summarizes the steps corresponding to the topology in Figure 8.4. First, the measured S-parameters S_{meas} are transformed into Y-parameters Y_{meas}, and then the two parallel admittances can be subtracted. Next, the resulting Y-parameters Y_A are transformed into Z-parameters Z_A and the series impedances, also the one connecting to the source if present, are subtracted. If the coupling admittance is considered, there is another transformation from the resulting Z-parameters Z_B into Y-parameters Y_B. Finally, the resulting Y-parameters Y_{DUT} are transformed into S-parameters S_{DUT}, which is the set of measurement data of the actual device.

$$1)\ S_{meas} \rightarrow Y_{meas}$$

$$Y_A = Y_{meas} - \begin{bmatrix} G_1 & 0 \\ 0 & G_2 \end{bmatrix}$$

$$2)\ Y_A \rightarrow Z_A$$

$$Z_B = Z_A - \begin{bmatrix} Z_1 + Z_3 & Z_3 \\ Z_3 & Z_2 + Z_3 \end{bmatrix} \tag{8.2}$$

$$3)\ Z_B \rightarrow Y_B$$

$$Y_{DUT} = Y_B - \begin{bmatrix} G_3 & -G_3 \\ -G_3 & G_3 \end{bmatrix}$$

$$4)\ Y_{DUT} \rightarrow S_{DUT}.$$

In the case of nonlinear measurements, the formulas are different because S-, Y-, Z-parameters are linear quantities and, as a consequence, these parameters cannot be used for representing the transistor under large-signal conditions. Therefore, the corresponding equations are now expressed in terms of port currents and voltages, and the de-embedding network is subtracted by applying Kirchoff's laws:

$$i_1' = i_1 - v_1 G_1$$

$$i_2' = i_2 - v_2 G_2$$

$$v_1' = v_1 - i_{gs}' Z_1$$

$$v_2' = v_2 - i_2' Z_2$$

$$v_{1DUT} = v_1' - \left(i_1' + i_2'\right) Z_3 \tag{8.3}$$

$$v_{2DUT} = v_2' - \left(i_1' + i_2'\right) Z_3$$

$$i_{1DUT} = i_1' - \left(v_{1DUT} - v_{2DUT}\right) G_3$$

$$i_{2DUT} = i_2' - \left(v_{2DUT} - v_{1DUT}\right) G_3.$$

Figure 8.5 illustrates the effect of de-embedding on linear measurements. The measured S-parameters clearly show a kink at low frequencies due to the lossy substrate. After de-embedding, the kink has disappeared. Also, we can notice that the gain (S_{21}) has increased after de-embedding, as we are subtracting the losses in the access transmission lines and pads. The gain after de-embedding is the actual gain of the device, so analyzing the gain as measured at the probe tips may be misleading.

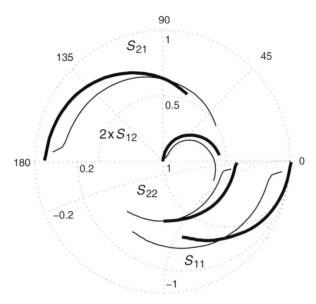

Figure 8.5 Measured S-parameters (0.3–50 GHz) before and after de-embedding; device is a FinFET (frequency range: 0.3 GHz – 50 GHz, $V_{gs} = 0.8$ V, $V_{ds} = 1.2$ V).

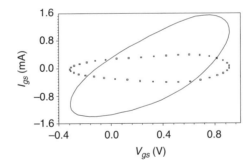

Figure 8.6 Measured (lines) and de-embedded (symbols) $I_{gs}(t)$ as function of $V_{gs}(t)$; device is a FinFET ($f_0 = 4$ GHz, $P_{in} = -0.24$ dBm, $V_{gs} = 0.3$ V, $V_{ds} = 0.9$ V).

Also, the nonlinear measurement data is clearly different after de-embedding, as illustrated by Figure 8.6. The figure shows the time-domain waveform of the gate–source current as a function of the time domain waveform of the gate–source voltage of a FinFET. A FinFET is an emerging MOSFET-type transistor technology [8]. We can observe that the de-embedding rotates the trajectory. The resulting trajectory is physically meaningful. The intrinsic large-signal model of a FET-type transistor consists in general of the parallel connection of a current source and a charge source (Figure 8.8). As the DC gate current can be neglected in the considered device, there is only a charge source at the gate, or in other words the behavior in the small-signal sense is capacitive. It is well known that a capacitor corresponds to a phase shift of 90° between the current through

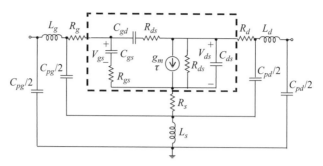

Figure 8.7 Small-signal equivalent circuit model of a FET-type device.

the capacitor and the voltage across it, and this is indeed confirmed by the de-embedded measurement. Note that this simple explanation ignores the device's parasitic elements. One should not overlook that the de-embedding process does not remove the transistor's extrinsic parasitic elements (Figure 8.7). Also here, the message is that analyzing the results at the reference plane of the probe tips may lead to wrong conclusions.

It is important to note that the de-embedding procedure allows the reference plane to be shifted closer to the actual transistor but the effect is different for linear and nonlinear measurements. In the former case, de-embedding allows one to completely remove the parasitic contributions associated to the dummy structures, while in the latter case these contributions should be included in the source and load impedances presented to the transistor at the new reference planes. We should bear in mind that the transistor behavior under small-signal approximation depends only on the properties of the device at the selected DC bias point. Hence, once the transistor have been fully characterized with linear measurements (e.g., S-parameter measurements), the device response can be predicted for any external excitation and termination at the fixed DC bias point. On the other hand, the transistor behavior under large-signal conditions depends on the properties of the device for the particular time-varying signal (i.e., DC and RF components) appearing at its ports. As a result, the transistor characterization based on nonlinear measurements (e.g., large-signal network analyzer measurements) for a given time-varying signal cannot be used for predicting its response for any external excitation and termination. Consequently, the shift of the reference plane for nonlinear measurements implies that the source and load impedances presented to the device at the new planes should include also the parasitic contributions associated to the dummy structures.

8.3   Measurements for linear model construction

The objective of constructing a linear model for microwave devices is to determine the component values of the small-signal equivalent circuit. A typical example of a small-signal equivalent circuit for a FET is depicted in Figure 8.7. Note that the topology of the schematic has to be tailored for each device type. For example, substrate effects should

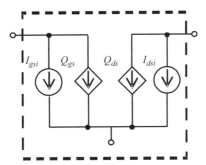

Figure 8.8 Large-signal quasi-static intrinsic equivalent circuit model of a FET-type device.

be included if the device is fabricated on a silicon substrate. The small-signal equivalent circuit consists of an extrinsic part, representing the physical layout (e.g., ohmic source and drain resistances) and the intrinsic part, which represents the actual device operation. As the behavior of a transistor depends on the applied DC bias voltages, the components of the intrinsic part of the small-signal equivalent circuit are bias-dependent. As the components of the extrinsic part have their origin in the device layout, which is passive, the component values are assumed to be bias-independent.

The small-signal equivalent circuit component values are extracted from S-parameter measurements. The S-parameter measurements have to be conducted under various DC bias conditions, to be able to identify the bias-dependency of the intrinsic model. The details of the extraction procedures are documented in other chapters within this book. In terms of experiment design, the focus of the present chapter, it is important that the S-parameter measurement data is collected under a set of DC bias voltages that cover the operating range of the device. Usually, a rectangular grid gets defined and the DC gate–source voltage and DC drain–source voltage are varied in equidistant steps. The grid is bounded by the maximal DC power dissipation limit, and by the gate–source and drain–source voltage combinations that are critical for breakdown. The size of the steps depends on the device type. It should be small enough to capture accurately the change in device behavior as a function of the DC operation. The reason is that the bias-dependent small-signal equivalent circuit is the cornerstone for nonlinear model construction.

The typical configuration for the intrinsic part of the nonlinear model for a FET is depicted in Figure 8.8. As the extrinsic part is bias-independent, its topology remains identical to the one in the small-signal model, and is therefore not reported here. The model is obtained by integrating the bias-dependent small-signal equivalent circuit parameters towards the intrinsic port voltages. Note again that the actual topology of the intrinsic nonlinear model may have to be adjusted depending on device type. For example, the shown simplified schematic implicitly ignores nonquasi-static effects, such as R_{gs}, R_{gd}, τ, and possible higher order effects.

If the step in DC operating point during the S-parameter measurements is too large compared to the change in device behavior, errors are accumulated during the integration step. The optimal step is strongly dependent on device technology. Typical values for a low-power GaAs HEMT are 50 mV for the step in DC gate–source voltage and 100 mV

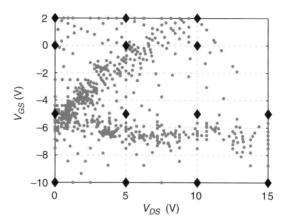

Figure 8.9 Illustration of bias points selected as a result of the iterative procedure. The initial bias points are marked with a diamond. Device is a GaN HEMT.

for the step in DC drain–source voltage. Depending on the device technology's supply voltage, this results in several hundred S-parameter measurements. In the case of high-power devices, such as GaN HEMTs, which have a much larger DC operating range, an equidistant grid would easily result in several thousand S-parameter measurements, which is practically no longer feasible to be collected. To optimize the experiment design, a nonuniform grid has to be adopted. The density of DC operating points under which S-parameter measurements are collected should be higher in regions of fast variation in device characteristics and can be less in regions with slow variation in device characteristics. Several approaches to identify the fast-changing regions have been reported in the literature [9, 10]. One starts with collecting S-parameter measurements on a so-called coarse grid, as represented by the diamond symbols in Figure 8.9. Next, the change in S-parameter characteristics between a grid point and its immediate neighbors is evaluated. If the change is above a predefined threshold, additional points (represented by the dots) are being added. In the example of a small-size GaN device shown in Figure 8.9, we observe a higher density in points at the transition to the pinch-off region, and also at the transition between the linear region at low drain-source voltages and the saturation region. In the shown example, there are no dots at the combination of high gate–source and high drain–source voltages due to the DC power dissipation limit. By adopting such an adaptive experiment design, the required number of measurements can be reduced to an acceptable number, even for high-power devices.

8.4   Measurements for model validation

The use of microwave measurements is not only for constructing device models but clearly also for evaluating their accuracy. In this section, we discuss how measurements can be selected for optimally evaluating a transistor's model accuracy. We make a distinction between validating linear models and nonlinear models.

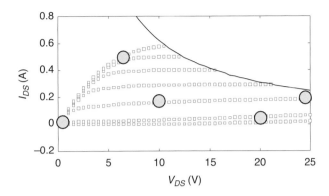

Figure 8.10 Representative bias points to check model accuracy. Device is a low-power GaN HEMT.

8.4.1 Linear model validation

The linear model is the simplest model for a microwave transistor, but nevertheless the importance of its validation should not be underestimated. The reason is that this model suffices to design various microwave circuit types (e.g., low-power high-gain amplifier, low-noise amplifier (in case the noise model is available as well)), but even more importantly, it is the basis for nonlinear model construction.

When considering S-parameter measurements for model validation, several degrees of freedom exist. One can vary the DC operating point, frequency range, environmental temperature, etc. It is practically not feasible to check the model under all possible conditions, so an adequate strategy should be adopted. In terms of the DC operating point, it is important to check the model under the operating conditions corresponding to the circuit design in which the model is going to be used. In the case of an amplifier design, a typical approach is to evaluate the model accuracy for a number of bias points along the load line corresponding to the intended amplifier design type (e.g., class A, class AB, etc.). When the small-signal model is planned to be used as the cornerstone for nonlinear model construction, the evaluation should include the extreme bias points that are along the integration path. An illustration is shown in Figure 8.10. It is suggested to test the linear model under cold conditions (drain–source voltage equal to 0 V), pinchoff condition, the knee region, near the power dissipation limit, and near the maximum drain–source voltage limit.

Another experiment design parameter is the frequency range. The recommendation is to test the model over a bandwidth that is about 20% broader than the bandwidth of interest for the intended application. An illustration is included in Figure 8.11 [11]. It compares measurements (symbols) to simulations. In the case of the simple model (indicated by a dashed line), there is a deviation between the measured and simulated S_{22} at higher frequencies. A more complex model, of which simulation results are represented by the solid line, yields accurate S_{22} results up to 90 GHz. The message is that if the target application is in the millimeter wave range, then obtaining good model accuracy at microwave frequencies does not guarantee good results at millimeter

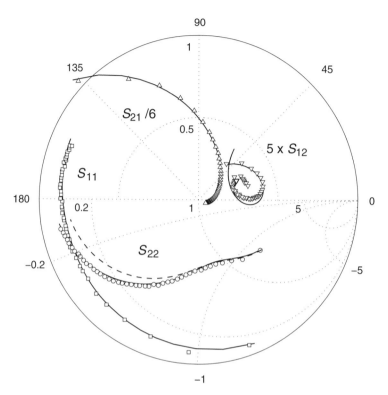

Figure 8.11 Measured (symbols) and simulated (solid lines) S-parameters. Dashed line represents S_{22} simulated by the model extracted without taking into account the output conductance time delay; device is a GaAs pHEMT (frequency range: 4 GHz – 90 GHz, $V_{gs} = 0.0$ V and $V_{ds} = 1.5$ V).

wave frequencies. The higher the frequency, the more distributed effects start playing a role, and such effects are not visible in measurement data in the lower frequency range. While it sounds obvious that a model's accuracy has to be checked across the frequency range of interest, this is often overlooked in practice. Designers rely on library models as offered by foundries and often do not pay close attention to the valid operating frequency range as specified by the foundry. This regularly results in fabricated circuits that do not perform as simulated.

Due to the spread in experimental conditions, metrics were formulated to summarize the overall quality of a linear model [12]. The formulas are summarized under equation (8.4). The first metric $M1\_SP_{xy}$ (equation (8.4a)) is a measure of how well the model agrees with the S-parameter measurements. The subscripts xy denote that there is such a result for each of the four S-parameters. The summations are over frequency range and over biases. The formula is in a rational form to express the error relative to the magnitude of the S-parameter. For example, a deviation of 0.1 dB has a different significance if it concerns the gain S_{21} or the low-valued S_{12}. As this metric results in four values, namely one for each of the *S*-parameters, other approaches have been proposed. The expression for $M2\_SP$ (equation (8.4b)) is in terms of the unilateral transducer gain G_{TU} (equation (8.4c)). The reasoning is that this power gain definition is of importance for amplifier design, a primary application area of transistor models, and, moreover, its expression (equation (8.4c)) groups three *S*-parameters. The missing S_{12} is the least important one in transistor modeling, owing to its low value, and therefore a metric based on the unilateral transducer gain is adequate. If there is interest to include S_{12} in the evaluation as well, one can make use of metric M3_SP (equation (8.4d)) that is based on the stability parameter K (equation (8.4e)).

The advantage of metrics is that they are compact, meaning that they summarize the model accuracy in a limited number of values, depending on the set of chosen metric expressions, but on the other hand this poses a drawback as well. By summarizing the results into a few numbers, one loses track of the origin of the model inaccuracy. So the recommendation is to use these metrics in combination with an in-depth analysis at a selected number of experimental conditions, as explained in the preceding paragraphs.

$$M1\_SP_{xy} = \sqrt{\frac{\sum_M \sum_N \left| S_{xy}^{meas} - S_{xy}^{sim} \right|^2}{\sum_M \sum_N \left| S_{xy}^{meas} \right|^2}}$$

(a)

$$M2\_SP = \sqrt{\frac{\sum_M \sum_N \left(G_{TU}^{meas} - G_{TU}^{sim} \right)^2}{\sum_M \sum_N \left(G_{TU}^{meas} \right)^2}} \qquad G_{TU}^{Max} = \frac{1}{1 - |S_{11}|^2} |S_{21}|^2 \frac{1}{1 - |S_{22}|^2}$$

(b)   (c)

$$M3\_SP = \sqrt{\frac{\sum_M \sum_N \left(K^{meas} - K^{sim} \right)^2}{\sum_M \sum_N \left(K^{meas} \right)^2}} \qquad K = \frac{1 - |S_{11}|^2 - |S_{22}|^2 + |\Delta|^2}{2 |S_{12} S_{21}|}$$

(d)   (e)

(8.4)

8.4.2   Nonlinear model validation

The range of possible experimental conditions to evaluate the accuracy of a transistor's nonlinear model is substantially higher than in the case of the linear model. There are not only the bias, frequency, and environmental temperature as parameters, but also input power, load presented at input and at the output and this at baseband, fundamental, and harmonic frequencies, and finally the type of excitation (ranging from a simple CW excitation to a complex modulated excitation). There is also a choice to be made in representing the data (time domain, frequency domain, or modulation domain).

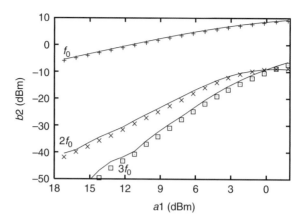

Figure 8.12 Measured and simulated ( ) magnitude of the first three harmonics of the output power; device is a MOSFET ($V_{gs} = 0.6$ V, $V_{ds} = 1.2$ V, $f_0 = 3.6$ GHz).

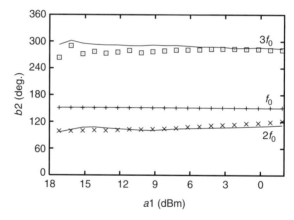

Figure 8.13 Measured and simulated ( ) phase of the first three harmonics of the output power; device is a MOSFET ($V_{gs} = 0.6$ V, $V_{ds} = 1.2$ V, $f_0 = 3.6$ GHz).

Figure 8.12 and Figure 8.13 show a model evaluation in the frequency domain [13]. The added value of the large-signal network analyzer instrumentation [2–4] is that not only the model's prediction of the magnitude but also that of the phase of the spectral components can be evaluated against measurements. When using spectrum analyzer measurements, only Figure 8.12 can be obtained.

To pinpoint the cause of model inaccuracies, namely whether it is related rather to the modeled nonlinear charge or nonlinear current sources (Figure 8.8), an evaluation in time domain is usually more instructive. To this purpose, the time-domain waveforms are not represented as a function of time, but are plotted against each other, as illustrated in Figure 8.14 [14]. The gate–source current waveform is plotted against the gate–source voltage waveform. As the input of a FET is capacitive (the gate leakage is small), we

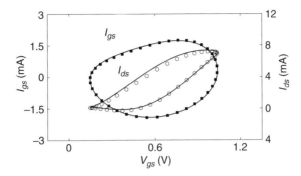

Figure 8.14 Measured (symbols) and simulated (lines) input (black squares, left axis) and transfer (white circles, right axis) loci; device is a FinFET ($f_0 = 15$ GHz, $V_{gs} = 0.6$ V, $V_{ds} = 0.6$ V, $P_{in} = 1.7$ dBm).

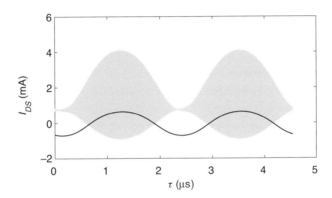

Figure 8.15 Measured RF envelope and baseband response (solid line) of instantaneous I_{ds} under two-tone excitation; device is a FinFET ($f_0 = 4$ GHz, tone spacing $= 440$ kHz, $V_{gs} = 0.3$ V, and $V_{ds} = 0.9$ V).

expect a 90° phase shift between the current and voltage. Therefore, the opening of the hysteresis curve is mainly a measure for the value of the input capacitor. If the model prediction deviates from the measurements, we can associate this with an over- or underestimation of the input capacitance value. Note that the shape is not a pure ellipse due to the fact that the input capacitor is bias-dependent and therefore not constant over the swing of the RF input signal. Similarly, we can plot the drain–source current time-domain waveform as a function of the gate–source time-domain waveform. In this case, we have a contribution from both the charge source and the current source. The opening of the shape is again related to the charge model, whereas the length and curvature of the shape can be related to the current source model.

The model can also be evaluated under experimental conditions using modulated excitations. An example using a two-tone excitation is presented in Figure 8.15 and Figure 8.16 [15]. In the measurement instrumentation, the RF response (represented in

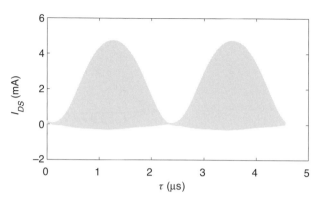

Figure 8.16 Simulated envelope of instantaneous I_{ds} under two-tone excitation; device is a FinFET ($f_0 = 4$ GHz, tone spacing $= 440$ kHz, $V_{gs} = 0.3$ V, and $V_{ds} = 0.9$ V).

gray in Figure 8.15) is measured separately from the response in the baseband (represented by a black solid line in Figure 8.15). By summing the two measured contributions, a comparison with model predictions (Figure 8.16) can be performed. Note that the standard instrumentation only measures the gray RF response [16], so comparing the RF response of Figure 8.15 to model predictions can lead to wrong conclusions. The evidence is that the simulated envelope in Figure 8.16 is physical as the current pinches off.

As in the case of using linear measurements for model validation, metrics have also been defined for nonlinear measurements-based evaluation [17]. As the degrees of freedom in experiment design are more extensive, there is also a larger variety in possible metrics. Depending on the application for which the model is targeted, only a subset of the metrics is considered and evaluated. Common metric expressions are listed under equation (8.5). Sums are made across ranges of fundamental frequencies, number of considered harmonics, and biases. Metrics can be expressed in frequency domain (equations (8.5a)–(8.5j)) or time domain (equation (8.5k)). In the former category, distinction is made according to whether the metrics are based on single-tone measurements (equations (8.5a)–(8.5d)) or two-tone measurements (equations (8.5e)–(8.5j)). The metric M1_LS_TD_Bp expresses the agreement between the measured and simulated complex envelopes of the scattered traveling voltage waves b_1 and b_2 at port p and around the hth harmonic of the carrier frequency f_0. In the equation, j refers to the time sample index. Weighted metrics (right column) have been introduced to emphasize that it is more important to have good model accuracy at conditions of interest for circuit design. For amplifiers, this corresponds to the spectral components with large amplitude levels (equation (8.5b)), the condition resulting in best power-added efficiency (PAE) (equation (8.5f)), and the condition with the worst, i.e., highest, third-order (equation (8.5h)) and fifth-order (equation (8.5j)) intermodulation levels. In practical cases, though, it has been demonstrated that the difference in results between the regular metric (left column) and the corresponding weighted metric (right column) is small [17].

$$M1\_LS\_1T = \sqrt{\frac{\sum_M \sum_N \sum_{f_0}^{H} |Pout^{meas} - Pout^{sim}|^2}{\sum_M \sum_N \sum_{f_0}^{H} |Pout^{meas}|^2}}$$

(a)

$$M1\_LS\_1T\_W = \sqrt{\frac{\sum_M \sum_N \sum_{f_0}^{H} \frac{|Pout^{meas}|}{\sum |Pout_h^{meas}|} |Pout^{meas} - Pout^{sim}|^2}{\sum_M \sum_N \sum_{f_0}^{H} |Pout^{meas}|^2}}$$

(b)

$$M2\_LS\_1T = \sqrt{\frac{\sum_M \sum_N \sum_{f_0}^{H} \left(Phase\left(Pout_{if_0}^{meas}\right) - Phase\left(Pout_{if_0}^{sim}\right)\right)^2}{\sum_M \sum_N \sum_{f}^{H} Phase\left(Pout_{if_0}^{meas}\right)^2}}$$

(c)

$$M3\_LS\_1T = \sqrt{\frac{\sum_M \sum_N |Gain^{meas} - Gain^{sim}|^2}{\sum_M \sum_N |Gain^{meas}|^2}}$$

(d)

$$M4\_LS\_1T = \sqrt{\frac{\sum_M \sum_N |PAE^{meas} - PAE^{sim}|^2}{\sum_M \sum_N |PAE^{meas}|^2}}$$

(e)

$$M4\_LS\_1T\_W = \sqrt{\frac{\sum_M \sum_N \frac{|PAE^{meas}|}{max|PAE_n^{meas}|} |PAE^{meas} - PAE^{sim}|^2}{\sum_M \sum_N |PAE^{meas}|^2}}$$

(f)

$$M1\_LS\_2T = \sqrt{\frac{\sum_M \sum_N |ImD3^{meas} - ImD3^{sim}|^2}{\sum_M \sum_N |ImD3^{meas}|^2}}$$

(g)

$$M1\_LS\_2T\_W = \sqrt{\frac{\sum_M \sum_N \frac{|IMD3^{meas}|}{max|imD3_n^{meas}|} |ImD3^{meas} - ImD3^{sim}|^2}{\sum_M \sum_N |ImD3^{meas}|^2}}$$

(h)

$$\text{M2\_LS\_2T} = \sqrt{\frac{\sum_M \sum_N \left| \text{Im}D5^{meas} - \text{Im}D5^{sim} \right|^2}{\sum_M \sum_N \left| \text{Im}D5^{meas} \right|^2}}$$

(i)

$$\text{M2\_LS\_2T\_W} = \sqrt{\frac{\sum_M \sum_N \frac{\left| \text{im}D5^{meas} \right|}{max \left| \text{im}D5_n^{meas} \right|} \left| \text{Im}D5^{meas} - \text{Im}D5^{sim} \right|^2}{\sum_M \sum_N \left| \text{Im}D5^{meas} \right|^2}}$$

(j)

$$\text{M1\_LS\_TD\_Bp} = \frac{1}{H} \sum_h \sqrt{\frac{\sum_M \sum_N \sum_j \left| b_p^{meas}(t_j) - b_p^{sim}(t_j) \right|^2}{\sum_M \sum_N \sum_j \left| b_p^{meas}(t_j) \right|^2}} .$$

(k)   (8.5)

8.5   Measurements for nonlinear model construction

An emerging research area is making use of nonlinear measurements to construct the nonlinear model of microwave devices. This is the most logical approach, as the procedure is direct, i.e., without detour via a bias-dependent small-signal equivalent circuit model and thus S-parameter measurements. This approach became possible with the introduction of the large-signal network analyzer [2]. This initiated the development of several modeling approaches, such as reference [18–22]. One can categorize the modeling approaches as either time-domain- or frequency-domain-based. In the next sections, we highlight one of the approaches in each domain, and discuss the corresponding experiment design. Both approaches have in common that they belong to the category of behavioral models. The difference with the equivalent circuit models, discussed in the previous sections, is that these models consider the device as a black box, or, in other words, no physical behavior is taken into account and thus has to be known. In the case of equivalent circuit models, their topology is associated with the physical behavior of the DUT. The practical interest in behavioral models originates from the fact that this type of model can equally well be applied to modeling microwave circuits, which is not feasible with an equivalent circuit representation.

8.5.1   Time-domain measurements-based model construction

The first approach is based on the time-domain representation of large-signal vectorial measurements. It finds its origin in state-space modeling, which is a widespread modeling concept. It has application not only in electrical engineering, but also in mechanical and chemical engineering.

The general formulation of the state equations is expressed by:

$$\dot{X}(t) = f_a \left(X(t), U(t) \right)$$

$$Y(t) = f_b(X(t), U(t)),$$ (8.6)

where $U(t)$ is the vector of inputs, $X(t)$ is the vector of state-variables, and $Y(t)$ is the vector of outputs. The superscript dot denotes time derivative. The functions $f_a(.)$ and $f_b(.)$ are analytical functions expressing the dependencies.

In applying the state-space modeling concept to microwave transistors, the state equations (equation (8.6)) are rewritten as expressed by equation (8.7) [20]. The port currents, considered as the outputs, are expressed as a function of the port voltages, and the time derivatives of port voltages and port currents up to an order dictated by the dynamics of the device under consideration.

$$I_1(t) = f_1\left(V_1(t), V_2(t), \dot{V}_1(t), \dot{V}_2(t), \ddot{V}_1(t), \ldots, \dot{I}_1(t), \dot{I}_2(t) \cdots\right)$$
$$I_2(t) = f_2\left(V_1(t), V_2(t), \dot{V}_1(t), \dot{V}_2(t), \ddot{V}_1(t), \ldots, \dot{I}_1(t), \dot{I}_2(t) \cdots\right)$$

$$(8.7)$$

with $I_i(t)$ the port currents and $V_i(t)$ the port voltages. The superscript dots denote (higher order) time derivatives.

An alternative formulation could be to express the scattered traveling voltage waves, b_1 and b_2, as functions of the incident traveling voltage waves, a_1 and a_2, and the necessary derived state variables (such as the time derivatives of a_1 and a_2).

Note that equation (8.7) is applicable only to devices that exhibit no long-term memory effects. In microwave electronics, one typically distinguishes between long-term and short-term memory effects. Short-term memory effects have time constants in the nanosecond range, in the order of the carrier-frequency period, while long-term memory effects have time constants in the millisecond or microsecond range. Physical phenomena giving cause to the latter are temperature effects and traps within the transistor. Also, the external circuitry of the measurement equipment can cause long-term memory effects, such as the low-frequency impedance of bias tees. The reader is referred to reference [23] for the extension of equation (8.7) to include the long-term memory effects.

In terms of experiment design, the aim is to cover the state space with measurement data, such that the functions $f_1(.)$ and $f_2(.)$ in equation (8.7) can be determined. A dense and complete coverage is a necessity to avoid inter- and extrapolation when using the resulting model in circuit design or other applications.

Measurements using a single-tone excitation do not suffice. Each measurement results in a trajectory such as represented by Figure 8.14. The acquired data concerns only the data points on the trajectory itself. This implies that a high amount of measurements at various experimental conditions (range of biases, carrier frequencies, and power levels) would be needed in order to obtain the required dense coverage of the state space. As the focus of this chapter is on making experiment design efficient, this type of measurement is not adequate.

The use of multitone excitations which tone spacing is small compared to the RF carrier frequency results in a high amount of data points. Take the example of a two-tone excitation, shown in Figure 8.15. One envelope period corresponds to almost 10 000 RF periods. When applying the Nyquist criterion to sample the measurement, one ends up with a multiple of this number in data points, and this in one measurement only.

If the spacing between the tones is smaller, the number of acquired data points rises. In such a way, one ends up with a situation by which the number of data points becomes too high to handle comfortably in data post processing. The solution is to

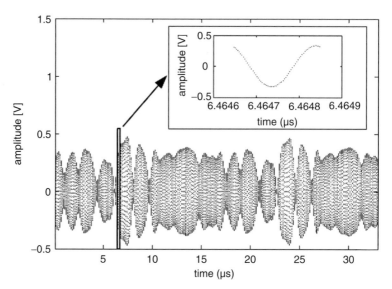

Figure 8.17 Time-domain waveform of 63-tone CDMA-like multisine; inset – zoom into one RF carrier frequency period. The device is a general-purpose PA on an evaluation board.

undersample the measurement. Two approaches have been proposed in the literature, namely uniform sampling in the time domain [24] and uniform sampling in the nonlinear metric domain [25].

The two approaches are discussed by means of an illustrative example. In the considered case, the excitation is more realistic with respect to practical applications than the two-tone excitation above, namely a code division multiple access (CDMA) excitation is considered. As the data is collected by large-signal vectorial measurements, the limitation is that signals should be periodic. For this reason, the modulated CDMA excitation gets approximated by a 63-tone multisine [26]. The amplitudes and phases of each of the tones in the multisine are determined in such a way that the resulting statistical properties of the multisine approximate those of the target digitally modulated excitation [27]. Figure 8.17 presents the envelope and zooms in to one RF period in the inset.

The time-domain uniform sampling of the modulation envelope as introduced in reference [24] is a straightforward approach, in analogy to the well-known sampling principle at a constant rate. It involves selecting N_{IF} RF carrier cycles along the period T_{IF} of the modulation envelope with the constant time interval $\Delta t_{IF} = T_{IF}/N_{IF}$ between consecutive sampling bins. To obtain a smooth representation of the RF waveform composed of maximum k_{max} harmonics, typically $N_{RF} = ovs \cdot 2k_{max}$ samples per period are required (with *ovs* denoting the oversampling factor). The data size reduction obtained through the modulation envelope sampling depends on the number N_{IF}. The approach is illustrated in Figure 8.18.

It was shown that the accuracy of the resulting model strongly depends on the actual shape of the envelope [28]. This can be intuitively understood by looking at Figure 8.20.

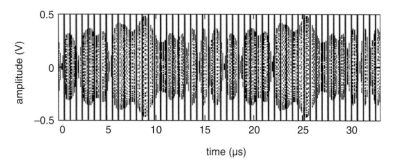

Figure 8.18 Time-domain uniform sampling of a CDMA-like multisine waveform; vertical lines – selected periods of the RF carrier frequency.

The plot shows the $(a_1(t), b_2(t))$ coverage for 18 RF periods, obtained by uniform sampling the modulation envelope period. It can be observed that the coverage is not uniform. There is a high amount of data points at small amplitudes, separated by a gap to a set of RF periods related to the larger amplitudes. The reason is that amplitudes are small for a large part of the envelope period, as a result of which only a few RF cycles at large amplitudes are selected. When constructing a model based on such data, the model is prone to be interpolated during its use in practical applications. As long as the sampling is uniform in time step, the obtained coverage strongly depends on the signal's characteristics, such as crest factor. In analogy to performing a denser grid of linear measurements in operating regions where the device's characteristics change quickly (Section 8.3), the aim is to look for an alternative method by which more data points are collected in regions of stronger nonlinearity. This gave rise to the development of the second approach for undersampling the envelope period, namely uniform sampling as according to the nonlinearity metric [25].

In the nonlinearity metric piecewise uniform sampling method [25], the RF carrier frequency periods are selected based on a nonlinearity metric calculated for many RF periods along the modulation envelope period. The nonlinearity metric σ is defined by equation (8.8) as the root-mean-square (RMS) of orthogonal distances between normalized trajectories $\tilde{b}_2(t_q, t_r)$ and the normalized reference (linear) trajectory $\tilde{b}_2(t_q, t_{ref})$:

$$\sigma_r = \sqrt{\frac{1}{N_{RF}} \sum_{t_q \in T_q} \left| \tilde{b}_2(t_q, t_r) - \tilde{b}_2(t_q, t_{ref}) \right|^2}, \forall t_r \in T_r \setminus t_{ref}. \tag{8.8}$$

The normalization involves dividing the instantaneous amplitudes of the output scattered traveling voltage wave $b_2(t)$ by the peak value of $a_1(t)$ waveform reached in each of the N_{IF} RF periods. The reference RF period $\tilde{b}_2(t_q, t_{ref})$ is chosen to be perturbed with a negligible nonlinear distortion. This usually occurs at low or moderate input signal levels driving the measured device into linear behavior. Next, RF periods are selected as according to uniform sampling of this nonlinearity metric, as illustrated by Figure 8.19. Such sampling results in a more uniform coverage, i.e., fewer gaps in between selected

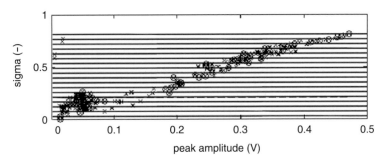

Figure 8.19 Piecewise equidistant sampling in the nonlinearity metric domain calculated for 512 RF cycles of the output multisine waveform (two times 256 RF periods corresponding to 63-tone multisine generated at -16 dBm and 0 dBm), and plotted as function of increasing input peak amplitudes; horizontal lines and circles denote 46 samples chosen piecewise equidistantly in the σ-domain and the dashed line marks σth parameter.

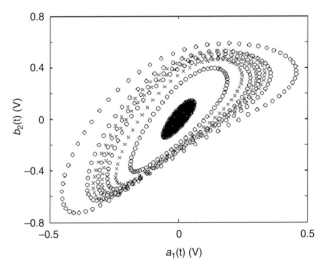

Figure 8.20 $(a_1(t), b_2(t))$ coverage (in (V)), represented by 18 RF periods (circles and crosses), corresponding to uniform sampling in time domain over the modulation envelope period.

RF periods, as presented in Figure 8.21. It has to be noted that the benefit of adopting this second approach as compared to the uniform sampling in time domain drops with increasing number of selected RF periods [29].

From the analysis above, one may conclude that using a complicated multisine excitation, such as one approximating a digitally modulated signal, always leads to the best model accuracy. The next example demonstrates that this is incorrect.

For the considered example, three behavioral models have been constructed [30]. The first model is based on power-swept single-tone measurements, called "Model 1", the second model is based on a measurement involving a 63-tone multisine excitation "Model 63", and the third model is based on the combined set of power-swept single-tone measurements and the measurement under 63-tone multisine excitation "Model 63+1". The three models are evaluated by applying the metrics introduced in Section 8.4.2 on

Table 8.1 Metric results for three behavioral models, constructed from, respectively, power-swept single-tone measurements 'Model 1', from a measurement with a 63-tone CDMA-like excitation 'Model 63', and from the combined set of measurements 'Model 63+1'.

Model Metric	M1_LS_1T	M3_LS_1T	M1_LS_TD_B1	M1_LS_TD_B2	M1_LS_2T	M2_LS_2T
Model 1	0.039	0.043	3.181	7.602	0.085	0.220
Model 63	0.669	0.734	0.813	2.837	0.805	0.904
Model 63+1	0.042	0.029	0.638	1.042	0.099	0.395

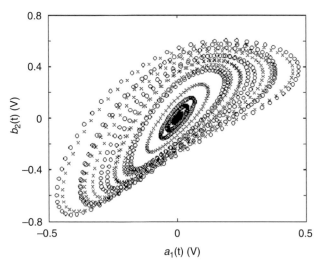

Figure 8.21 $(a_1(t), b_2(t))$ coverage (in (V)), represented by 18 RF periods (circles and crosses), corresponding to uniform sampling of σ nonlinearity metric over the modulation envelope period.

measurements that differ from the ones used for model construction. The results are summarized in Table 8.1.

It can be deduced from Table 8.1 that the model based on the multisine excitation, i.e., Model 63, does not always yield the lowest metric value, or, in other words, is not always the most accurate. As regards metrics based on single-tone excitation, M1_LS_1T and M3_LS_1T, Model 1 performs better than Model 63. The model based on the combined set of measurements, i.e., Model 63+1, yields similar results as Model 1. When analyzing the model predictions for a multisine excitation and adapting the metric expressions in the modulation domain, M1_LS_TD_B1 and M1_LS_TD_B2, Model 63+1 yields again the best results, and is even better than the model based on a – different – multisine excitation, Model 63. Model 1 results in a significant error. Finally, we compare the models for a two-tone excitation. Intuitively, one may expect that Model 63, based on a multisine, would be able to predict the two-tone measurement results, which is also a multisine, the best. The results show that Model 1 performs clearly the best, followed by Model 63+1. The explanation can be deduced from Figure 8.22. The figure shows the coverage of the measurement data used for the Model 1 and Model 63 constructions, respectively. It can be observed that the coverage of the former is very similar to the

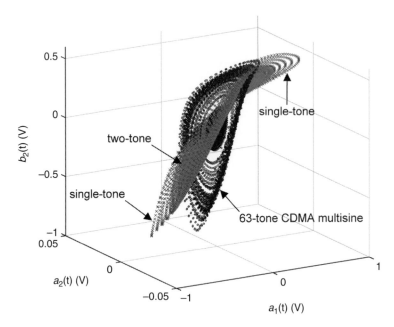

Figure 8.22 $(a_1(t), a_2(t), b_2(t))$ coverage for a power-swept single-tone measurement, a two-tone measurement, and a 63-tone CDMA-like multisine measurement.

coverage of the two-tone measurement, which explains the metric results. In the case of Model 63, extrapolation is required to predict the response for the two-tone excitation, and extrapolation usually results in larger inaccuracies. This example demonstrates once more that the coverage of the state space is primordial for constructing good state-space behavioral models.

8.5.2   Frequency domain measurements-based model construction

The second example of using large-signal vectorial measurements in model construction is a behavioral model expressed in the frequency domain. The scattered traveling voltage waves b_1 and b_2 are expressed as functions of the incident traveling voltage waves a_1 and a_2, as follows:

$$ b'_{ph} = Sf_{ph11} |a_{11}| + \sum_{\substack{ij \\ ij \neq 11}} \left(Sf_{phij} a'_{ij} + Sfc_{phij} \left(a'_{ij} \right)^* \right), \tag{8.9} $$

where p, h denote the output port and harmonic indices, respectively, and i, j stand for the input port and harmonic indices, respectively, at which the probing signal is applied. The superscript * stands for the conjugate operator, and the superscript prime denotes a phase shift relative to the phase of a_1 at the carrier frequency:

$$ x'_{ph} = x \exp\left(-jh\phi\left(a_{11}\right)\right). \tag{8.10} $$

The complex coefficients Sf_{phij} and Sfc_{phij} are called S-functions [22]. Note that for simplicity in equation (8.9) the DC part of the device's response is neglected. A more complete description can be found in reference [30].

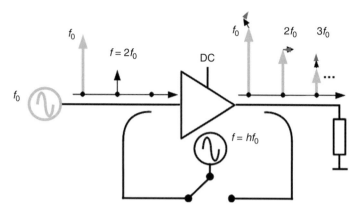

Figure 8.23 A simplified diagram of a setup to collect data for S-functions behavioral model extraction.

The meaning of the summation in equation (8.9) can best be explained by Figure 8.23. The principle is that the response of the DUT is linearized around a large-signal operating condition. The result at port 2 of applying a single-tone excitation at port 1 at frequency f_0 is indicated by the spectral components in gray. This state determines the large-signal operating condition. When a small probing signal is applied at a harmonic frequency (or at the fundamental frequency at port 2), the result is such that there is a linear response at each of the other spectral components. When the power of the small probing signal increases, the corresponding change in the device response cannot be described by a simple linear relationship with the incident phasors. To improve this and to keep the simple linear combination framework, new model terms are added. These additional terms are related to the conjugate of the incident traveling wave phasors [21].

In terms of experiment design, the first approach to determine the unknowns, the S-functions parameters, is by putting the device under a large-signal operating condition of interest, i.e., fixing bias condition, carrier frequency, and input power level, and then to sequentially inject a small probing signal at each of the harmonic frequencies at port 1 and at port 2, and also adding a measurement by which a small probing signal at the fundamental frequency f_0 is injected at port 2. For each of these excitations, the corresponding responses at all fundamental and harmonic spectral components are measured. To be able to distinguish between Sf_{phij} and Sfc_{phij}, at least two measurements, namely two different phase values for the probing signal, are required for each of the settings. Once all measurements have been collected, the S-functions parameters can be determined by solving a linear set of equations, starting from equation (8.9). Due to measurement imperfections (Figure 8.24), such as noise, the presence of harmonics generated by the sources, nonlinear interaction between source and DUT, and imperfect terminations, an optimization procedure is usually required. Finally, to have a general model for the device, this procedure has to be repeated for other large-signal operating conditions.

An alternative way to obtain the S-functions parameters is by applying the small probing signals at a frequency slightly offset. The method has been introduced in reference [31] and demonstrated by simulations, and later on by measurements [22]. This approach is illustrated by Figure 8.25.

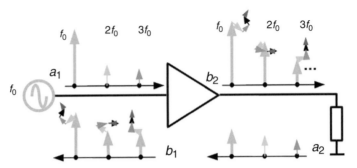

Figure 8.24 A simplified frequency-domain representation of the incident and scattered traveling voltage waves present at the ports of a two-port nonlinear component excited by a single-tone large-signal excitation at f_0 and a number of small-signals due to source harmonics, nonlinear interactions between source and DUT, and imperfect terminations.

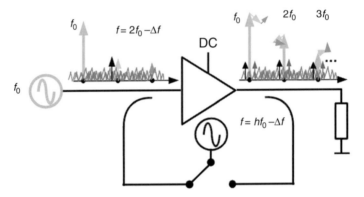

Figure 8.25 A simplified diagram of a setup to collect modulated data for S-functions behavioral model direct extraction.

The advantage of this approach is that no longer do all measurements have to be collected before the S-functions parameters can be determined by solving the set of linear equations. Instead, the S-functions parameters can be directly determined from a partial set of measurements, i.e., up to the harmonic of interest, making the overall measurement time shorter. The expressions to calculate the S-functions Sf_{phij} and Sfc_{phij} are given by the following formulas:

$$Sf_{phij} = \frac{b'_p\,(hf_0 - \Delta f)}{a'_1\,(jf_0 - \Delta f)}$$

$$Sfc_{phij} = \frac{b'_p\,(hf_0 + \Delta f)}{\left(a'_1\,(jf_0 - \Delta f)\right)^*}. \tag{8.11}$$

Note that one cannot assume an ideal single-tone large-signal excitation driving the DUT. In the real-life situation, the incident traveling voltage waves will always contain some small but nonzero harmonic components, contributing to the output phasors. As a result, the large-signal coefficients $Sfph_{11}$ cannot be directly calculated from the ratio of the output and the input phasors. Instead, one needs first to subtract the contribution of harmonics, as shown by equation (8.12), using the previously identified small-signal S-functions parameters.

$$Sf_{ph11} = \frac{\beta'_p (hf_0)}{a'_1 (f_0)}$$

$$\beta'_p (hf_0) = b'_p (hf_0) - \sum_{\substack{ij \\ ij \neq 11}} \left(Sf_{phij} a'_i (jf_0) + Sfc_{phij} \left(a'_i (jf_0) \right)^* \right). \qquad (8.12)$$

It is important to consider that the parameter values obtained may vary due to additive noise and phase noise affecting the measured quantities (Figure 8.25). Also, long-term memory effects may play a role. As a result, the choice of offset frequency is not arbitrary, but should be selected carefully, taking into account the presence of phase noise and, possibly, long-term memory effects.

To estimate the true values of the S-functions parameters and to assess the variability of the results, the measurements can be repeated a number of times. Based on this repeated set of measurements and assuming a zero-mean normal distribution of the measurement noise, one can estimate the values of the amplitudes and phases of the S-functions parameters using the errors-in-variables technique, as described in reference [22].

The knowledge of the uncertainties of the S-functions parameters can be of use to simplify the S-functions behavioral model [32]. The most dominant model parameters are selected based on the relative uncertainty of their estimated values evaluated against a threshold value. Relative uncertainty stands for the ratio of the uncertainty of the estimated parameter and its expected value. All S-functions parameters which have a relative uncertainty larger than the threshold are removed from equation (8.9). The fact that the relative uncertainty is a good metric to evaluate for model-order reduction can intuitively be understood as follows. In general, the relative uncertainty is largest for S-functions parameters that have small expected values. Namely, the latter indicates that the output phasor response to the probing input signal is small, resulting in a decreased signal-to-noise ratio and therefore an increased measurement uncertainty which directly affects the uncertainty on the estimated parameter. As a result, the contribution of such S-functions parameter to equation (8.9) is small, and therefore can be neglected. A threshold value of 1% or -40 dB has been experimentally demonstrated to result in a high level of model-order reduction without substantial loss in prediction accuracy [32].

The final point of attention is the fact that S-functions and other frequency-domain behavioral models based on describing functions rely on the linearization of the response of a nonlinear microwave device around a set of large-signal operating points. To ensure the validity of the linearization principle, and thus the validity of the model, one has to select appropriate stimuli levels during the experiments which are performed to extract the model parameters.

The idea behind the procedure proposed in reference [33] is to increase the amplitude of the small probing signal (Figure 8.26) and to observe the resulting change in corresponding scattered traveling voltage wave phasors. This is possible only when the small-signal and conjugate small-signal contributions are separated in frequency by

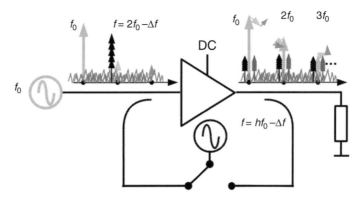

Figure 8.26 A simplified diagram of a setup to collect modulated data for S-functions behavioral model direct extraction; the probing tone power is swept to identify the linear validity range of the model.

applying the small signal slightly offset by Δf with respect to the fundamental frequency grid hf_0, as described above. For the linearization assumption to hold, it is required that the magnitude of the response at the offset frequencies $hf_0 - \Delta f$ and their conjugates $hf_0 + \Delta f$ change linearly, while the magnitude of the response on the fundamental frequency grid hf_0 remains unchanged. At the same time, the phase relationship between the response at hf_0 and at $hf_0 \pm \Delta f$ and the respective large- and small-signal inputs has to remain constant. Note that in the case of the response at the conjugate frequencies $hf_0 + \Delta f$, the phase differences are calculated with respect to the inverted phase of the probe tone. Finally, the probing signal level must not cause any response at the frequencies corresponding to higher order intermods of hf_0 and $hf_0 - \Delta f$ components. In general, one seeks the maximum amplitude of the probing signal at which all these assumptions are valid.

To select the appropriate levels in case of data that is corrupted by measurement noise, one can fit respectively linear and constant value models to the frequency components corresponding to such input–output relationships. As previously mentioned in connection with dealing with measurement noise, one can analyze the cost functions of an errors-in-variable (EIV) estimator while increasing the amplitude of the probing signal [33]. As long as the assumed relationship holds, the cost function value stays within the 95% confidence bounds of the expected value. The moment when the assumption is violated, the actual cost function value moves outside the confidence bounds. The maximum amplitude before that happens determines the desired maximum allowed probing signal level.

References

[1] A. Rumiantsev and N. Ridler, "VNA calibration," *IEEE Microwave*, vol. 9, no. 3, pp. 86–99, 2008.

[2] J. Verspecht, P. Debie, A. Barel, and L. Martens, "Accurate on wafer measurement of phase and amplitude of the spectral components of incident and scattered voltage waves at the signal ports of a nonlinear microwave device," *IEEE MTT-S Int. Microw. Symp.*, 1995, pp. 1029–1032.

[3] P. Blockley, D. Gunyan, and J. B. Scott, "Mixer-based, vector-corrected, vector signal/network analyzer offering 300 kHz-20 GHz bandwidth and traceable phase response," *IEEE MTT-S Int. Microw. Symp.*, pp. 1497–1500, 2005.

[4] D. Williams, P. Hale, and K.A. Remley, "The sampling oscilloscope as a microwave instrument," *IEEE Microwave*, vol. 8, no. 4, pp. 59–68, 2007.

[5] H. Cho and D. E. Burk, "A three-step method for the de-embedding of high-frequency *S*-parameter measurements," *IEEE Trans. Electron Devices*, vol. 38, no. 6, 1371–1375, 1991.

[6] M. C. A. M. Koolen, J. A. M. Geelen, and M. P. J. G. Versleijen, "An improved de-embedding technique for on-wafer high frequency characterization," *IEEE Bipolar Circuits and Technology Meeting*, pp. 188–191, 1991.

[7] E. P. Vandamme, D. Schreurs, and C. van Dinther, "Improved three-step de-embedding method to accurately account for the influence of pad parasitics in silicon on-wafer RF test-structures," *IEEE Trans. Electron Devices*, vol. 48, no. 4, pp. 737–742, 2001.

[8] V. Subramanian, B. Parvais, J. Borremans, A. Mercha, D. Linten, P. Wambacq, *et al.*, "Planar bulk MOSFETs versus FinFETs: an analog/RF perspective," *IEEE Trans. Electron Devices*, vol. 53, no. 12, pp. 3071–3079, 2006.

[9] S. Fan, D. E. Root, and J. Meyer, "Automated data acquisition system for FET measurement and its application," *ARFTG Conf.*, pp. 107–119, Fall 1991.

[10] C. van Niekerk and D. Schreurs, "A new adaptive multi-bias S-Parameter measurement algorithm for transistor characterisation," *Int. J. Numerical Modelling: Electron. Networks, Devices and Fields*, vol. 18, no. 4, pp. 267–281, 2005.

[11] G. Crupi, D. Schreurs, A. Raffo, A. Caddemi, and G. Vannini, "A new millimeter wave small-signal modeling approach for pHEMTs accounting for the output conductance time delay," *IEEE Trans. Microw. Theory Tech.*, vol. 56, no. 4, pp. 741–746, 2008.

[12] M. Pirazzini, G. Fernández, A. Alabadelah, G. Vannini, M. Barciela, E. Sánchez, and D. Schreurs, "A preliminary study of different metrics for the validation of device and behavioral models," *Spring Autom. RF Tech. Group Conf.* (ARFTG), 2005.

[13] D. Schreurs, "Capabilities of vectorial large-signal measurements to validate RF large-signal device models," Automatic RF Techniques Group Conference (ARFTG), Fall 2001.

[14] G. Crupi, D. Schreurs, and A. Caddemi, "Accurate silicon dummy structure model for nonlinear microwave FinFET modeling," *Microelectron. J.*, vol. 41, no 9, pp. 574–578, 2010.

[15] G. Crupi, G. Avolio, D. Schreurs, G. Pailloncy, A. Caddemi, and B. Nauwelaers, "Vector two-tone measurements for validation of nonlinear microwave FinFET model," *Microelectron. Eng.*, vol. 87, no. 10, pp. 2008–2013, 2010.

[16] G. Avolio, G. Pailloncy, D. Schreurs, M. Vanden Bossche, and B. Nauwelaers, "On-wafer LSNA measurements including dynamic bias," *Eur. Microw. Conf. (EuMC)*, 2009, pp. 930–933.

[17] D. Schreurs, "Systematic evaluation of non-linear microwave device and amplifier models," *Eur. Microw. Integrated Circuits Conf. (EuMiC)*, 2006, pp. 261–264.

[18] D. Schreurs, J. Verspecht, B. Nauwelaers, A. Van de Capelle, and M. Van Rossum, "Direct extraction of the non-linear model for two-port devices from vectorial non-linear network analyzer measurements," *Eur. Microw. Conf.*, 1997, pp. 921–926.

[19] D. Schreurs, J. Verspecht, S. Vandenberghe, and E. Vandamme, "Straightforward and accurate nonlinear device model parameter estimation method based on vectorial large-signal measurements," *IEEE Trans. Microw. Theory Tech.*, vol. 50, no. 10, pp. 2315–2319, 2002.

[20] D. Schreurs, J. Wood, N. Tufillaro, L. Barford, and D. Root, "Construction of behavioural models for microwave devices from time-domain large-signal measurements to speed-up high-level design simulations," *Int. J. RF Microw. Computer Aided Eng.*, vol. 13, no. 1, pp. 54–61, 2003.

[21] J. Verspecht and D. Root, "Polyharmonic distortion modeling," *IEEE Microwave*, vol. 7, no. 3, pp. 44–57, 2006.

[22] M. Myslinski, F. Verbeyst, M. Vanden Bossche, and D. Schreurs, "S-functions extracted from narrow-band modulated large-signal network analyzer measurements," *Autom. RF Techniques Group Conf.* (ARFTG), Fall, 2009.

[23] D. Schreurs, K. Remley, M. Myslinski, and R. Vandersmissen, "State-space modelling of slow-memory effects based on multisine vector measurements," *Autom. RF Techniques Group Conf. (ARFTG)*, Fall, 2003, pp. 81–87.

[24] D. Schreurs and K. Remley, "Use of multisine signals for efficient behavioural modelling of RF circuits with short-memory effects," *Autom. RF Techniques Group Conf. (ARFTG)*, Spring, 2003, pp. 65–72.

[25] D. Schreurs, K.A. Remley, and D. F. Williams, "A metric for assessing the degree of device nonlinearity and improving experimental design," *IEEE Int. Microw. Symp.*, pp. 795–798, 2004.

[26] M. Myslinski, "Using modulated signals for the validation and construction of non-linear models for microwave components," Ph.D. thesis, K. U. Leuven, Belgium, 2008.

[27] J. C. Pedro and N. B. Carvalho, "Designing multisine excitations for nonlinear model testing," *IEEE Trans. Microw. Theory Tech.*, vol. 53, no. 1, pp. 45–54, 2005.

[28] D. Schreurs, M. Myslinski, and K. A. Remley, "RF behavioural modelling from multisine measurements: influence of excitation type," *Eur. Microw. Conf. (EuMC)*, 2003, pp. 1011–1014.

[29] M. Myslinski, D. Schreurs, and B. Nauwelaers, "Impact of sampling domain and number of samples on the accuracy of large-signal multisine measurement-based behavioral model," *Int. J. RF and Microw. Computer-Aided Eng. (RFMICAE)*, vol. 20, no. 4, pp. 374–380, 2010.

[30] NMDG newsletter IMS2009 special edition, 2009. [Available online: http://www.nmdg.be/newsletters/Newsletter_IMS09.html].

[31] D. E. Root, J. Verspecht, D. Sharrit, J. Wood, and A. Cognata, "Broad-band poly-harmonic distortion (PHD) behavioral models from fast automated simulations and large-signal vectorial network measurements," *IEEE Trans. Microw. Theory Tech.*, vol. 53, no. 11, pp. 3656–3664, 2005.

[32] M. Myslinski, F. Verbeyst, M. Vanden Bossche, and D. Schreurs, "S-functions behavioral model order reduction based on narrow-band modulated large-signal network analyzer measurements," *Autom. RF Techniques Group Conf. (ARFTG)*, 5 p., Spring, 2010.

[33] M. Myslinski, F. Verbeyst, M. Vanden Bossche, D. Schreurs, "A method to select correct stimuli levels for S-functions behavioral model extraction," IEEE MTT-S *Int. Microw. Symp.*, pp. 1170–1173, 2010.

9 Practical statistical simulation for efficient circuit design

Peter Zampardi, Yingying Yang, Juntao Hu, Bin Li, Mats Fredriksson, Kai Kwok, and Hongxiao Shao

Skyworks Solutions, Inc.

9.1   Introduction

In wireless handset design, specifically power amplifiers (PAs), there is constant pressure to improve time-to-market while maintaining high yields. To meet these demands, designers need to evaluate current design practices and identify areas for improvement. Presently, some PA designers spend a great deal of time bench-tuning to optimize circuits. Because this is very time consuming, the main focus is obtaining the best "nominal" performance, and process variation is generally an afterthought. Frequently, new circuit topologies are implemented and minimal sample sizes are evaluated (often on a single wafer) leading to "one-wafer wonder" results.

Unfortunately, as the design is run over many wafers, normal process variations take their toll degrading the initial "hero" performance and, in the extreme case, lead to unacceptable yields. These variations are often blamed on the starting material or the fabrication process but, in reality, are due to *expected* process variations.

Including process statistics in the simulation phase can greatly reduce the occurrence of these frustrating events. To date, the implementation of statistical simulations in microwave designs (and III–V designs, specifically) has been limited, even though it is commonplace in silicon (Si) digital or analog-mixed signal design [1–6].

What are the barriers? The first is that methodology used in the Si design community is usually centered on inherently time-consuming Monte Carlo (MC) simulations [4–7]. While necessary for most Si designs, where neighboring device mismatches are critical, the additional complexity and added simulation time makes it "unfit" for III–V designs where devices are large and wafer turnaround time is short (weeks compared to months). Some Si foundries provide "corner" models, but these are derived by driving figures of merit (like f_T) that are not necessarily important for RF design. Most foundries provide customers with an option to fabricate "spread" wafer lots that capture the expected process variation (by changing process variables) [8], but do not provide a way to easily simulate that set of wafers to allow designers to close the loop with simulation.

Another barrier is that many approaches for modeling of GaAs devices are curve-fitting-based rather than being physics- and scaling-based. Curve fitting makes it cumbersome, if not impossible, to provide a set of models that accurately tracks real-life process variations.

The final impediment is that most statistical analysis training focuses on using a particular software package, separate from the circuit simulator [9]. This creates a large barrier, since designers do not have the time to learn another piece of software (or do not want to fragment the design flow further). To make the statistical simulation useful to designers, it should be accessible in the design environment.

To overcome these barriers, the development of a statistical-simulation capability embedded in a designer-friendly design flow has several key requirements. The approach should:

- be predictive and represent real-life examples (no nonphysical variations are allowed);
- be simple, convenient, and faster than "trial-and-error." Otherwise, it is viewed as an extra burden or nice "window dressing" for design reviews;
- provide insight into what can be changed to make a better design, not just indicate how "poor" the design is. The simulation approach should be intuitive enough that designers can easily assess layout or design changes to reduce variation.
- allow closure of the simulation loop by comparison with measurement of similar process spread wafers.

We have implemented a design flow that takes advantage of the attributes of III–V HBT technology (but is applicable to other technologies) by: including a "unified" core modeling approach and statistical simulation based on design-of-experiments (DOE), selecting orthogonal only material/process/operational variables, and implementing these in an advanced design system (ADS). This allows the high-level integration of design, simulation, and statistical analysis of a PA in a single tool. Greater process tolerance is achieved because this DOE-based flow makes designers aware of process variation and allows circuit redesign before committing the design to GaAs (before final tape-out). The method has resulted in the following benefits for our development teams and our customers:

1. Resulting designs are more robust and vary less over process, allowing customers to "set it and forget it" after these parts are used in their phones.
2. Verification of the design topology is a result of including these simulations in the design review process and provides a foundation for failure-modes effects and analysis (FMEA).
3. The method has eliminated numerous circuit topologies in the early stages of development that were found to have unacceptable process variation.
4. It provided guidance on process development directions and led to refinement of process-control monitors for our application.
5. It provided a tool to determine if requests for tighter control on device parameters, such as beta, are reasonable, or if there are other root causes for circuit variation.
6. It serves as a de-bugging tool (allowing "what-if" simulation). The resulting insight guides (and is faster than) physical wafer fabrication.

In this chapter, we describe the key elements of our approach to include processing statistics into a III–V PA design process. This is a three-tier approach where we begin by determining the fundamental material/process parameters that affect the device and, in turn, the product performance. This first tier is necessary for two very important reasons: (1) experimentally running these variants allows important correlations to be determined (or incorrectly assumed ones to be disregarded) which later simplifies the circuit level simulations, and (2) it allows us to close the loop by comparing our statistical simulations to products fabricated on these variant wafers when we are done. The second tier is implementing a "unified" model where all devices in the technology are physically coupled together so that when we do statistical simulation, all the states are physically realizable. This means that we use physics, geometry, and process parameters to develop the models for each device rather than curve-fitting devices individually. The third tier is to integrate the "unified" model and the circuit-level statistical simulation, in ADS, based on tiers 1 and 2, into the design flow.

In Section 9.2, we present the objective and key elements of this approach and compare/contrast them with more commonly used methodologies – for example, MC simulations and corner models. We will then discuss in detail the three tiers of our approach. Some key features and advantages of this approach versus others are highlighted. In Section 9.3, we present two examples that demonstrate the power of this design flow. Both examples show significantly reduced performance variation while maintaining excellent nominal performance. These results demonstrate the impact of design topology and layout on performance variation. In Section 9.4, we summarize our work.

9.2　Approach, model development, design flow

9.2.1　Objective and key elements of this approach

Simply stated, the main objective of this approach is to capture as much process variation (and product) with as few variables as possible. The first two key elements of this approach are the physics-based "unified" modeling and DOE circuit simulation. Both are based on the assumption that PA designs use large devices and that device mismatch is negligible (so each device does not need to be unique). The orthogonal-only-variables element focuses on reducing the number of simulation runs. Pareto-driven analysis (a Pareto chart is a graph showing the importance of the factors as a bar chart in descending order of importance), within the simulator, provides easily accessible insight to designers. This high-level single-tool integration is the third key element to make this approach a powerful and practical tool for the III–V design community.

9.2.1.1　Physics-based "unified" modeling

Our unified modeling approach is the foundation of this design flow. This geometrical and physical modeling approach (similar to that used in some Si bipolar modeling) is described in detail in a later section. The term "unified" means that devices fabricated from the same junctions or layers are forced to share not only the same variation, but the same model parameters where possible, thus "unifying" the behavior. This is important since real-world devices behave this way!

In a traditional curve-fitting approach, different devices are modeled independently and little thought is given to device consistency, for example, HBTs do not share model parameters with junction diodes or semiconductor layer resistors. In the extreme, model parameters for the devices of the same type, say HBTs, of different size or geometries were not linked. As a result, it takes many parameters to vary each of these devices statistically on an individual basis, and some nonphysical statistical states can occur.

A simulation approach based on individual device corner modeling has been reported [10]. That approach is useful if there is a single design performance criterion, for characterizing a system in which a single component dominates the system behavior, or when all the components vary independently (as if the parts were all independent discrete components). The possible statistical system responses, in those cases, can be obtained by simulating all combinations of corners for each device. However, for *integrated* circuits (MMICs), the assumption that circuit components vary independently or that one device changes while other devices are constant is clearly incorrect. Adopting that approach would result in nonphysical states and wasting effort worrying about variations that can never really occur. A good example of what we mean by unified modeling is shown in Figure 9.1a for a semiconductor resistor. The model for this device is based only on process control monitor (PCM) data values, geometrical calculations, and the BC diode model (which is also consistent with the HBT). So, rather than individually extracting parameters for this device, we only have to take some limited measurements for validation (Figure 9.1b). This reduces the workload for model generation (it is not just for HBTs) and ensures that the devices all respond to process changes in the same way.

Our unified modeling approach is a physical corner modeling approach. It automatically generates corner models, by inputting statistical DOE parameters which control all the onchip devices together rather than individually, and naturally guarantees physically possible circuit corner responses.

The parameters used for DOE simulations can be fundamental epi parameters (such as doping and thickness) or can be abstracted to a higher level of process control measurement parameters. At various stages of development we have used both, but for circuit simulations, we typically use the PCM parameters. This method reduces the simulation iterations, as opposed to running through different individual device corner models.

9.2.1.2   DOE circuit simulation

While statistical models based on MC simulation are the norm in the silicon industry, they may be too cumbersome for designing in III-V technologies. In particular, a DOE approach allows a faster mapping of circuit performance and is more easily coupled to DOE lots fabricated in our manufacturing line.

DOE is heavily used in the semiconductor industry but has gained an undeservedly bad reputation due to misuse in six-sigma implementations. Six-sigma is a quality management methodology that relies heavily on the use of DOE [11]. Unfortunately, the misuses include trying to apply it where it does not make sense (it is great for crank-turning optimization problems, but not good for finding the crank) and trying to apply

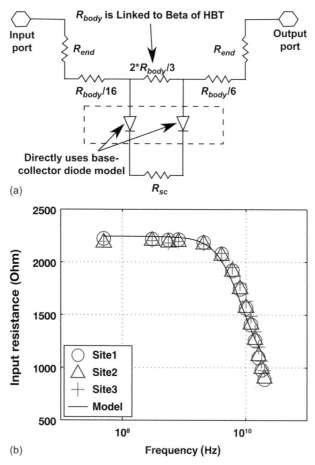

(a)

(b)

Figure 9.1 An example of unified device modeling approach. Instead of modeling the base layer resistor as a separate entity, parameters and models for other devices sharing the same physical layers are incorporated in its model topology and model parameters. (a) Base layer resistor model consists of 1 sheet resistance, which is directly linked with a HBT parameter, β; 2 base–collector junction diode model, which is directly "borrowed" from the diode device models to describe the underneath layer. (b) The unified modeled base resistor model accurately predicts the impedance drop at higher frequency contributed by junction capacitance of the underneath junction diode.

the methodology blindly to replace thinking and understanding the problem. However, when used correctly, it provides a powerful tool and is the cornerstone of this modeling methodology. While a discussion of DOE is beyond the scope of this chapter, the reader can find an excellent tutorial on DOE methodology in reference [12] and more detailed treatments in reference [10, 14]. Essentially, a DOE approach changes multiple factors (variables) in a controlled fashion and allows the impact of these factors, and different orders of interaction among the factors, to be determined with a minimal number of experiments.

Table 9.1 Independent epi material, process, and circuit operational variables which cover all the statistical changes of all the onchip devices and their corresponding circuit operational changes. Most designs do not involve all the onchip devices; therefore, the statistical variables needed in simulations are less than 10 in most cases.

PCM parameters	Description
PCMVt	Threshold voltage for FET in BiFET process
PCMRef	Emitter resistance
PCMbeta	DC gain for HBT
PCMvbe	Turn on voltage for diode (base-emitter) and HBT
PCMvbc	Turn on voltage for diode (base-collector) and BC junction in HBT
PCMRbcontTLM	Base contact resistance
PCMRscsh	Subcollector sheet resistance
PCMRscCont	Subcollector contact resistance
PCMdWt	Thin-film resistor width variation
PCMrhot	Thin-film resistor sheet resistance
PCMvsc	Schottky diode turn-on voltage

9.2.1.3   High-level single-tool integration and implementation

We implemented these models in Agilent ADS so that a simple DOE box could be placed in the schematic that would automatically include states corresponding to process corners. The distinction between this approach and simple corner or MC models is that the results can be quickly analyzed using a statistical software package like RS/1®, JMP®, or in this case, ADS® itself. We also point out that it is extremely difficult or close to impossible to obtain the Pareto information from MC data – at least in ADS. You would have to export it to a statistical analysis package such as JMP or write some post-processing functions in ADS to figure out which parameters or interactions are the most sensitive.

For a GaAs HBT chip, there are many variables that can potentially vary based on starting material or fab variation. For epi-variation, the models were implemented to allow individual material parameters (like doping and thickness) to be varied. However, for this work, we use model parameters such as beta that are responses to the doping and thickness [15]. This link is necessary to help understand the circuit response (in other words, it is important to know which parameter caused beta to change). As a result, only independent (orthogonal) epi and process variables, represented by model parameters, are accessible for statistical circuit simulations, even though orthogonality is not generally required for statistical simulation. The benefits are twofold: it minimizes the simulation time to observe the same circuit response obtained with more correlated variables; and it eliminates any nonphysical circuit responses of correlate parameters going in incorrect directions. The accessible parameters are listed in Table 9.1. The parameters are for our BiFET process (which includes a MESFET). While we have applied the exact same methodology to the FET, we restrict our discussion here to the HBT and related devices. The variations of the independent parameters are obtained from PCM data, which also provides the correlation among parameters (based on material DOE runs [16]) and validation of the orthogonality of the parameters [17]. Reviewing the semiconductor resistor example again, instead of using both R_{bsh} (also called R_b) and β variables to describe the resistor and HBT variation, only β is accessible (but R_{bsh} is linked to β inside the model code) to simulation. Our strategy is to catch as much variation as possible with as few parameters as possible.

For III–V HBT technologies, all frontend devices are formed by reusing junctions (base–emitter or base–collector junctions, etc.) or layers (emitter, base, subcollector), and backend devices (such as thin-film resistors, MIM capacitors, and inductors) are independently formed but may share metal layers. We separate model parameters into two sets: layout (geometry)-dependent and material (epi)-dependent. The geometry parameters describe the layout (configuration and size) of a given type device. When varied, they affect all devices that share a particular set of layers/junctions. In addition to the sheet resistance, the geometrical dependence/variations are important for thin-film resistor and epi-resistor simulations. The material-dependent (epi) parameters describe the differences in parameters due to the starting material. This allows us to (a) model the same geometry set of devices on different materials (or their variation) by changing only a few parameters related to the differences in material design; (b) drive variations across devices that share the same material layers (for example, the base, emitter, or collector layers), greatly reducing the number of parameters needed to describe all device variations in a given epi material. The following section describes the selection of material parameters and the impact of material variation on the devices so that the epi-dependent parameters for modeling (including correlations) can be determined.

9.2.2 Three-tier approach

9.2.2.1 Tier one: parameter (factor) selection

Epitaxial wafer parameter selection and impact on PCMs

Many different parameters can affect the performance of an HBT and circuit fabricated using them. One practical constraint we needed to impose was that any epi material DOE used for product qualification fits within a single process lot (in our case fewer than 20 wafers). This basically allows for three independent epi parameters (eight corners plus center) to be used with duplicates of each (allowing for some indication of process variation) experiment. Our electrical parametric control monitor set already includes orthogonal determination of many important parameters (R_e, C_{bc}, C_{be}, etc.) so we focused on parameters that (1) had a potentially large impact on the circuit performance and (2) could not be uniquely determined from DC PCM data.

For bipolar transistors, it is well known that the properties of the base play a large role in determining the major device characteristics [18]. Because carbon doped III–V HBTs are grown by well-controlled epitaxy, we can accurately specify the base thickness and doping. The base thickness (BT or w_b) and base doping (BD or N_b) have a substantial impact on the DC current gain (β), turn-on voltage (V_{be}), base resistance (R_{bsh}), and – depending on the bias region – RF gain (f_T or f_{MAX}). The base thickness and doping occur as a product in both β and R_{bsh}. This makes it impossible to understand (using nondestructive methods) whether an observed change in DC gain or base-sheet resistance is due to a thickness change or a doping change. Therefore, base doping and thickness

Table 9.2 DOE material matrix

Run Number	BT	BD	CT
1	Nom	Nom	Nom
2	Nom + 11%	Nom+12.5%	Nom+12.5%
3	Nom+11%	Nom+12.5%	Nom–12.5%
4	Nom+11%	Nom–12.5%	Nom+12.5%
5	Nom+11%	Nom–12.5%	Nom–12.5%
6	Nom–11%	Nom+12.5%	Nom+12.5%
7	Nom–11%	Nom+12.5%	Nom–12.5%
8	Nom–11%	Nom–12.5%	Nom+12.5%
9	Nom–11%	Nom–12.5%	Nom–12.5%

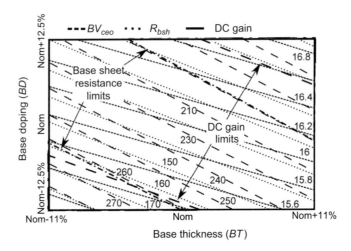

Figure 9.2 Response of several important device parameters to base doping and thickness.

were selected as two essential parameters for our epi-material DOE. Another important circuit parameter is the base-collector capacitance, C_{bc}. Since PAs are operated with large voltage swings on the base–collector junction, the collector thickness (CT) determines the C_{bc} for a significant portion of the operating range. An interaction between the DC gain and the breakdown voltage, BV_{ceo} was also expected (and breakdown also depends on the collector thickness). Therefore, CT was selected as the third parameter. Using RS1 software, we developed the experimental matrix shown in Table 9.2. The parameters were varied well outside the expected range to ensure that we captured the normally expected variations *within* the DOE range.

Figure 9.2 shows the measured dependence of several key parameters (β, R_{bsh} (R_b) and BV_{ceo}) on the base thickness and doping with a nominal collector thickness. Analyzing this data, we found several interesting results. First, the β and R_{bsh} are linearly related to one another (slope = 1.06), as shown in Figure 9.3. This was unexpected since the formulas for DC gain of HBTs limited by bulk recombination have a different doping

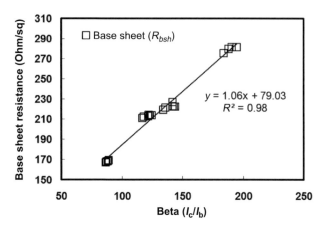

Figure 9.3 Relationship between DC gain (β) and base sheet resistance. Note that they are linearly related with a slope of 1.06.

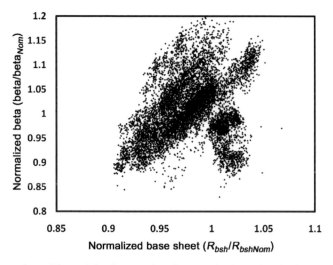

Figure 9.4 Same plot as Figure 9.3 using raw data from production PCM database. Notice that it does not look linear but is instead a fuzzball.

dependence for the β and R_{bsh} [19]. We expected β to follow:

$$\beta = \upsilon\tau/w_b = \upsilon \cdot 2.8x10^{38}/w_b N_b^2 \tag{9.1}$$

where υ is the average carrier velocity in the base, τ is the minority carrier lifetime, w_b and N_b the base thickness and doping, respectively. The fact that R_{bsh}/β is a constant implies that the β varies as $1/N_b$, rather than $1/N_b^2$. This is actually beneficial, since it allows a simpler implementation of statistical process models, since knowing R_b or β implies the other. This provides a good example of how the DOE approach is appropriate for determining these relationships. Figure 9.4 shows a similar plot made from data pulled directly from the manufacturing database. In this case, one observes

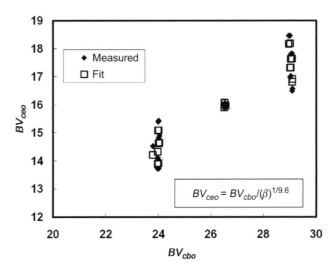

Figure 9.5 Open base breakdown voltage, BV_{ceo}, versus open emitter breakdown voltage, BV_{cbo}. Circles are measured data, squares are fit to equation (9.2) using $n = 9.6$.

a "fuzz-ball" that makes it impossible to determine the proper correlation of these variables. This happens for two reasons. The first is that, in production, the accepted data is much tighter than what we have used in the DOE. This results in short "lever arms" for the DOE and makes it more difficult to see the correlations. The second is that there are other factors that have a second-order impact on the gain and sheet resistance that are outside of the DOE. While these factors could be included, we are interested in capturing the first-order relationship between the parameters and those factors are not related to fundamental material changes.

Another unexpected relationship uncovered from this experiment was that we found that the BV_{ceo} is only *weakly* coupled to β. The typical values found in the literature for the relationship between BV_{cbo} and BV_{ceo} follow [20]:

$$BV_{CEO} = BV_{CBO}/\beta^{1/n}, \qquad (9.2)$$

where BV_{ceo} is the open-base breakdown, BV_{cbo} is the open emitter breakdown, and β is the DC gain. However, many modern bipolar transistors (including Si/SiGe) tend to deviate from the $n = 2$ found discussed in most textbooks. For the HBTs measured in this experiment, a value of 9.6 was determined by fitting to the data. A comparison of BV_{ceo} versus BV_{cbo} to equation (9.2) with $n = 9.6$ is shown in Figure 9.5. One reason for the possible discrepancy is that β for the HBT is typically measured at high currents (near peak β), while the breakdown is measured at low currents. Parasitic resistance effects also cause deviation from this theory, as discussed by Yeats *et al.* [21].

For variations in the collector thickness, we find that the breakdown and the reverse bias C_{bc} are well correlated to the thickness, as expected. This correlation of PCM behavior is a very important link in relating the circuit results back to the PCM data. As we have shown, this is a critical step, since it can often uncover some unexpected relationships that

Table 9.3 Ranges of RF parameter from epi DOE material matrix.

Part	Gain	ACPR1	ACPR2	PAE
Cell-1	28.9−30.7	45.6−53.4	57−61.4	NA
Cell-2	26.3−29	49.6−53.3	60.3−63.2	36.4−39.2
Cell-3	24.2−28	42.5−49.6	56.5−58	39.8−42
PCS-4	28.4−31.6	45.3−47.3	53.5−55.6	37.3−39.3

must be understood for proper model implementation. Having evaluated the dependence of the PCM parameters on the epi material variations (for brevity, we have only shown a limited subset), we next proceed to product level testing to confirm that we have selected the important material parameters for the circuit.

PA module level validation of parameter selection

To demonstrate this method – and validate that we have selected the correct material parameters that are important for our circuits – several different PA modules were evaluated. The first three circuits in Table 9.3 are 900 MHz cellular PA modules. The fourth part is a 1.8 GHz PCS module. Dies from each wafer were built in distinct assembly runs and the data collected was segregated for each specified epi material variant. The test results allowed process- and material-dependent PCM parameters to be separated. As anticipated, we found the key RF performance parameters to be strongly dependent on the starting epi material parameters we varied. Table 9.3 also shows the range of key parameters achievable with this DOE for the measured parts. Because of design differences, not all of the parts respond in the same way to variations in the parameters we used. As a result, conclusions about how a given parameter affects circuit performance must be evaluated on a circuit-by-circuit basis. As previously stated, the parameter ranges we selected are purposely wider than is acceptable for our customer specification, allowing interpolation of the data to the region of interest (i.e., our incoming material specifications). We evaluate the circuit yields and performance for the epi material variations and use this information to set incoming material specifications for given product families. As a result, the circuit yields and performance variations due to epi material variability are quickly established with this approach. From the table we note that the linearity, as measured by the first adjacent channel power ratio (ACPR1), can be substantially changed due to the material variations. Examples of how to tie this back to the PCM results are shown in Figures 9.6–9.8. By plotting a given parameter over the variation space, and comparing it to the PCM data over the same space, we can determine exactly what material parameters (and resulting PCM space) will give an acceptable performance. Figure 9.6 shows the RF gain under IS-95 modulation as a function of the base thickness and doping for Cell-3 (the data in table is for CDMA-2K). Comparing this figure to Figure 9.2, we note that the RF gain for this circuit follows the same general trend as the DC current gain. Figure 9.7 shows the linearity characteristics (first- and second-channel adjacent power ratios – ACPR1 and ACPR2, respectively)

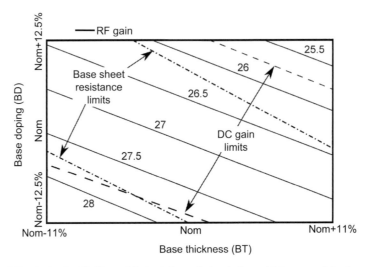

Figure 9.6 RF gain (under IS-95 modulation) for variation in base thickness and doping.

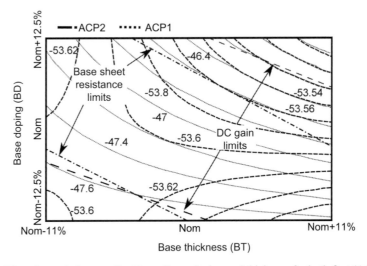

Figure 9.7 Linearity variation as a function of base doping and thickness for both first (ACP1) and second (ACP2) channel linearities.

for this amplifier. Note that in this case, the dependence of these parameters can deviate significantly from just the DC current gain. This shows the importance of separating out the base thickness and doping contributions to the DC gain and RF performance parameters. Finally, Figure 9.8 shows the power-added efficiency (PAE) for the circuit. The PAE is also independent of DC gain. The collector thickness did not have a significant effect on this module for this modulation scheme but can play a role in other aspects of PA performance such as ruggedness. When the collector thickness does play a role,

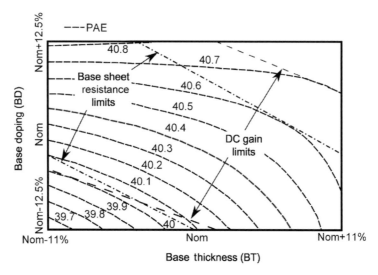

Figure 9.8 PAE as a function of base thickness and doping.

plots similar to these are generated for each collector thickness and compared in order to see where parts fall out of spec.

This methodology can be applied to any measured parameter from these circuits, including yield. If required, this early feedback on parameter sensitivity allows design fixes to be implemented that result in much improved circuit yield that would otherwise go undetected until full production ramp-up. This type of systematic DOE approach also allows FMEA analysis for product performance so that if wafer yield varies on future lots, a simple comparison can help to identify the root cause. As an example, if a circuit exhibits performance failure at some combination of base doping, base thickness, and collector thickness, but was within PCM pass/fail criteria, we can easily compare it to the parameters in Figure 9.2 (or a similar plot) to determine what PCMs should be looked at to identify the cause of the yield loss and tie it back to the material characteristics. Combining this performance data with full characterization measurements over temperature, bias, and frequency allows us to confidently set part specifications, and develop data sheets, for our products early in the product development life cycle. This also allows the required sampling of multiple material lots to be achieved within this minimal number of fabrication lots, resulting in time and money savings. Most importantly, this methodology allows us to provide our customers with a data sheet, detailing expected product variations, due to material, in the pre-production phase of the product instead of waiting for high volume production data.

From analyzing the PCM data collected from these wafers, we correlate important parameters to one another (such as the beta and R_b as described before) within and across devices. While evaluating DOE wafers is a very useful exercise in itself, using data from them to generate statistical models for simulation allows consideration of variation even earlier in the design cycle. This approach enabled us to uncover some unexpected relationships between device parameters that otherwise would have gone unknown. Continuing to run these DOE wafers provides ongoing data for validation and refinement of the model statistics and provides greater insight into possible yield pitfalls. As an example, comparing the variation in measured product performance from each DOE wafer to the wafer-to-wafer variation of our experiment (driven by the variables we purposely changed), we observed that measured variation could not be fully explained by the DOE factors we used. This led to investigation of other areas for improvement in the product yield related to packaging and assembly. This direction for improvement would have gone unexplored without applying our DOE approach.

In this section, we developed and presented a DOE approach to epi material selection for new product qualification and statistical model generation. This approach provides us with greater insight into the device characteristics from analyzing the PCM data and comparing to published data. In particular, the relationships between β and R_{bsh}, as well as the effect of β on BV_{ceo} were evaluated and showed behavior that was unexpected compared to common belief in the literature. Simply comparing the data from these wafers also allowed us to eliminate variation of many parameters that simply do not change with material structure. This greatly simplifies the modeling task. This approach allows greater insight into the causes of product variation when the contributing factors are not easily determined from PCM measurements and provides product-based standards for acceptable material specifications. In particular, it allows the effects of base doping and thickness to be independently determined, since they cannot be easily separated from normal DC PCM measurements (how many times have you heard someone ask why the circuit behaves differently but beta is the same?). This approach also provides a critical link between material parameters and PCMs, which then allows the link between PCM and product performance to be made (via the material parameters), as we have demonstrated. It is also very efficient, since the specific structures fit in a single-wafer lot and can be "banked" and used for any new product in development, making sure that there is always a good spread of material available to quickly evaluate module sensitivity and for comparison with simulation using the statistical models we develop later in this chapter.

9.2.2.2   Tier two: "unified" statistical model development

Model parameter selection for statistical simulation

In the previous subsection, we explored and identified the fundamental material changes that effected device and circuit performance. Having identified those factors, we now consider how they impact the device model parameters. There are several key overall considerations to be taken into account in the selection of model parameters that may be varied for statistical simulation. First, the number of user-controlled parameters should be minimized. This keeps simulation setup relatively simple and does not put an undue burden on the designer. Second, the correlations between important parameters are placed within the model code, hidden from the user. This helps to minimize the inputs while making sure that unrealistic sets of parameters are not entered by the designer. Besides HBTs, other devices also have variations that are important to include in the simulations. For devices that are fabricated from "front-end" or active layers, physical

Table 9.4 Measured parameters, modeled parameters, and observed change on curves.

Device parameters	Impacted model parameters	Observations
V_{be}	I_s, I_{bei}, I_{ben}	Controls turn-on voltage and shifts both I_c and I_b
β	I_{bei}, I_{ben}	Controls I_b, R_b and changes DC current gain
R_e	R_e	Tracks emitter resistance, affects G_m and RF gain
R_b	R_{bi}, R_{bx}	Highly correlated to β, affects RF power gain
R_c	R_{ci}, R_{cx}	R_{ci} is insignificant, R_{cx} is correlated to R_{csh}
f_T	T_f	Tracks transit time, depends slightly on base and strongly on collector design
C_{be}, C_{bc}	$C_{je}, M_e, P_e, C_{jc}, C_{jen}, M_c, P_c$	Affect total transit time, impedance, and RF gain

parameters are correlated to the HBT input parameters. For backend devices (precision resistors, MIM caps, and inductors) the PCM data is used, since these devices are not correlated to the frontend devices. The selection criteria for variable model parameters were conducted by considering (1) the important parameters for circuit design, (2) what data is available from the process control measurements, and (3) what impacts (1) and (2) based on the knowledge of the HBT device physics.

For RF-circuit design, some of the important device parameters are turn-on voltage (V_{be}), DC current gain or β (I_c/I_b), device parasitic resistances (R_e, R_b, and R_c), junction capacitances (C_{be} and C_{bc}), and the RF gain (related to transit time). All of these parameters are also measured in our fab process control monitoring. The next step is to look at the device physics, and compare these parameters to the model parameters. In this case, we considered the parameters of the VBIC model, but this technique can be applied to the Gummel–Poon model, HiCUM, MEXTRAM, or AHBT as well. What we observed is that the turn-on voltage, DC gain, base resistance, and transit time are all functions of the base thickness and doping (as expected from device physics). DC gain and R_b are actually correlated in a well-controlled process, so only one of them is needed as an input. Table 9.4 shows the key device parameters and the major model parameters that depend on them. The dependence of the model parameters on the fundamental material changes (base thickness, W_b, base doping, N_b, and collector thickness, W_c) was experimentally verified using a DOE approach discussed in the previous section. The transistor electrical parameters from the PCM measurements of these wafers were then correlated back to the physical parameter changes. For example, only β or R_b needs to be allowed as an input in statistical modeling, since they are directly correlated. Knowing the physical correlations, we then look at what PCMs can be correlated to each other (minimizing variables) so that the PCM data can drive the simulation. What we also discovered by doing the material DOE is that many parameters simply do not change because of normal material variations [18] so they do not need to be included in the statistical simulation. From an understanding of the device physics (Berkner [22] provides a very nice description of how basic device curves respond to model parameter changes), we also make the critical link between PCM parameters and model parameters, discussed in the next section.

Connecting PCM (technology variations) to model parameters
using device physics

Keeping in mind that we want to vary as few parameters as possible and keep them physically consistent, we considered the device parameters most important to PA circuit design and their relationships with the measured PCM and model parameters (shown in Table 9.4). In order to accommodate the turn-on shift of the HBT, it is useful and convenient for designers simply to enter the measured V_{be} of the device from the PCM data (or in a DOE control box). In extensive measurement data analysis, we noted that the ideality factors of the collector and base currents are independent of the base doping, base thickness, and collector thickness. As a result, to shift the turn-on voltage properly, we only need to shift the base and collector saturation currents. For a shift in V_{be} (ΔV_{be}), we have:

$$I_{s,new} = I_{s,old} e^{q\delta V_{be}/N_F kT} \tag{9.3}$$

$$I_{bei,new} = I_{bei,old} e^{q\delta V_{be}/N_{EI} kT} \tag{9.4}$$

$$I_{ben,new} = I_{ben,old} e^{q\delta V_{be}/N_{EN} kT}, \tag{9.5}$$

where we take the parameters I_s, I_{bei}, I_{ben}, N_f, N_{ei}, N_{en} from the given model parameter deck. As an example, if we have $n_F = 1.079$ and a ΔV_{be} of -10 mV, we get:

$$I_{s,new} = I_{s,old}/1.43. \tag{9.6}$$

The shift in the DC current gain arises almost solely from changes in the base current (again, validated empirically). As a result, to account for the variation in β, we need to modify only I_{bei} and I_{ben}. We do this by forming a ratio between the nominal β and the "PCM β", and applying the shifting factor to the above model parameters. If β is higher on the given wafer, then I_{bei} and I_{ben} must be lower. These two simple modifications account for most of the DC current variations. From the discussion in the earlier section, R_b is scaled in the same way as the DC gain (it must increase if β increases). Next, we consider the emitter resistance, R_e. The emitter resistance measurements for modeling naturally require that self-heating be accounted for (and removed). However, for PCM data collection, such detailed measurements are too long and time consuming. An open-collector measurement (aka fly-back) is often used for process control. Fortunately, there is good correlation between the fly-back measurement and the more detailed Re measurement. It is also useful to note that the emitter resistance scales as area, so that the resistance of various devices can be calculated/scaled from the PCM measurement of a minimum area device. Finally, we will discuss the variation of τ_F (i.e., f_T) based on these parameters. Unfortunately, τ_F is not "well" correlated to β or R_b over the allowed range of our material variation. However, there is a loose relationship, since for a constant doping, if W_b decreases, β increases and τ_F decreases. The weakness of this correlation is also due to our PCM measurement conditions. Because of this weak correlation, we simply use the PCM measurement for f_T to adjust the τ_F, but with the caveat to designers entering parameters by hand that there are some physical relationships they need to make sure are adhered to (β and f_T must go in the same

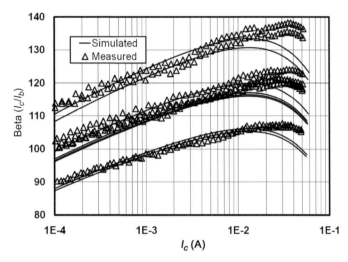

Figure 9.9 DC gain (β) simulation (red lines) versus DOE wafer measurements (green symbols).

direction). This problem is circumvented if a set of DOE states is preprogrammed for the simulation. It should also be mentioned that this implementation can be performed even without knowing the exact parameter relationships to allow first-order simulation until those relationships are known (i.e., can be added to already existing models).

While there are many other parameters that can affect the performance of circuits, some of these parameters are very difficult to measure accurately in a manufacturing environment, so are not included. For example, C_{be} and C_{bc} can be important, and are measured at PCM, but the accuracy and number of data points for accurate use in modeling are insufficient for statistical simulations. From the previous section, we have found, empirically, that the base parameters account for most of the variation in our PA circuits, so collector thickness (as discussed earlier) is not considered in our primary simulations. In addition, we have found it useful to include the variation of precision resistors and metal-insulator-metal (MIM) caps to get accurate simulation of circuits with on-chip bias circuits. For the thin-film resistor, a model similar to reference [23] is used that accounts for the changes in sheet resistance and effective resistor width (both measured at PCM). For the MIM, the capacitance value per unit area measured at PCM is used for simulation. The variations account for a significant portion of the product level variation that we have experienced.

Device model level validation and recentering

To validate the statistical model, the characteristics of an HBT were simulated using the model with selected PCM parameters measured from the DOE wafers as the statistical control input. The comparison of the simulated results with the measurements shows the good model tracking of the DOE variations. Figure 9.9 shows the DC β simulation of an HBT over the DOE variations compared to corresponding measurements from the DOE wafers. Figure 9.10 shows the comparison between the simulated and measured

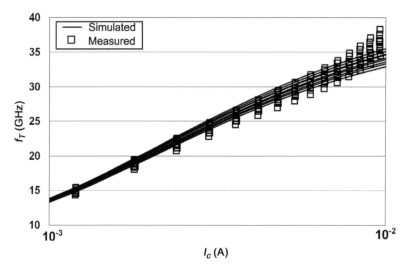

Figure 9.10 Collector current dependence of cutoff frequency with DOE varied PCM inputs. The symbols are measured data and the lines are the simulation.

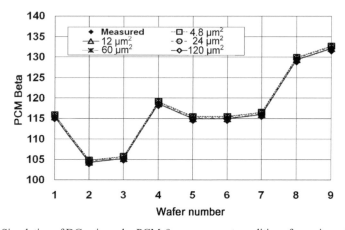

Figure 9.11 Simulation of DC gain under PCM β measurement conditions for various sized HBTs.

cutoff frequency (f_T) dependence on the collector current of the HBT over the DOE variations.

Further validation was carried out by evaluating the selected PCM parameters using the model simulation under the same conditions employed in the measurement. The simulated PCM parameters were then checked against measured values. An example of such validation results is given in Figure 9.11, which shows the PCM β simulation for various sized HBTs and their comparison with the corresponding PCM measurement result. It can be clearly seen that the process variations represented by the PCM parameters are well tracked by the statistical model simulation.

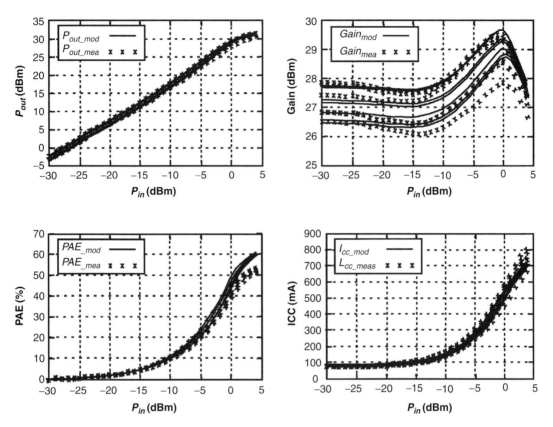

Figure 9.12 Simulation of output performance of a PA circuit with DOE variations (lines) compared to measured data (symbols).

Module level statistical model validation

The ultimate goal of the statistical modeling is to track and simulate the PA circuit performance variation due to the fluctuation in wafer process and epi material properties. In this section, we demonstrate some results from simulation, using this simple statistical model, of a power amplifier circuit built on the DOE wafers with intentional variations that represent the corners of our manufacturing process. As shown in Figure 9.12, good agreement between simulation and measurement in both power performance and material variation, for output power, has been achieved with our simple statistical modeling approach without any "tweaking" for large-signal performance.

Figure 9.13a shows the comparison between the simulated and measured quiescent currents (I_{cq}) of the PA over the DOE variations. The measurement data was from our production test database. It can be seen that the simulation can quite reasonably track the I_{cq} variations resulting from the controlled variations. Figure 9.13b shows the measured RF Gain across the material DOE variations. The variation is predicted well and the slight offset to nominal can be easily attributed to other factors. For example, the simulations shown here only include HBT variations (no variation on thin-film resistor or MIM capacitor) which partially accounts for the scatter in the data. It is also common

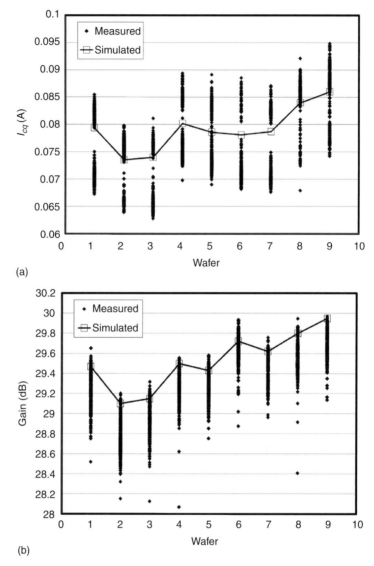

(a)

(b)

Figure 9.13 Comparison of (a) measured quiescent current (I_{cq}) and (b) RF power gain across a fabrication lot using DOE material that varied the base doping and thickness.

that other factors besides the die (such as packaging/assembly) can impact both of these parameters. However, the models track the changes in the gain, even if there is a slight offset in nominal value.

In this section, we have outlined and demonstrated a simple method for providing statistical models for an InGaP-GaAs HBT process. This method has been applied to a handset power amplifier circuit simulation with good results tracking process variations. The model can be used to help de-bug design issues and make PA designs more robust to process variations.

9.2.2.3   Tier three: integration into design flow

Model implementation in simulation environment

Having developed and verified the device level statistical models, we now consider how these are included in the design flow. The emphasis is on keeping it simple and painless enough that designers would want to use it. We also contrast and compare this approach to other commonly used implementations.

Another feature of this implementation is that all the statistical parameters can be easily fixed to their nominal values or to values set by the designer for wafer-specific simulations. This allows designers to skip simulations of parameters that are not included or critical for their particular design (for example, why simulate Schottky diode variation if it is not in your circuit?) which greatly reduces the total simulation time. The number of statistical parameters in a typical circuit simulation, after such selection considerations, is less than 10.

Numerical performance of DOE implementation compared to
MC implementation

The predicted mean and distribution ranges from MC depend on the number of variables and number of simulations. The higher the ratio of simulation runs to number of variables, the more accurate the predictions. In reality, one normally does not know which of the selected variables has the most impact on a particular design. Running hours of simulation to find this out is not very appealing when you are up against a project deadline. Worse, it can result in the critically important parameters not getting selected if not enough iterations are run! Both cases can lead to unrealistic distribution range predictions and inaccurate mean predictions. As mentioned earlier, DOE is widely used in the semiconductor industry but is not often taught to circuit designers.

When neighboring device mismatches are negligible, our study indicates that DOE is a superior simulation choice. Figure 9.14 shows that for a simulation of four variables, 64 000 MC simulations lead to much wider I_{cq} ranges compared to only 240 MC simulations. Since typical products ship in the millions, the distribution range using 64 000 MC simulations is more representative of reality. However, these 64 000 MC simulations take 18.5 hours to finish! Figure 9.14 also shows that a very large number of MC simulations are required to approach the results of running a full factorial (2 kmp) DOE simulation. The difference is that the DOE simulation only took 28 seconds. To further compare these methods, we analyzed the importance of each of the five variables using 250 MC runs and full factorial DOE simulation ($2^5 = 32$ runs) of another design (Figure 9.15) using JMP statistical software. The analyzed effects are expressed by equations (9.7) (MC) and (9.8) (DOE). Comparing these equations, it is obvious that the weight and direction of each variable from both results are identical. The major difference is the predicted means. As we indicated earlier, this can be caused by using too few MC runs for the number of variables.

$$I_{cq}(mA) = 235 - 0.768 beta - 34.5VT - 1.69TaNRho + 0.336Ref \quad (9.7)$$

$$Icq(mA) = 240 - 0.782 beta - 34.6VT - 1.69TaNRho + 0.310Ref. \quad (9.8)$$

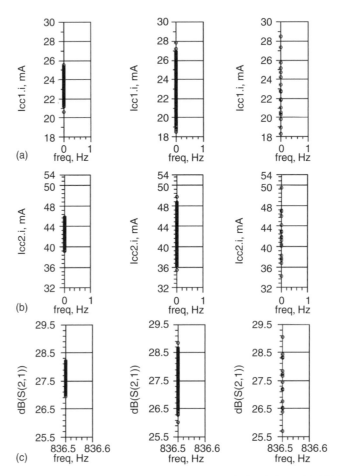

Figure 9.14 Circuit response comparison between DOE and MC simulations: (a) 240 MC simulations, (b) 64 000 MC simulations, and (c) full-factorial DOE.

To refine our DOE implementation, different DOE designs were evaluated. The full factorial method (2 kmp) uses maximum and minimum values (only two levels) of each factor, so it requires 2^n simulation runs for n parameters and allows subsets of parameters to be considered. For example, if we varied three factors, the simulation points would be represented by the corners on a cube. If we want to look at only two of these factors, we would just look at the faces of the cube. Other experimental designs sacrifice some information (such as interaction information) to reduce the number of required experimental runs by assuming that higher order interactions are unimportant or by allowing them to be "aliased" with main factors. These alternative designs may not project properly and represent various alternative placement of the points or different numbers of levels for the factors [24, 25]. Plackett-Burman experimental designs: interactions between factors are considered negligible, and because of the design, some main effects and interactions cannot be separated (they are confounded or aliased). Box-Behnken experimental designs: three levels for each factor and a center point and can

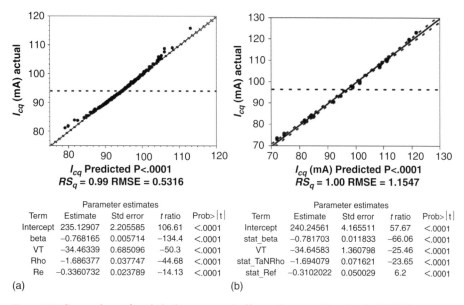

Figure 9.15 Comparison of statistical parameters' effects of an auto bias circuit: (a) MC simulations (250 runs) and (b) full-factorial DOE (32 runs).

be used for quadratic models. The simulation points are at the mid points of the cube example given above for three factors, central composite design (CCD) contains the same points as the full-factorial and additional "star" points. For a two-factor design, the star points are basically at 45° rotation of the box corners and represent extreme values of the parameters. This type of design is used to model curvature in the modeling of the responses (output variables). The 3k design is a full factorial with three levels for each factor (so it includes the corners of the cube and all the center points). For more information on the implementation within ADS, the reader is referred to reference [26].

As shown in Table 9.5, the full factorial DOE method gives consistent results with the much denser sampled 3k method, but is far more time efficient. The results of other DOE methods are not consistent with each other, even though they are slightly faster than the full factorial. This inconsistency among DOE methods is partly due to our orthogonal only (at device level) variable selection. The full factorial method is optimal considering accuracy and simulation time, and was selected to be the method implemented in our design flow.

DOE implementation compared to "sensitivity analysis"

"Sensitivity analysis" simulation, another popular analysis technique for understanding circuit variability, was also evaluated. It predicts significantly different results than the full factorial DOE or the dense sampled DOE (3k) (Table 9.5). The reason is that the "sensitivity" method only considers small perturbations around a nominal condition with one parameter changed at a time. This assumption does not apply to our particular products; therefore, it is not a recommended method for improving design robustness.

Table 9.5 DOE methods comparisons. The full factorial DOE method (2kmp) gives consistent results, in terms of performances ranges and variable Pareto orders, with the much denser sampled 3k method, but takes much less time to run. "Sensitivity Analysis" predicts totally different results.

| Method | Sim time (s) | # Samples | Icq1(mA) at 25°C | | | Variable order in 25°C Pareto chart | | | | | | | | | |
			I_{cq1_Min}	I_{cq1_Max}	I_{cq1} range	VT	Beta	V_{be3}	Ref	Rho	dw	dl	SCdV	BCdV	V_{cc}
2kmp	488	1024	19	31	12	1	4	5	2	2	3	6	6		
Plackett-Burman	8	12	19	27	8	1	5	4	2	2	3		6	7	
Box-Behnken	33	181	21	27	6	1	4	5	2	2	3	6	6		
CCD	493	1045	17	32	15	1	4	5	2	2	3	6	6		
3k	16070	59049	19	31	12	1	4	5	2	3	3	6	6		
Sensitivity							4	1	3			2			

| Method | Sim time (s) | I_{cq2}(mA) at 25°C | | | Variable order in 25°C Pareto chart | | | | | | | | | |
		I_{cq2_Min}	I_{cq2_Max}	I_{cq2} range	VT	Beta	V_{be3}	Ref	Rho	dw	dl	SCdV	BCdV	V_{cc}
2kmp	488	34	54	20	1	3	5	2	2	4	6	7		
Plackett-Burman	8	35	48	13	1	3	5	2	2	4	6	7	8	
Box-Behnken	33	39	48	9	1	3	5	2	2	4	6	7		
CCD	493	33	64	31	1	3	5	2	2	4	6	7		
3k	16070	34	54	20	1	3	5	2	2	4	6	7		
Sensitivity						3	4	2			1			

| Method | Sim time (s) | dB(S21) at 25°C | | | Variable order in 25°C Pareto chart | | | | | | | | | |
		$dB(S_{21})\_Min$	$dB(S_{21})\_Max$	$dB(S_{21})\_range$	VT	Beta	V_{be3}	Ref	Rho	dw	dl	SCdV	BCdV	V_{cc}
2kmp	488	25.7	29.5	3.8	1	4	5	5	3	2	6	7	7	
Plackett-Burman	8	26.1	28.7	2.6	1	4	5	5	3	2	6	7	7	
Box-Behnken	33	26.8	28.7	1.9	1	4	5	5	3	2	6	7	7	
CCD	493	24.2	29.7	5.5	1	4	5	5	3	2	6	6	7	
3k	16070	25.7	29.5	3.8	1	4	5	5	3	2	6	7	7	
Sensitivity						3	2	5	2	1				

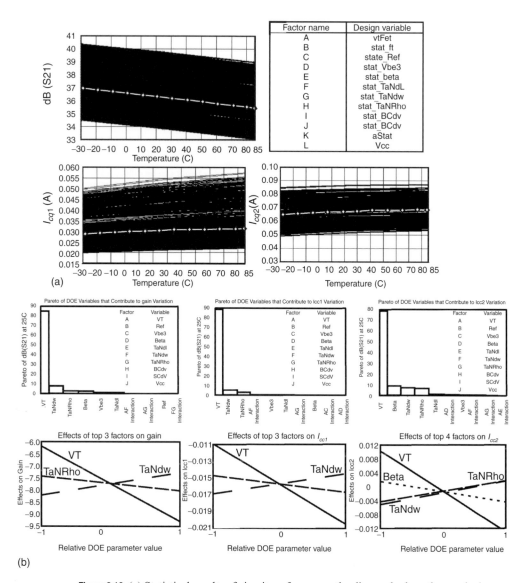

Figure 9.16 (a) Statistical results of circuit performance, the diamonds show the nominal simulation, and (b) statistical Pareto chart.

Integrated design flow in ADS

Because models, design schematics, DOE simulations, and simulation analysis (Figure 9.16) are all integrated into ADS, a DOE design flow is practical, even for designers who are not highly trained in statistical analysis. Statistical analysis has been intimidating to many designers in the past due to the need for special training in the subject. This has been one of the major barriers to its use in PA design.

As shown in Figure 9.17, our integrated design flow consists of iterating DOE simulations, reviewing circuit performances, reviewing Pareto charts, identifying circuit

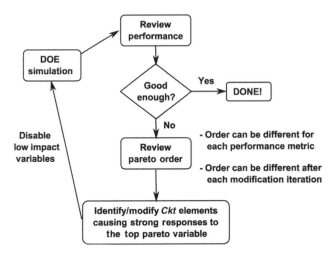

Figure 9.17 The DOE-Pareto-driven design flow chart.

elements that drive top Pareto factors, and modifying the circuit. Once modified, the loop is followed until satisfactory results are obtained. The Pareto charts provide useful information about which variables contribute most to the performance variation of the circuit. Such information is powerful for identifying which part of a design or which specific element should be modified.

9.3   Examples of application to real circuits

9.3.1   Dual band PA

In this example, the first-cut design had wide I_{cq} and RF gain variations (Figure 9.18a). Pareto analysis showed that threshold voltage variations of the FET devices, width variations of some critical thin-film resistors, and DC gain variation of HBT devices were the dominant factors for the circuit performance variations (Figure 9.18c). To address these top-order effects, the internal reference voltage and values of three resistors were identified as critical components to change. Changing these elements drastically reduced the performance variations while the nominal performance was maintained (Figures 9.18a and 9.18b).

9.3.2   WCDMA FEM

For our second example, battery voltage (V_{cc}) variation caused large performance variation in the initial design (Figure 9.19). Through the DOE-Pareto driven design flow, insufficient ballasting and rising voltage on a particular node were identified as the root cause. As a result, clamping diodes and increased ballasting were implemented in the circuit. Measured results (Figure 9.19) verified that as V_{cc} changed from 3.2 to 4.5 V, I_{cq}, after the design improvements, is relatively constant and the standard deviation at

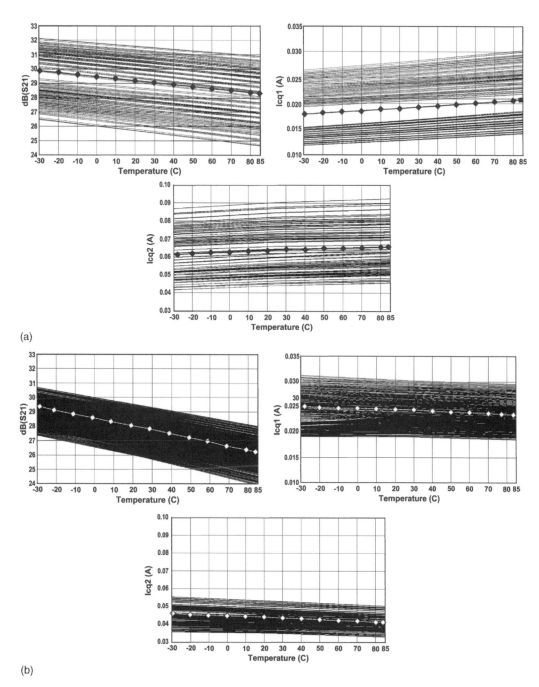

Figure 9.18 Results and analysis of DOE simulations of a dual band PA design: (a) performance results (rf gain, I_{cq1}, and I_{cq2}) of first-cut design, the diamonds show the nominal simulation; (b) performance after modifications targeting reduction of top-order variables, the diamonds show the nominal simulation; (c) Pareto chart for initial design.

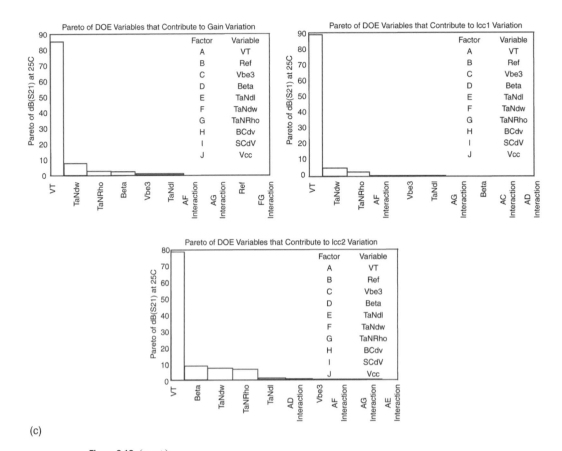

(c)

Figure 9.18 (*cont.*)

high V_{cc} is significantly reduced. This example further illustrates the effectiveness of our DOE-Pareto driven flow.

9.4   Summary

This chapter demonstrates the utility of modeling/design flow integrating a unified statistical model, DOE statistical circuit simulations (of both epi/process and circuit operational variables), Pareto analysis, and design schematics into a single tool for a PA design. The design iterations are driven by the highest ranked Pareto variable, and the interactive process of design modification, simulation, and analysis was accomplished in a matter of minutes. We have presented how each key element of our three-tier approach is determined and distinguished from previous statistical simulation or design work. Our examples, using this approach, show that DOE-based simulations are an effective tool for guiding circuit design and reducing performance variation for a given process. We started with a discussion of wafer-level material parameters and related those to observed PCM and circuit variations. We also showed how to correlate parameters to

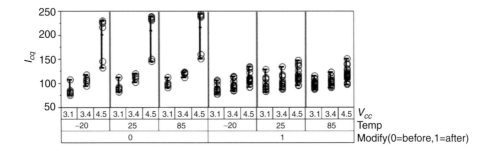

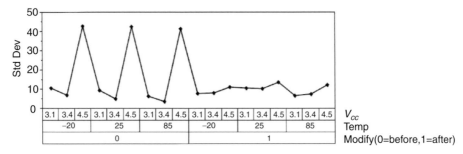

Figure 9.19 Measured data for a WCDMA FEM design. It confirms the simulation results from before and after design modifications. The design change, targeted at eliminating V_{cc} impact, significantly tightened design performances.

one another and use the information collected to simplify modeling. We showed how these changes related to model parameters and presented the "unified" model foundation. These models were validated by simulating the PCM measurement and comparing to measured circuit data from DOE wafers. Finally, we discussed the advantages of DOE implementation and concluded with two circuit examples.

Acknowledgments

We would like to acknowledge the Kopin Corporation for growing the epi material used in this work, especially K. Stevens and R. Welser. We would also like to thank K. Weller, K. Buehring, and R. Ramanathan for their support of this project. The Skyworks design team, especially Mary Ann Abarientos, Gary Zhang, Ede Enobakhare, Peter Tran, and Nick Cheng are also gratefully acknowledged. The authors wish to thank Bac Tran for providing circuit test data, Jing Li for providing production line PCM test data, and for his early support of this work, and continuous support from the design groups (especially Jane Xu and Sunny Chen) at Skyworks' Newbury Park facility. We are also very thankful to Jack Sifri (Agilent) for kindly sharing his knowledge and some of the original templates for Pareto plots. Robin Accomazzo's help on the artwork and tables is much appreciated. Finally, we would like to thank the brave designers who acted as guinea pigs and helped us to refine this flow.

DEDICATION

The authors would like to dedicate this work to our friend and colleague Juntao Hu who passed away before the completion of this chapter. Without his dedication and commitment to this activity, this work would not have been possible.

References

[1] W. F. Davis *et al.*, "Statistical IC simulation based on independent wafer extracted process parameters and experimental designs," *Proc. Bipolar Circuits and Technol. Meeting, 1989*, pp. 262–265, 1989.

[2] M. Schroter *et al.*, "Statistical modeling of high-frequency bipolar transistors," *IEEE Bipolar/BicMoS Circuits Technol. Meeting (BCTM)*, pp. 54–61, 2005.

[3] K. M. Walter *et al.*, "A scalable, statistical SPICE Gummel–Poon model for SiGe HBT's," *J. of Solid-State Circuits*, vol. 33, no. 9, 1998.

[4] W. Schneider, M. Schroter, W. Kraus, and H. Wittkopf, "Statistical simulation of high-frequency bipolar circuits," *Des., Autom. Test in Eur. Conf. Exhibition*, 2007, pp. 1397–1402.

[5] C. C. McAndrew, "Statistical modeling for circuit simulation," *Proc. 4th Int. Symp. Quality Electron. Des.*, 2003.

[6] S. W. Director *et al.*, "Statistical integrated circuit design," *IEEE J. Solid-State Circuits*, vol. 28, no. 3, March 1993, pp. 193–202.

[7] D. L. Harame *et al.* "Design automation methodology and RF/analog modeling for RF CMOS and Si Ge BiCMOS technologies," *IBM J. Res. Dev.* vol. 47, no. 2/3, Mar/May 2003.

[8] W. Rensch, S. Williams, and R. J. Bergeron, "BiCMOS Fab Yield Optimization," *IBM Microelectron.*, pp. 23–25, First Quarter 2000.

[9] SAS Institute Inc. 2008. JMP® 8 *Design of Experiments Guide*. Cary, NC: SAS Institute Inc.

[10] J. Carroll and K. Chang, "Statistical computer-aided design for microwave circuits," *IEEE Trans. Microw. Theory Tech.*, vol. 44, no. 1, Jan. 1996.

[11] M. Harry and R. Schroeder, *Six Sigma: The Breakthrough Management Strategy Revolutionizing the World's Top Corporations*. New York: Currency/Doubleday, 2000.

[12] *http://www.agilent.com/find/eesof-doe*

[13] G. E. P. Box, J. S. Hunter, and W. G. Hunter, *Statistics for Experimenters*, Wiley & Sons, 1978.

[14] NIST/SEMATECH *e-Handbook of Statistical Methods*, http://www.itl.nist.gov/div898/handbook/, 2011.

[15] J. Hu, P. J. Zampardi, C. Cismaru, K. Kwok, and Y. Yang, "Physics-based scalable modeling of GaAs HBTs," *Proc. 2007 Bipolar Circuits and Technol. Meeting*, pp. 176–179.

[16] P. J. Zampardi, D. Nelson, P. Zhu, C. Luo, S. Rohlfing, and A. Jayapalan, "A DOE approach to product qualification for linear handset power amplifiers," *2004 Compound Semiconductor Mantech Conf.*, Miami, FL, pp. 91–94.

[17] J. Hu, P. J. Zampardi, H. Shao, K. H. Kwok, C. Cismaru, "InGaP-GaAs HBT statistical modeling for RF power amplifier designs," *Dig. Compound Semiconductor Integrated Circuit Symp.* 2006, San Antonio, 2006, pp. 219-222 and P. J. Zampardi, "Modeling challenges for future III–V technologies," *Workshop on Compact Modeling for RF-Microwave Applications (CMRF)*, Long Wharf, Boston, October 3, 2007.

[18] H. K. Gummel, "Measurement of the Number of Impurities in the Base Layer of a Transistor," *Proc. IRE*, vol. 49, no. 4, p. 834, April 1961.

[19] R. Welser *et al.*, "Role of neutral base recombination in high gain AlGaAs/GaAs HBT's," *IEEE Trans. Electron Devices*, vol. 46 , no. 8, pp. 1599–1607, Aug. 1999.

[20] R. Gray and P. Meyer, *Analysis and Design of Analog Integrated Circuits*. New York: Wiley & Sons, 1984, p. 27.

[21] B. Yeats *et al.*, "Reliability of InGaP emitter HBTs at high collector voltage," *GaAs IC Symp., 2002. 24th Annual Tech. Dig.*, Oct. 20–23, 2002, pp. 73–76.

[22] J. Berkner, *BCTM Shortcourse, Bipolar Model Parameter Extraction*, 2001.

[23] F. Larsen, M. Ismail, and C. Abel, "A versatile structure for on-chip extraction of resistance matching properties," *IEEE Trans. Semicond. Manuf.*, vol. 19, no. 2, May, 1996.

[24] SEMATECH "Intermediate design of experiments (DOE)," *Technology Transfer* #97033268A-GEN, 1997

[25] *http://edocs.soco.agilent.com/display/ads2009/Using+Design+of+Experiments+%28DOE%29*

[26] ADS 2009 Update 1, October 2009, *Tuning, Optimization, and Statistical Design*, Agilent Technologies, Inc. 2000–2009.

Trademarks

RS/1 is a registered trademark of Brooks Software

JMP is a registered trademark of SAS Institute Inc

Agilent Advanced Design System (ADS) is a registered trademark of Agilent

Six-Sigma is a registered trademark of Motorola, Inc

10 Noise modeling

Manfred Berroth

University of Stuttgart

Spontaneous fluctuations of electrical signals are usually called noise. Especially for sensitive amplifiers, noise is a challenging problem for the design engineer. Sensitive amplifiers are not only required in receivers of communications systems, but also for radar as well as measurement systems. The range of any communication system is limited by the acceptable signal-to-noise ratio. In electronic circuits, three types of noise are present:

- Thermal noise, caused by random motion of current carriers;
- Shot noise, caused by drift current of an applied electric field;
- Flicker noise, caused by slow fluctuations in conductivity.

Noise can be described by statistical mathematical equations. Some fundamentals of transistors are discussed in Section 10.3 with the extensions to include noise. Section 10.4 describes noise measurement setups. Finally the parameter extraction procedure is presented.

10.1 Fundamentals

10.1.1 Probability distribution function

Noise is described by the mathematical probability theory. In electrical engineering, usually current or voltage signals are described as functions of time, e.g., $i(t)$ or $v(t)$.

Those signals are continuous in time and in their values. If those signals are noisy, they become a continuous random variable $P(x)$. We define the probability distribution function

$$P(x \le x_\mathrm{o}) = \int_{-\infty}^{x_o} f(x)dx \qquad (10.1)$$

We can derive the probability density function $f(x)$ as the derivative of the probability distribution function

$$dP(x) = f(x)dx \qquad (10.2)$$

indicating that the continuously fluctuating variable has the probability dP having a value between x and $x + dx$.

The expectation value $E(x)$ is given by

$$E(x) = \int_{-\infty}^{\infty} x f(x) dx \tag{10.3}$$

In any experiment, the instantaneous values of the fluctuating quantities change from observation to observation, but average values remain constant.

The definition of the nth moment of the average value is given by

$$\overline{x^n} = \int x^n dP, \tag{10.4}$$

where the integration is to be extended over the range of allowed values of x. For a true fluctuating quantity, the average value of the first moment is zero

$$\overline{x} = \int x dP \overset{!}{=} 0 \tag{10.5}$$

In this case the most important average value is the mean square value

$$\overline{x^2} = \int x^2 dP. \tag{10.6}$$

There are also central moments defined:

$$\overline{(x - \overline{x})^n} = \int (x - \overline{x})^n dP. \tag{10.7}$$

The most important one is called standard deviation σ,

$$\sigma^2 = \overline{(x - \overline{x})^2} = \int (x - \overline{x})^2 dP = \int (x - \overline{x})^2 f(x) dx \tag{10.8}$$

There are two methods of averaging. First, a single system subjected to spontaneous fluctuations is investigated and the averages are determined over a sufficiently long time interval.

The noise measurements of electronic devices and circuits are performed with this method.

The second method considers an ensemble of a large number of identical systems and the average value is determined by taking instantaneous values over all elements at a certain instant.

For an ergodic process, the values of the fluctuating quantity produced by a single system over a long time interval have the same probability density function as the values of an ensemble of identical systems at a certain instance.

For continuous random variables the normal distribution, often named Gaussian, has the following probability density function as shown in Figure 10.1:

$$f(x) = \frac{1}{\sigma \sqrt{2\pi}} e^{-\frac{(x - \overline{x})^2}{2\sigma^2}}. \tag{10.9}$$

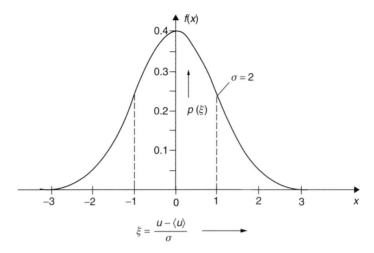

Figure 10.1 Gaussian probability density function.

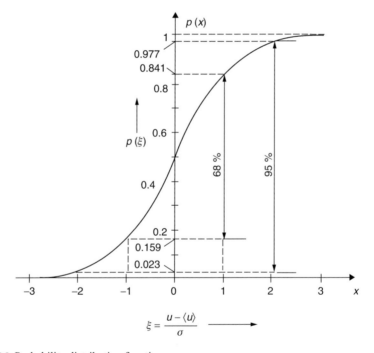

Figure 10.2 Probability distribution function.

The pdf is presented in Figure 10.2 and is given by:

$$P(x_0) = \frac{1}{2}\left(1 + \varphi\left(\frac{x_0 - \bar{x}}{\sigma\sqrt{2}}\right)\right)$$

(10.10)

with

$$\varphi(z) = \frac{2}{\sqrt{\pi}} \int_0^z e^{-u^2} du. \tag{10.11}$$

10.1.2 Correlation of fluctuating quantities

Two fluctuating quantities x and y are called independent or uncorrelated if a given value of x in no way restricts the possible values of y. Then it is

$$\overline{xy} = \overline{x\overline{y^x}} = 0 \tag{10.12}$$

because $\overline{y^x} = 0$ as x is a constant in this averaging process and y is independent of x.

However, if there is a relation between the fluctuating variables x and y, for example $y = ax$ where a is a constant, then, for a given value of x, also y is known too,

$$\overline{xy} = a\overline{x^2}. \tag{10.13}$$

In this case, the quantities x and y are completely correlated.
We define the correlation coefficient by

$$c = \frac{\overline{xy}}{\sqrt{\overline{x^2 y^2}}}. \tag{10.14}$$

If $c = 0$, then x and y are uncorrelated. If $|c| = 1$, they are completely correlated. In the case $0 < |c| < 1$, the quantities are partly correlated.
A general case of partial correlation is

$$y = ax + z. \tag{10.15}$$

With a constant and z as a fluctuating quantity, then x and y are linearly correlated.
If two fluctuating quantities are added and the mean square value of this sum is calculated, the correlation coefficient has to be taken into account.

$$\overline{(x+y)^2} = \overline{x^2} + \overline{2xy} + \overline{y^2} = \overline{x^2} + 2c\sqrt{\overline{x^2 y^2}} + \overline{y^2}. \tag{10.16}$$

10.1.3 Correlation functions

Instead of two random variables x and y, two values of the function $x(t)$ can also be taken at two different time instants t_0 and $t_0 + \tau$. We now define a correlation function which describes the relation between different time intervals of the function $x(t)$.

For the two random variables x and y for two different time instants t_1 and t_2, we define the cross-correlation function $R_{xy}(t_1, t_2)$

$$R_{xy}(t_1, t_2) = \iint x_1 y_2 f(x_1, y_2, t_1, t_2) dx_1 dy_2$$

$$= \overline{x(t_1) y(t_2)} \tag{10.17}$$

with $x_1 = x(t_2)$

$$y_2 = y(t_2).$$

For a stationary process, f and R_{xy} depending only on the time difference

$$\tau = t_1 - t_2$$

$$R_{xy}(\tau) = \iint x_1 y_2 f(x_1, y_2, \tau) dx_1 dy_2 \qquad (10.18)$$

$$= \overline{x(t_0)\, y(t_0 + \tau)}.$$

The cross-correlation function reduces to the auto-correlation function, if $x = y$.

$$R_x(\tau) = \iint x_1 x_2 f(x_1, x_2, \tau) dx_1 dx_2 \qquad (10.19)$$

$$= \overline{x(t_0)\, x(t_0 + \tau)}.$$

10.1.4   Fourier analysis of fluctuating quantities

Any periodic signal $x(t)$ can be developed into a Fourier series for the time interval:

$$0 \le t \le T_0$$

$$x(t) = \sum_{\nu=-\infty}^{\infty} c_\nu e^{j\nu\omega_1 t} \qquad (10.20)$$

with

$$c_\nu = \frac{1}{T_1} \int_0^{T_1} x(t) e^{-j\nu\omega_1 t} dt \qquad (10.21)$$

and

$$\omega_1 = \frac{2\pi}{T_1}. \qquad (10.22)$$

The amplitude of the harmonic frequency $\nu\omega_1$ is:

$$A_\nu = c_\nu e^{j\nu\omega_1 t} + c_{-\nu} e^{-j\nu\omega_1 t} = 2|c_\nu|\cos(\nu\omega_1 t + \varphi_\nu). \qquad (10.23)$$

For the mean square value, $\overline{A_\nu^2}$ can be calculated:

$$\overline{A_\nu^2} = \overline{2a_\nu a_{-\nu}} = \frac{2}{T_1^2} \int_0^{T_1} \int_0^{T_1} x(t) x(t') e^{-j\nu\omega_1(t-t')} dt dt'. \qquad (10.24)$$

We can define a time autocorrelation function R_x^t:

$$R_x^t(\tau) = \lim_{T\to\infty} \frac{1}{2T} \int_{-T}^{T} x(t+\tau) x(t)\, dt. \qquad (10.25)$$

If the random process is ergodic, the time auto-correlation function R_x^t is equal to the statistical autocorrelation function R_x. Then we get:

$$R_x^t(\tau) = \lim_{T \to \infty} \frac{1}{2T} \int_{-T}^{T} \sum_{\nu=-\infty}^{\infty} c_\nu^* e^{-j\nu\omega_1 t} \left(\sum_{\mu=-\infty}^{\infty} c_\mu e^{j\mu\omega_1(t+\tau)} \right) dt. \tag{10.26}$$

Using the orthogonality:

$$\lim_{T \to \infty} \frac{1}{2T} \int_{-T}^{T} e^{-j(\nu-\mu)\omega_1 t} dt = \lim_{T \to \infty} \frac{\sin(\nu-\mu)\omega_1 T}{(\nu-\mu)\omega_1 T} = \begin{cases} 1 \text{ when } \nu = \mu \\ 0 \text{ when } \nu \neq \mu \end{cases}, \tag{10.27}$$

we get:

$$R_x(\tau) = \sum_{\nu=-\infty}^{\infty} |c_\nu|^2 e^{j\nu\omega_1\tau}. \tag{10.28}$$

As $x(t)$ is usually a real function of time, this simplifies to:

$$R_x(\tau) = c_0^2 + 2 \sum_{\nu=1}^{+\infty} |c_\nu|^2 \cos(\nu\omega_1\tau). \tag{10.29}$$

The Fourier transform $V(f)$ of a function $x(t)$ is defined by:

$$V(f) = \int_{-\infty}^{\infty} x(t) e^{-j\omega t} dt \quad \text{with } \omega = 2\pi f. \tag{10.30}$$

The inverse Fourier transform is given by:

$$x(t) = \int_{-\infty}^{\infty} V(f) e^{j\omega t} df. \tag{10.31}$$

If we apply the Fourier transform to the autocorrelation function, we get:

$$\int_{-\infty}^{\infty} R_x(\tau) e^{-j\omega\tau} d\tau = \int_{-\infty}^{\infty} \sum_{\nu=-\infty}^{\infty} |c_\nu|^2 e^{j\nu\omega_1\tau} e^{-j\omega\tau} d\tau \tag{10.32}$$

$$= \sum_{\nu=-\infty}^{\infty} |c_\nu|^2 \delta(f - \nu f_1) = S(f).$$

The power spectral density $S(f)$ of the periodic signal $x(t)$ is obviously the Fourier transform of the auto-correlation function. Then, the inverse Fourier transform can be used to define the autocorrelation function:

$$R_x(\tau) = \int_{-\infty}^{\infty} S(f) e^{j\omega t} df. \tag{10.33}$$

For real variables, we get $A_v = 2|c_v|$ and therefore,

$$S_A(f) = 4 \int_0^\infty R_x(\tau)\cos(2\pi f\tau)d\tau \qquad (10.34)$$

and

$$R_x(\tau) = \int_0^\infty S_A(f)\cos(2\pi f\tau)df. \qquad (10.35)$$

For a stationary process, the auto-correlation function exhibits the following properties

1.
$$R_x(0) = \overline{x(t_0)}^2 = \int_0^\infty S_A(f)df. \qquad (10.36)$$

This equation also describes the total power of the random signal and is called the theorem of Parseval.

2.
$$R_x(\tau) \leq R_x(0). \qquad (10.37)$$

3.
$$R_x(-\tau) = R_x(\tau). \qquad (10.38)$$

4.
$$R_x(\pm\infty) = (\overline{x})^2. \qquad (10.39)$$

As the probability density function and the probability distribution function can describe a random variable quantity in the time domain, the power spectral density function is a useful description of noise signals in the frequency domain.

10.1.5   Noise response of noiseless linear time-invariant circuits

If a random signal $x_e(t)$ is applied to the input of a noiseless linear and time-invariant circuit, the output response of the circuit is given by:

$$x_a(t) = \int_{-\infty}^\infty h(t-\tau)x_e(\tau)d\tau = h(t) * x_e(t) \qquad (10.40)$$

with impulse response $h(t)$ of the circuit.

The transfer function in the frequency domain is given by:

$$H(f) = \int_{-\infty}^\infty h(t)e^{-j2\pi ft}dt. \qquad (10.41)$$

For the output power spectral density of the random input signal, we get:

$$S_a(f) = S_e(f)|H(f)|^2. \qquad (10.42)$$

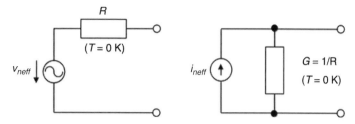

Figure 10.3 Two equivalent circuits for the description of noise in resistors.

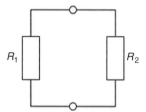

Figure 10.4 Thermal noise in resistive circuit with two identical resistors R.

This can be used for network analysis as a noise signal can be interpreted within a frequency interval Δf as a harmonic sinusoidal signal with the equivalent root mean square amplitude:

$$V_{neff} = \sqrt{S(f)\,\Delta f}. \tag{10.43}$$

10.2 Noise sources

10.2.1 Thermal noise

The charge carriers in any conductions material at temperatures T above 0 K show random motions similar to Brownian motion [1]. The energy per degree of freedom is given by kT. The thermal noise of resistors can be described either by a voltage or current noise source in the equivalent circuits shown in Figure 10.3.

As derived first from Nyquist [2] the noise generated by one resistor is forcing a current in any load resistance. If, as in Figure 10.4, two identical resistors R, which are both at temperature T, are connected, the electromotive force due to thermal agitation in resistor R_1 generates a current whose value is obtained by dividing by $2R$. The noise power P available in any load resistance is then

$$P = \frac{\sqrt{\overline{v^2}} \cdot i}{2} = \frac{\overline{v^2}}{4R} = \frac{\overline{i^2} \cdot R}{4}. \tag{10.44}$$

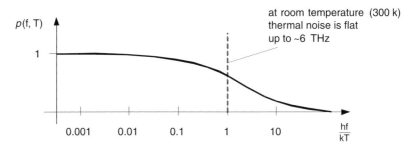

Figure 10.5 Thermal noise power spectral density.

From statistical thermodynamics, we know that

$$S(f) = \frac{hf}{e^{\frac{hf}{kT}} - 1}. \tag{10.45}$$

The total noise power is then

$$P_v = \int_0^\infty S(f)df, \tag{10.46}$$

as circuit theory is limited to physical dimensions which are much smaller than the wavelength of the maximum signal frequency. In integrated circuits, this is valid even up to the transit frequencies of modern transistors of several hundreds of GHz. The power spectral density of the thermal noise according to equation (10.45) is valid for frequencies f much less than kT/h.

At room temperature, this transition frequency is about 6 THz, as shown in Figure 10.5; therefore, we can assume a constant spectral density.

$$S(f) = kT = S_0. \tag{10.47}$$

Therefore, thermal noise is also called white noise as it shows constant spectral density in the interesting frequency range.

For the noise equivalent circuits of Figure 10.3, we get:

$$\overline{v_{neff}^2} = \int_0^\infty v_{neff}(f)^2 df \tag{10.48}$$

$$\overline{i_{neff}^2} = \int_0^\infty i_{neff}(f)^2 df. \tag{10.49}$$

The spectral densities of the thermal noise are:

$$v_{neff}(f) = \sqrt{4RS(f)}. \tag{10.50}$$

or

$$i_{neff}(f) = \sqrt{\frac{4S(f)}{R}}. \tag{10.51}$$

The effective thermal noise voltage v_{neff} across a resistor R is therefore given by:

$$v_{neff} = \sqrt{v_{neff}(f)^2 \, \Delta f} = \sqrt{4kTR\Delta f} \qquad (10.52)$$

or

$$i_{neff} = \sqrt{i_{neff}(f)^2 \, \Delta f} = \sqrt{4kT\Delta f / R}. \qquad (10.53)$$

If several resistors are connected in series, we get:

$$v_{neff}(f)^2 = 4k \sum_\nu R_\nu T_\nu. \qquad (10.54)$$

In the case of parallel connected resistors, the total effective noise current spectrum is given by:

$$i_{neff}(f)^2 = 4k \sum \frac{T_\nu}{R_\nu}. \qquad (10.55)$$

The electronic circuit may also include energy storage elements such as capacitors or inductors. However, those elements do not add further thermal noise, but they influence the thermal noise spectrum:

$$v_{neff}(f)^2 = 4S(f)\mathrm{Re}\,\{\underline{Z}\} \qquad (10.56)$$

$$i_{neff}(f)^2 = 4S(f)\,\mathrm{Re}\,\{\underline{Y}\}. \qquad (10.57)$$

10.2.2   Shot noise

Thermal noise is present without any external induced current in the device. Shot noise, however, is generated when charge carriers cross an energy barrier according to the statistics of this process [3]. In the transistors, this can happen at any $p-n$ junction. Therefore, a DC current $I \neq 0A$ is associated with a noise component due to the statistical nature of the current and only the average value of $\langle i(f) \rangle = I$.

The noise component is then:

$$i_n(t) = i(t) - I, \qquad (10.58)$$

so the noise power is described by:

$$\langle i_n(t)^2 \rangle = \int_0^\infty i_{eff}(f)^2 \, df. \qquad (10.59)$$

The power statistics is derived with the assumption that the charge carriers are independent from each other and a large number of carriers are crossing the barrier with an average rate of:

$$y(t) = \sum_\nu z(t - t_\nu) \qquad (10.60)$$

with $z(t - t_\nu) = 0$ for $t < t_\nu$.

The spectral density of $y(t)$ is then:

$$S_y(f) = 2qI|\Psi(f)|^2, \tag{10.61}$$

with

$$\Psi(f) = \sum_\nu \int_{-\infty}^{\infty} z(t - t_\nu)e^{j\omega t}dt. \tag{10.62}$$

For the effective noise current, we get:

$$i_{eff}(f)^2 = 2qI|\Psi(f)/\Psi(0)|^2. \tag{10.63}$$

For low frequencies, this can be simplified to:

$$i_{eff}(f)^2 = 2qI. \tag{10.64}$$

The shot noise is unavoidable in diodes and bipolar transistors, but also in field effect transistors (FETs) with junctions or Schottky metal gate electrodes. However, if the current across the barrier is small, other noise sources dominate in those devices.

10.2.3   Low-frequency noise

The first observation of low-frequency noise was made in thermionic tubes [4]. It is observed in all electronic materials and devices, including field effect or bipolar transistors. The measured spectral density of the noise current shows a $1/f$ dependence at low frequencies. This is attributed to a reciprocal frequency. This can simply be described by the term:

$$S(f) = \frac{C}{f}, \tag{10.65}$$

with C any constant. The noise power in any frequency band between the lower frequency f_1 and the upper frequency f_2 is given by:

$$P(f_1, f_2) = \int_{f_1}^{f_2} S(f)df = C\ln\left(\frac{f_2}{f_1}\right). \tag{10.66}$$

The problem of infinite power arises if either f_1 approaches zero or f_2 becomes infinite. In electronic circuits, there is always an upper frequency limit which also causes a decay of at least $1/f_2$ at some very high frequency. The lower frequency limit is given by the observation time.

The first model of the $1/f$ noise was proposed by McWorther as summing a distribution of time constants and integrating the Lorentzian characteristic over all possible time constants τ [5].

The wave function of an electron decays exponentially into the oxide barrier of a MOSFET, the time constant τ associated with the trapping effect is given by:

$$\tau = \tau_0 e^{\gamma x}, \tag{10.67}$$

where x is the direction into the oxide and y the attenuation coefficient. This coefficient can be determined by:

$$\gamma = \frac{4\pi}{h}\sqrt{2m_e\varphi_B},$$ (10.68)

where m_e is the effective mass of the electron in the oxide and φ_B is the tunneling barrier height seen by the electron at the interface. For the $Si - SiO_2$ System $\gamma \simeq 10^8$ cm^{-1} and $\tau_0 \simeq 10^{-10}s$. The traps are distributed in energy and can be accessed by tunneling electrons if the quasi Fermi level in the silicon lines up with oxide trap energy density. The power spectral density of the fluctuations in the number of trapped electrons is:

$$S'_{N_t} = \frac{4\tau}{1+\omega^2\tau^2}N_t(x, E)\, p_t(1 - p_t)\Delta V\Delta E,$$ (10.69)

where $N_t(x, E)$ is the trap density at position x into the oxide at energy E and p_t is the probability that the trap is filled. ΔV is the element volume and ΔE is the energy element. The probability p_t is given by:

$$p_t = \left(1 + e^{\frac{E_t-E_{fn}}{kT}}\right)^{-1}.$$ (10.70)

The spectral density of the induced fluctuations in the channel charge can be written as:

$$S'_{V_g} = \frac{S_{QN}}{(\text{WLC}_{\text{OX}})^2}$$ (10.71)

with $\Delta V = WLdx$ we get the total gate voltage noise spectrum S_{Vg} by integrating over space and energy:

$$S_{Vg} = \frac{4q^2}{(WLC_{ox})^2}\int_0^{dm}\int_{E_V}^{E_C}\frac{N_t(x, E)\,\tau(x)}{1+\omega^2\tau^2(x)}p_t(1 - p_t)dxdE.$$ (10.72)

For uniform distributions of the traps in space and energy, we get:

$$S_{Vg} = \frac{2kTq^2}{\pi\left(WLC_{ox}^2\gamma\right)}\frac{N_t\left(E_{fn}\right)}{f}$$ (10.73)

as the integral over E for the term $N_t(x, E)P_t(1 - P_t)$ can be approximated by $kTN_t(E_{fn})$.

The simplified notation of the $1/f$ noise is then:

$$S_{v_g} = \frac{K_f}{WLC_{ox}^2 f}$$ (10.74)

with

$$K_f \equiv \frac{2kTq^2N_t\left(E_{fn}\right)}{\pi\gamma}.$$ (10.75)

In SPICE this is used by the drain current noise equivalent:

$$S_{id}(f) = \frac{K_f I_d^{A_f}}{C'_{ox} L^2 f^{E_f}} \qquad (10.76)$$

with $A_f \approx E_f \approx 1$.

Beside the number fluctuation model by McWorther, there is a mobility fluctuation model by Hooge, which fits quite well the $1/f$-noise measured on III–V devices [6]. The spectral density for the mobility is:

$$S_\mu(f) = \frac{\alpha(0)\overline{\mu}^2}{N_{tot}} \qquad (10.77)$$

$$\alpha(0) = \frac{\alpha_H}{e_n(\tau_1/\tau_0)} \qquad (10.78)$$

where α_H is a measured constant with τ_1, and τ_2 are the bounds on the time constant. This results in a power density spectrum:

$$S_{id}(f) = \frac{\alpha_H I_{d^2}}{N_{tot} f}. \qquad (10.79)$$

N_{tot} is the total number of charge carriers and γ is a flicker noise parameter. A unified model for MOSFETs from Hung is implemented in recent SPICE models that is given by:

$$S_{id}(f) = \frac{kТq^2\mu_{\text{eff}} I_d}{\alpha\gamma f C'_{ox} L^2} \left[A \ln\left(\frac{N_0 + N^*}{N_L + N^*}\right) + B(N_0 - N_L) + \frac{C}{2}\left(N_0^2 - N_L^2\right)\right]$$
$$+ \Delta L \frac{kТI_d^2}{8 f W L^2} \frac{A + BN_L + CN_L^2}{(N_L + N^*)^2}. \qquad (10.80)$$

For the unified model, μ_{eff} is determined at the given bias condition, k is the Boltzmann constant, T is the absolute temperature, α is the scattering coefficient, and q is the element charge. There are further flicker noise parameter A, B, and C, and various charge densities included, N_0 is the charge carrier density at the source side of the channel, N_L is the charge density at the drain side, and N^* is given by:

$$N^* = kТ\frac{(Cox + C_d + C_{it})}{q^2} \qquad (10.81)$$

with C_d the depletion capacitance and C_{it} a further model parameter.

As there are several noise sources present in electronic devices and circuits, the measured noise power spectral density at the output of an amplifier is the aggregation of all. Figure 10.6 shows the regions in the noise power spectral density of an amplifier. It is assumed that in region I the noise spectral density is limited below the minimum observed frequency $f_{obs} = \frac{1}{T_{obs}}$. In region II the flicker noise dominates up to the corner frequency which might be in the kHz regime for some bipolar devices or in the MHz regime for FET devices. In region III the thermal noise sources are dominating. At the upper frequency limit of the amplifier gain bandwidth, the gain decreases and the noise power spectral density increases in region IV.

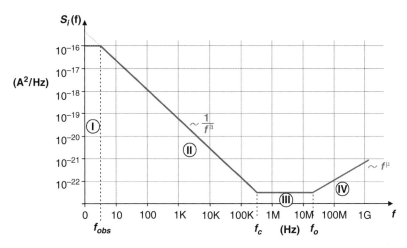

Figure 10.6 Noise power spectral density of transistor amplifier.

10.3 Noise analysis in linear network theory

The noise analysis of linear networks is mostly done in the frequency domain with two-port theories. In general, the noise signal at the input is related to the signal at the input by the signal to noise ratio. The definition of the noise factor F of a two-port is given by:

$$F = \frac{S_i / N_i}{S_O / N_O}.$$
(10.82)

Often, the following definition of noise figure NF is used.

$$NF = 10\log(F).$$
(10.83)

For linear network elements, the noise sources are related to the real part of the admittance or impedance respectively.

$$v_{eff}(f)^2 = 4S(f)\operatorname{Re}\{\underline{Z}\}$$
(10.84)

$$i_{eff}(f)^2 = 4S(f)\operatorname{Re}\{\underline{Y}\}.$$
(10.85)

Therefore, an equivalent noise resistance or conductance can be defined with a reference temperature $T_o(e.q. T_o = 290\,\text{K})$ and using the temperature noise equivalence.

$$v_{eff}(f)^2 = 4kT_o R_{eq}$$
(10.86)

$$i_{eff}(f)^2 = 4kT_o G_{eq}.$$
(10.87)

Often, an equivalent noise temperature T_r is used to which a resistor with the value of $\operatorname{Re}\{\underline{Z}\}$ must be heated to generate the same noise power spectral density as the network element.

$$T_r = \frac{S(f)}{k}.$$
(10.88)

For noisy linear two-ports, there is a theory to describe the noise by separate sources outside a noiseless two-port. Depending on the notation of the two-port parameters, different equivalent notations are used. In the impedance notation, noise voltage sources are added at the input and output port in series as shown in Figure 10.12.

$$\underline{V}_1 = \underline{Z}_{11}\underline{I}_1 + \underline{Z}_{12}\underline{I}_2 + \underline{V}_{N1} \tag{10.89}$$

$$\underline{V}_2 = \underline{Z}_{21}\underline{I}_1 + \underline{Z}_{22}\underline{I}_2 + \underline{V}_{N2}. \tag{10.90}$$

Using the admittance notation, noise current sources are added at the input and output part in parallel.

$$\underline{I}_1 = \underline{Y}_{11}\underline{V}_1 + \underline{Y}_{12}\underline{V}_2 + \underline{I}_{N1} \tag{10.91}$$

$$\underline{I}_2 = \underline{Y}_{21}\underline{V}_1 + \underline{Y}_{22}\underline{V}_2 + \underline{I}_{N2}. \tag{10.92}$$

For amplifiers, often the ABCD matrix is used. A noise voltage source is then added in series and a noise current source in parallel to the input port.

$$\underline{V}_1 = \underline{A}_{11}\underline{U}_2 + \underline{A}_{12}\underline{I}_2 + \underline{V}_N \tag{10.93}$$

$$\underline{I}_1 = \underline{A}_{21}\underline{V}_2 + \underline{A}_{22}\underline{I}_2 + \underline{I}_N, \tag{10.94}$$

as the noise figure is defined by

$$F = \frac{S_i N_o}{N_i S_o} = 1 + \frac{|\underline{V}_N + \underline{Z}_S\underline{I}_N|^2}{8kT\Delta f \mathrm{Re}\left\{\underline{Z}_S\right\}} \tag{10.95}$$

with source impedance $\underline{Z}_S$.

At the output of a noisy two-port, the noise figure can be described as:

$$F = \left(1 + {}^{N_{2p}}/_{G_a} \cdot N_i\right) \tag{10.96}$$

with G_a the available gain and N_{2p} the noise generated by the noise sources of the two-port.

As the noise figure at the output port of the two-port depends on the matching of the source impedance $\underline{Z}_s$ to the optimal impedance $\underline{Z}_{opt}$ of the two-port for minimum noise figure F_{min}, we can write:

$$F = F_{min} + \frac{|\underline{I}|^2|\underline{Z}_s - \underline{Z}_{opt}|^2}{8kT\Delta f \mathrm{Re}\left\{\underline{Z}_s\right\}}. \tag{10.97}$$

The minimum noise figure is achieved for $\underline{Z}_s = \underline{Z}_{opt}$:

$$F_{min} = 1 + \sqrt{|\underline{V}_N|^2\underline{I}_{N2}} \cdot \frac{(\mathrm{Re}\left\{\rho\right\}) + \sqrt{1 - (\mathrm{Im}\left\{\rho\right\})^2}}{4kT\Delta f}. \tag{10.98}$$

The optimum impedance $\underline{Z}_{opt}$ is given by:

$$|\underline{Z}_{opt}| = \sqrt{|\underline{V}_N|^2/|\underline{I}_N|^2} \tag{10.99}$$

and

$$\rho = \frac{\underline{I}_N \cdot \overline{\underline{V}}_N}{\sqrt{|\underline{V}_N|^2 |\underline{I}_N|^2}}.$$ (10.100)

Most often, equivalent parameters are given to characterize the noise of two-ports:

$$R_n = \frac{|\underline{V}_N|^2}{8kT\Delta f}$$ (10.101)

$$G_n = |\underline{I}_N|^2 \frac{(1 - |\rho|^2)}{8kT\Delta f}$$ (10.102)

$$\underline{Y}_{corr} = \rho \sqrt{|\underline{I}_N|^2 / |\underline{V}_N|^2}.$$ (10.103)

In a measurement system, a reference impedance $\underline{Z}_0$ is used; therefore, the reflection factor Γ_{opt} is defined:

$$\underline{\Gamma}_{opt} = \frac{\underline{Z}_{opt} - \underline{Z}_0}{\underline{Z}_{opt} + \underline{Z}_0}.$$ (10.104)

Then, the noise figure can be described as:

$$F = 1 + \frac{G_n + R_n |\underline{Y}_s + \underline{Y}_{corr}|^2}{\text{Re}\{\underline{Y}_s\}},$$ (10.105)

or

$$F = F_{min} + (F_0 - F_{min}) \frac{|\underline{\Gamma} - \underline{\Gamma}_{opt}|^2}{(1 + |\underline{\Gamma}|^2) |\Gamma_{opt}|^2}.$$ (10.106)

For measurements, usually the following notation is used:

$$F = F_{min} + \frac{R_n}{G_s} \left[(G_s - G_{opt})^2 + (B_s - B_{opt})^2 \right].$$ (10.107)

Basically, any noisy two-port exhibits a minimum noise figure F_{min} if the source impedance $\underline{Z}_s = G_s + jB_s$ is exactly equal to the optimum impedance $\underline{Z}_{opt} = G_{opt} + jB_{opt}$. The noise resistance R_n is a measure of the steepness of the increase of the noise figure for deviations of the source impedance from the optimum one. Those are the low noise parameters describing the noise of any two-port and those noise parameters are applied to the transistor models. An example of the noise figure dependence on the source impedance for a MOSFET is shown in Figure 10.7. The minimum of the noise figure F_{min} is achieved at $\underline{Z}_s = \underline{Z}_{opt}$.

An efficient method for computer-aided noise analysis of linear networks is based on the correlation matrix representation of noisy two-ports [7]. Any noisy two-port can be represented by a noise-equivalent circuit which consists of the original, but noiseless two-port and two additional noise sources. As the two-port theory has six different forms, there are six different noise equivalent circuits, depending upon the type of the additional noise sources and their arrangement relative to the noiseless two-port. Most often used

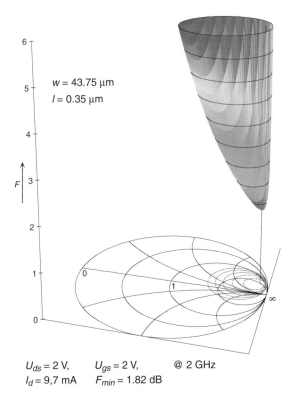

$w = 43.75\ \mu\text{m}$
$l = 0.35\ \mu\text{m}$

$U_{ds} = 2$ V, $U_{gs} = 2$ V,   @ 2 GHz
$I_d = 9{,}7$ mA $F_{min} = 1.82$ dB

Figure 10.7 Noise figure of a MOSFET depending on source impedance.

are the admittance, impedance, and chain representations. The three representations
are shown in Figure 10.12 with the corresponding electrical two-port matrix and their
correlation matrix. The admittance matrix uses two current noise sources i_1 and i_2 in
parallel at the input and the output of the noiseless two-port, respectively. The impedance
representation uses two voltage noise sources V_1 and V_2 in series at the input and the
output port, respectively. In the chain representation, a voltage noise source $V1$ is added
in the series and current noise source i_1 in parallel at the input port. As the noise is
assumed to be stationary random processes, a physically significant description of these
sources is given by their self- and cross-power spectral densities which are defined as the
Fourier transform of their auto- and cross-correlation functions. Arranging these spectral
densities in matrix form leads to the so-called correlation matrices. The elements of the
matrices are denoted by $C_{S_1 S_{2*}}$ where the subscript indicates that the spectral density
refers to the noise sources S_1 and S_2. The matrices themselves are denoted by C and by
a subscript which specifies the representation.

   Noise sources are usually characterized by their mean fluctuations in bandwidth Δf
centered on frequency f. The mean fluctuations are closely related to power spectral

densities by the following term:

$$\langle S_i S_j^* \rangle = 2\Delta f C_{S_i S_j^*}. \tag{10.108}$$

The factor 2 results from the frequency range from $-\infty$ to $+\infty$. The correlation matrix C belonging to the noise sources S_1 and S_2 can then be written as:

$$C = \frac{1}{2\Delta f} \begin{bmatrix} \langle S_1 S_1^* \rangle & \langle S_1 S_2^* \rangle \\ \langle S_2 S_1^* \rangle & \langle S_2 S_2^* \rangle \end{bmatrix}. \tag{10.109}$$

Each representation can be transformed into each other by simple transformation operations. A derivation is based on linear network theory. The noise signal of the new representation can be expressed in terms of the noise signals of the old representation by the convolution integral:

$$x'(t) = \int_{-\infty}^{\infty} H(p) x(t-p) dp, \tag{10.110}$$

where the transformation is characterized by the weighting matrix $H(p)$. Using this relation, the auto- and cross-correlation functions are calculated. Fourier transforming them leads to the transformation formula:

$$C' = TCT^{\dagger}. \tag{10.111}$$

Here, C and C' denote the correlation matrix of the original and resulting representation and T is the Fourier transform of $H(p)$. The plus sign denotes the Hermitian conjugation.

In Figure 10.12, the most often used representations of the transformation matrices are presented.

Interconnection of noisy two-ports can be described with operations using the correlation matrices. Of particular interest are two two-ports in parallel, in series, or cascaded. The resulting correlation matrix is related to the correlation matrices of the original two-ports by:

$$C_y = C_{y1} + C_{y2} \quad (parallel) \tag{10.112}$$

$$C_Z = C_{Z1} + C_{Z2} \quad (series) \tag{10.113}$$

$$C_A = A_1 C_{A2} A_1^{\dagger} + C_{A1} \quad (cascade). \tag{10.114}$$

For transistor modeling, the derivation of the correlation matrix from the measured noise parameters is required. Using the chain representation, the correlation matrix is given by:

$$C_A = 2kT \begin{bmatrix} R_n & \frac{F_{min}-1}{2} - R_n Y_{opt} \\ \frac{F_{min}-1}{2} - R_n Y_{opt}^* & R_n |Y_{opt}|^2 \end{bmatrix}. \tag{10.115}$$

A typical noise figure test setup is shown in Figure 10.8.

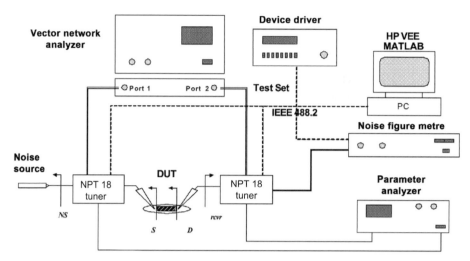

Figure 10.8 Noise figure test setup.

10.4   Noise measurement setups

Different noise measurement setups are used for low-frequency flicker noise and high-frequency noise characterization. As thermal noise of transmission lines, contact probes, and any parasitic resistances of the DUT deteriorates the noise figure, a precise equivalent-circuit model can be derived from broadband S-parameter measurements at the required operating points of the DUT. Appropriate extraction procedures are available for FETs [8] as well as for bipolar transistors [9]. As an example, all four S-parameters of a MOSFET with a gate length of 0.3 μm and a gate width of 394 μm are presented in Figure 10.9 in the frequency range from 445 MHz up to 40 GHz. There is an excellent agreement for all four S-parameters between the extracted equivalent circuit model and measurements at ht operating point of $V_{gs} = 1.5$ V and $V_{ds} = 1.2$ V.

To determine the noise spectral densities of the DUT, the equivalent circuit model of the DUT has to be completed for all noise sources as shown in Figure 10.9 of the MOSFET.

All thermal noise sources are related to the resistance R and the absolute temperature T according to (10.52) or (10.53), so only the gate and drain current noise contributions are unknown, according to Figure 10.10. The cross-correlation of gate and drain current can be determined by the noise parameter extraction procedure. The high-frequency noise figure measurement setup is shown in Figure 10.8. First, calibration and S-parameter measurements of the DUT are performed with a vector network analyzer and appropriate DC-bias.

The noise figure meter determines the output noise power for a dedicated bandwidth in the frequency range from 0.1 GHz up to 18 GHz. As there are four noise parameters, at least four measurements are required. Therefore, impedance tuners are used at the input to modify the noise source impedance, which has to be characterized for each setting also by S-parameter measurements. To improve accuracy, statistically distributed

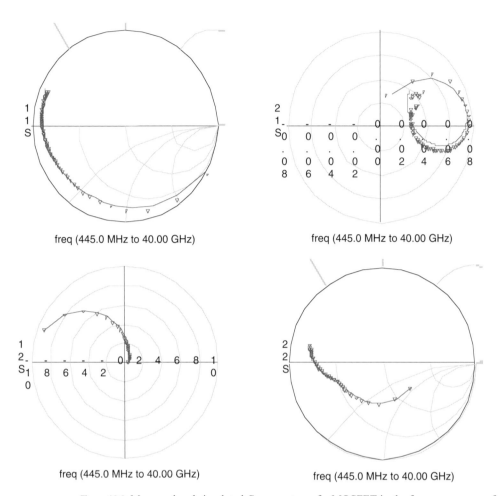

freq (445.0 MHz to 40.00 GHz)

freq (445.0 MHz to 40.00 GHz)

freq (445.0 MHz to 40.00 GHz)

freq (445.0 MHz to 40.00 GHz)

Figure 10.9 Measured and simulated S-parameters of a MOSFET in the frequency range of 445 MHz to 40 GHz.

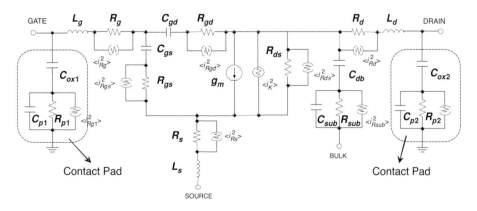

All resistors generate thermal noise with the spectral density $<i_R^2> = \dfrac{4kT_0}{R}$

Figure 10.10 Equivalent circuit and noise sources of a MOSFET.

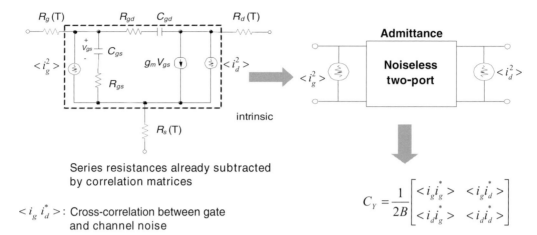

R_g (T)
R_{gd}
C_{gd}
R_d (T)
V_{gs}
C_{gs}
$g_m V_{gs}$
R_{gs}
$< i_g^2 >$
$< i_d^2 >$
intrinsic
R_s (T)
Series resistances already subtracted by correlation matrices
$< i_g i_d^* >$: Cross-correlation between gate and channel noise
Admittance
Noiseless two-port
$C_Y = \frac{1}{2B}\begin{bmatrix} < i_g i_g^* > & < i_g i_d^* > \\ < i_d i_g^* > & < i_d i_d^* > \end{bmatrix}$

Figure 10.11 Gate and drain noise current sources of a MOSFET.

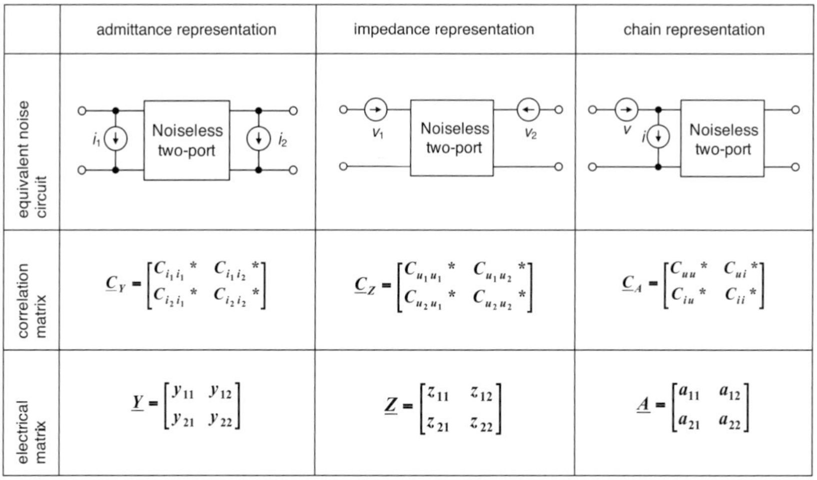

admittance representation
impedance representation
chain representation
equivalent noise circuit
Noiseless two-port
i_1
i_2
v_1
v_2
v
i
correlation matrix
$\underline{C}_Y = \begin{bmatrix} C_{i_1 i_1}{}^* & C_{i_1 i_2}{}^* \\ C_{i_2 i_1}{}^* & C_{i_2 i_2}{}^* \end{bmatrix}$
$\underline{C}_Z = \begin{bmatrix} C_{u_1 u_1}{}^* & C_{u_1 u_2}{}^* \\ C_{u_2 u_1}{}^* & C_{u_2 u_2}{}^* \end{bmatrix}$
$\underline{C}_A = \begin{bmatrix} C_{uu}{}^* & C_{ui}{}^* \\ C_{iu}{}^* & C_{ii}{}^* \end{bmatrix}$
electrical matrix
$\underline{Y} = \begin{bmatrix} y_{11} & y_{12} \\ y_{21} & y_{22} \end{bmatrix}$
$\underline{Z} = \begin{bmatrix} z_{11} & z_{12} \\ z_{21} & z_{22} \end{bmatrix}$
$\underline{A} = \begin{bmatrix} a_{11} & a_{12} \\ a_{21} & a_{22} \end{bmatrix}$

Figure 10.12 Three representations of noise equivalent circuits.

source impedances are applied at the input and a least-square algorithm delivers the four noise parameters at a given operating point and frequency.

The noise paraboloids of three different types of transistor are shown in Figure 10.13.

The noise figure for various input impedances can be presented as a parabolic area as shown in Figure 10.13. At the optimum source impedance, the curve exhibits its lowest point. Examples for MOSFET, HEMT, and bipolar transistor are compared with HEMT devices showing a flat minimum, while MOSFETs exhibit a sharp increase of the noise figure for deviations from the optimum source impedance Z_{opt}. The minimum noise figure of any transistor also depends on the bias voltages. Figure 10.14 shows the

Noise hyperboloids

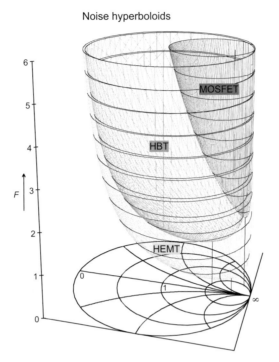

Figure 10.13 Noise figure paraboloids for a MOSFET, an HBT, and an HEMT.

measured minimum noise figure for drain and gate voltage scans, indicating a strong increase of the noise in the linear region of the MOSFET where the gain drops.

10.5   Transistor noise parameter extraction

10.5.1   RF noise extracted using correlation matrices

The design of high-frequency devices and circuits is based on compact transistor models. Each device is composed of an intrinsic transistor with detailed scaling rules and external parasitic elements. There are detailed extraction routines to determine all elements of the transistor equivalent circuits by S-parameter measurements up to beyond 100 GHz. An example of such a high-frequency equivalent circuit model for a MOSFET is shown in Figure 10.10.

As all resistors generate thermal noise, an equivalent noise current source is connected in parallel. The circuit simulators include a noise analysis option, which includes the noise sources for the resistive elements. The temperature can be set by default or an individually specified value. Here, the gate noise is assumed to be induced only by capacitive coupling, so the main noise source of the intrinsic device is the channel noise represented by the noise current source $\langle i_K^2 \rangle$.

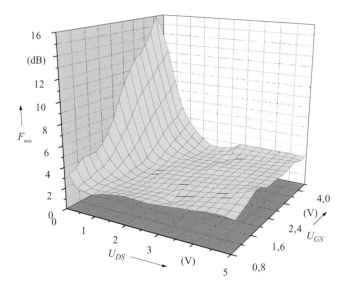

Figure 10.14 Noise figure of a MOSFET depending on gate-source and drain-source voltages.

To determine this channel current noise, the following extraction procedure can be applied to broadband noise parameter measurements, knowing the equivalent circuit elements from S-parameter measurements. The noise parameter extraction procedure is similar to analytic equivalent circuit parameter extraction procedures, using linear network theory in combination with the correlation matrices.

The noise parameter measurements deliver the four noise parameters Fmin, Rn, Gopt and Bopt as shown, for example, in Figure 10.15 in the frequency range from 2 to 18 GHz. The correlation matrix C_A^{Trans} of the measured device can be calculated according to equation (10.114). This correlation matrix includes the total noise caused by the intrinsic device as well as all parasitic elements.

First, the parasitic pad capacitances and resistances are de-embedded. The correlation matrix of the pads is given by:

$$\left[C_y^{pad}\right] = 2kT_0\text{Re}\left\{\left[Y^{pad}\right]\right\}. \tag{10.116}$$

The measured correlation matrix has to be converted to the Y representation:

$$\left[C_y^{Trans}\right] = \left[T_{A\to y}\right]\left[C_A^{Trans}\right]\left[T_{A\to y}\right]^+. \tag{10.117}$$

Then, the pad parasitics are subtracted from the measured correlation matrix data.

$$\left[C_y^{Tp}\right] = \left[C_y^{Trans}\right] - \left[C_y^{pad}\right]. \tag{10.118}$$

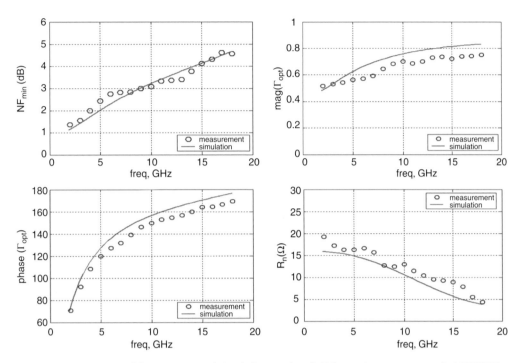

Figure 10.15 Measurement and simulation results of all four noise parameters of a MOSFET.

The external parasitic inductances and series resistances R_g and R_d can be de-embedded in the Z representation:

$$\left[C_Z^{ser} \right] = 2kT \mathrm{Re} \left[Z^{ser} \right] \tag{10.119}$$

$$\left[C_Z^{T_p} \right] = \left[T_{y \to z} \right] \left[C_A^{T_p} \right] \left[T_{y \to z} \right]^{+} \tag{10.120}$$

$$\left[C_Z^{T_s} \right] = \left[C_Z^{T_p} \right] - \left[C_Z^{ser} \right]. \tag{10.121}$$

The intrinsic device also includes a substrate admittance, which can be de-embedded in the Y representation:

$$\left[C_{y,T}^{1} \right] = \left[T_{Z \to y} \right] \left[C_Z^{T_s} \right] \left[T_{Z \to y} \right]^{+} \tag{10.122}$$

$$\left[C_{y,T}^{2} \right] = \left[C_{y,T}^{1} \right] - 2kT \mathrm{Re} \left\{ \begin{bmatrix} 0 & 0 \\ 0 & y_{sub} \end{bmatrix} \right\}. \tag{10.123}$$

The source resistance has to be de-embedded in the impedance representation:

$$\left[C_{Z,T}^{2} \right] = \left[T_{y \to z} \right] \left[C_{y,T}^{2} \right] \left[T_{y \to z} \right]^{+} \tag{10.124}$$

$$\left[C_{Z,T}^{int} \right] = \left[C_{Z,T}^{2} \right] - 2kT \mathrm{Re} \left\{ \begin{bmatrix} 0 & R_s \\ R_s & 0 \end{bmatrix} \right\}. \tag{10.125}$$

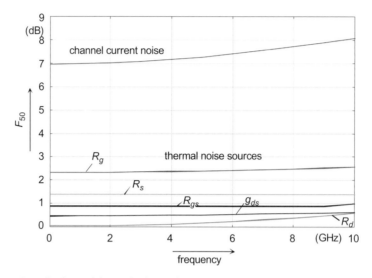

Figure 10.16 Contributions of the equivalent noise sources of a MOSFET.

After transformation to the admittance representation, the noise correlation matrix directly indicates the noise current spectral densities of the intrinsic transistor:

$$\left[C_{y,T}^{int}\right] = \left[T_{Z \to y}\right] \left[C_{Z,T}^{int}\right] \left[T_{Z \to y}\right]^+ \tag{10.126}$$

$$\left[C_{y,T}^{int}\right] = \frac{1}{2\Delta f} \begin{bmatrix} \langle i_g i_g^* \rangle \langle i_g i_d^* \rangle \\ \langle i_d i_g^* \rangle \langle i_d i_d^* \rangle \end{bmatrix} \tag{10.127}$$

$$S_{ig} = \frac{\langle i_g^2 \rangle}{\Delta f} = \frac{\langle i_g i_g^* \rangle}{\Delta f} \tag{10.128}$$

$$S_{id} = \frac{\langle i_d^2 \rangle}{\Delta f} = \frac{\langle i_d i_d^* \rangle}{\Delta f}. \tag{10.129}$$

In Figure 10.16 the measured noise spectral density of the gate and drain current sources are presented for a MOSFET with a gate length of 0.35 μm. As all elements of the equivalent circuits can be determined from S-parameter measurements and the noise contribution can be extracted according to the presented procedure, the noise contribution of each element to the total noise figure of the device can be easily investigated by simulations where only selected noise sources are considered while all others are neglected. An example is given in Figure 10.16 indicating that the channel current noise is the dominant source and the gate and source, resistances.

Finally, the measured and simulated noise parameters of a MOSFET with the associated gain are given in Figure 10.17.

10.5.2 $1/f$ noise sources and 50 Ohm noise measurement

Noise correlation matrices allow for a very elegant and systematic extraction of the transistor noise source parameters, as these allow the parasitics to be de-embedded in a

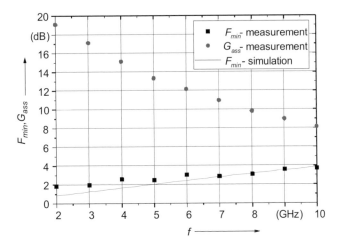

Figure 10.17 Measured and simulated noise figure of a MOSFET.

very easy way. However, in certain cases, it is impossible to use correlation matrices, if the four noise parameters have not been determined. For example, in the low-frequency range, instead of excessive source-pull measurement, the noise powers are measured for only a few source resistance values. In the microwave region, it might be desirable to determine the transistor noise model from standard 50 Ohm measurements. A microwave tuner might not be available, or one might want to speed up measurement time.

The task is therefore to determine the noise model parameters from a measured noise power or noise figure, respectively, while equivalent circuit elements, and source and load impedances are known. For a system with two sources, $\langle |v_1|^2 \rangle$ and $\langle |v_2|^2 \rangle$, for which the short-circuit noise current at the output, $\langle |i_{out}|^2 \rangle$, was measured, for example, the following equation will have to be solved:

$$\langle |i_{out}|^2 \rangle = A\langle |v_1|^2 \rangle + B\langle |v_2|^2 \rangle + C\langle v_1 v_2^* \rangle + D, \tag{10.130}$$

where A, B, C describe how the noise is transmitted through the circuit, and D gives the noise contribution of the source to the output noise. Depending on the number of unknowns, and on the formulas for $\langle |v_1|^2 \rangle$ and $\langle |v_2|^2 \rangle$, a certain number of independent measurements are required. Therefore, the measurement needs to be repeated for different source impedances, or at different frequencies. But fitting a parametrized function to measured points is not the tough part of this task. It is to determine the equation itself.

The traditional way would be to take advantage of superposition laws in the small-signal domain, and to calculate the contribution of each of the noise sources to the output short-circuit noise current. This is quite elaborate, especially if the noise sources are correlated.

Even in this case, applying a correlation-matrix-based approach significantly simplifies the analytic determination of noise figure or output noise power. In the following, the algorithm will be explained with the example of a GaAs HBT $1/f$ noise model. Restriction to the $1/f$ region significantly simplifies the equivalent circuit, but the complexity of

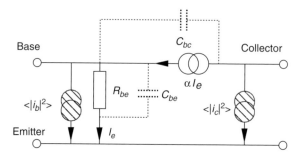

Figure 10.18 Intrinsic HBT equivalent circuit in the low-frequency region.

the intrinsic equivalent circuit is not really relevant. Only minor adjustments are required to obtain the respective formula valid in the microwave region. The same method has been applied, e.g., to derive the formula to determine Pucel's PRC and Pospieszalski's temperature-based HEMT noise models from 50 Ohm noise figure measurements [10].

The starting point is the intrinsic equivalent circuit of the HBT that is shown in Figure 10.18. The capacitances are drawn with dashed lines, since the frequency range is so low that they can be omitted without compromising accuracy. The noise of the intrinsic transistor is determined by the two short-circuit noise currents at base $\langle |i_1|^2 \rangle$ and collector side $\langle |i_2|^2 \rangle$, respectively.

$$\langle |i_1|^2 \rangle = 2q \Delta f I_b + KF \frac{I_b^{AF}}{f^{FB}} + KL \frac{I_b^{AL}}{1 + (f/FL)^2} \tag{10.131}$$

$$\langle |i_2|^2 \rangle = 2q \Delta f I_c \tag{10.132}$$

$$\langle i_1 i_2^* \rangle = 0. \tag{10.133}$$

While the electron charge q and the base- and collector currents I_b and I_c are known a priori, the parameters KF, AF, FB, KL, AL, and FL need to be determined from frequency-dependent measurements at different base currents.

In order to simplify the calculation of the noise current delivered to the load, we first transform the short-circuit noise currents of equations (10.131)–(10.133) into open-circuit voltage sources. Transformation into voltage sources is easily achieved using the transformation matrices, since the elements of the C_y matrix are the short-circuit noise currents given above, while the open-circuit noise voltages are given by the elements of the Z-type matrix C_z. The resulting voltage noise sources read:

$$\langle |v_{i1}|^2 \rangle = |Z_{i,11}|^2 \langle |i_1|^2 \rangle + |Z_{i,21}|^2 \langle |i_2|^2 \rangle \tag{10.134}$$

$$\langle |v_{i2}|^2 \rangle = |Z_{i,21}|^2 \langle |i_1|^2 \rangle + |Z_{i,22}|^2 \langle |i_2|^2 \rangle \tag{10.135}$$

$$\langle v_{i1} v_{i2}^* \rangle = Z_{i,11} Z_{i,21}^* \langle |i_1|^2 \rangle + Z_{i,21} Z_{i,22}^* \langle |i_2|^2 \rangle, \tag{10.136}$$

where $Z_{i,11} \ldots Z_{i,22}$ denote the Z-parameters of the intrinsic HBT. At this point already, the benefits of this approach are already obvious. First, the details of the intrinsic equivalent circuit are not necessarily relevant, as it is described through its two-port Z-parameters. One can even rely on measured data after de-embedding of the extrinsics.

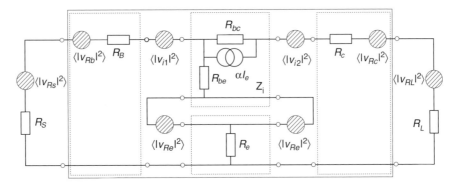

Figure 10.19 Low-frequency equivalent circuit of the HBT connected to source and load.

Second, through all these transformations, the equation remains a linear combination of the original noise currents, as assumed from the start in equation (10.130).

An equivalent circuit for the measurement setup is shown in Figure 10.19. Transformation of the intrinsic HBT's noise sources into voltage sources $\langle|v_{i1}|^2\rangle$, $\langle|v_{i2}|^2\rangle$ is already anticipated in the schematic.

The next step is to account for the noise of the emitter resistance, $\langle|v_{Re}|^2\rangle$. The emitter resistance is understood to be a two-port that is in series–series connection to the intrinsic transistor. Hence, their noise voltage sources can be added, according to the rules of correlation-matrix algorithm. The resulting sources describing the noise sources of the intrinsic HBT with emitter resistance R_e read:

$$\langle|v_{i1e}|^2\rangle = \langle|v_{i1e}|^2\rangle + \langle|v_{Re}|^2\rangle \tag{10.137}$$

$$\langle|v_{i2e}|^2\rangle = \langle|v_{i2e}|^2\rangle + \langle|v_{Re}|^2\rangle \tag{10.138}$$

$$\langle v_{i1e}v_{i2e}^*\rangle = \langle v_{i2}v_{i2}^*\rangle + \langle|v_{Re}|^2\rangle. \tag{10.139}$$

Looking at the equivalent circuit, Figure 10.19, it is observed that the noise voltage source $\langle|v_{i1e}|^2\rangle$ at the base is in series with the base resistance's noise source $\langle|v_{Rb}|^2\rangle$ and the noise source of the source, $\langle|v_{Rs}|^2\rangle$. The same holds true for the collector side, where $\langle|v_{i2e}|^2\rangle$ is in series with the collector resistance's noise source $\langle|v_{Rc}|^2\rangle$ and the noise source of the load, $\langle|v_{Rl}|^2\rangle$. These noise sources along a branch can simply be added, which yields:

$$\langle|v_1|^2\rangle = \langle|v_{i1e}|^2\rangle + \langle|v_{Re}|^2\rangle + \langle|v_{Rb}|^2\rangle + \langle|v_{Rs}|^2\rangle \tag{10.140}$$

$$\langle|v_2|^2\rangle = \langle|v_{i2e}|^2\rangle + \langle|v_{Re}|^2\rangle + \langle|v_{Rc}|^2\rangle + \langle|v_{Rl}|^2\rangle \tag{10.141}$$

$$\langle v_1v_2^*\rangle = \langle v_{i2}v_{i2}^*\rangle + \langle|v_{Re}|^2\rangle. \tag{10.142}$$

These two noise sources account for the total noise of the system. Source, load, and HBT are now completely noise-free. $\langle|v_1|^2\rangle$ is located at the base port of the HBT, and $\langle|v_2|^2\rangle$ is located at its output port.

The output noise current $\langle|i_{csc}|^2\rangle$ that is measured in $1/f$ measurements can now be derived on the basis of the abstract Z-parameters of the HBT, without even considering

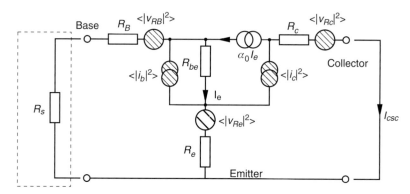

Figure 10.20 Measurement setup for the $1/f$ noise measurement of the HBT, with source-pull option.

the details of the intrinsic equivalent circuit.

$$\langle |i_{csc}|^2 \rangle = \frac{|A|^2}{R_L^2} \left[|B|^2 \langle |v_1|^2 \rangle + \langle |v_2|^2 \rangle + 2\mathrm{Re}\left(B \langle v_1 v_2^* \rangle \right) \right] \tag{10.143}$$

with

$$A = \left(\frac{Z_{21} Z_{12}}{(R_s + Z_{11}) R_L} - 1 - \frac{Z_{22}}{R_L} \right)^{-1} \tag{10.144}$$

$$B = \frac{-Z_{21}}{R_s + Z_{11}}. \tag{10.145}$$

It is now necessary to replace the terms $\langle |v_1|^2 \rangle$, $\langle |v_2|^2 \rangle$, and $\langle v_1 v_2^* \rangle$ by the original noise source formulas. This yields a linear equation in the form of equation (10.130). Performing a sufficient number of measurements varying frequency, current, and source impedance, it can be solved for the unknowns relying on a least-squares algorithm.

In the present case, the equivalent circuit was rather simple, due to the low-frequency range considered. But is also applicable to RF equivalent circuits that account for capacitances and transit times in the intrinsic circuit, and where the lead inductances and extrinsic capacitances cannot be neglected. Basically, the complexity of the intrinsic equivalent circuit is unimportant, since it is only described by its Z-parameters. For example, an algorithm for determining the Pospieszalski noise model from 50 Ohm noise sources was derived in a similar manner [10]

To conclude the chapter, a few results will be shown that were obtained for a $3 \times 30\,\mu m^2$ InGaP/GaAs HBT from the foundry of the Ferdinand-Braun-Insitut, Leibniz-Institut für Höchstfrequenztechnik, Berlin, Germany [11]. The measurement setup is shown in Figure 10.20, and the results, measurement, and model, are shown in Figure 10.21. The setup allows the source resistance to be varied, basically by changing it from $R_s = 10\,\mathrm{Ohm}$ to $R_s = 10\,k\mathrm{Ohm}$.

The measurement with $R_s = 10\,k\mathrm{Ohm}$ shows that the measured noise is dominated by the base-emitter $1/f$ noise source described by equation (10.131). The $1/f$ part, as

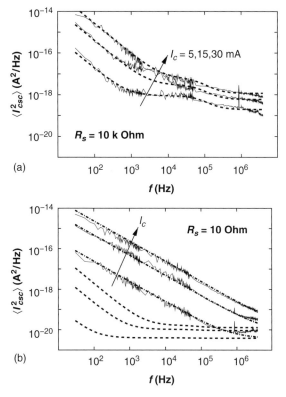

Figure 10.21 Measurement and simulation of $1/f$ noise of the HBT, parameter is the base current. Solid line: measurement, dashed line: simulation with only base-emitter $1/f$ noise source, dash-dotted line: full $1/f$ noise model.

well as the Lorentz-type spectrum showing a distinct cutoff frequency, are observed, in addition to the white noise floor. Also, the dependence of the noise spectrum on current is described well.

However, if only this source is considered in the model, it fails to predict the low-impedance case, $R_s = 10\,\Omega$, as shown by the dashed line in Figure 10.21b. The noise is drastically underestimated. What happens in this case can be explained when looking at the measurement setup, Figure 10.20. The condition $R_s + R_b + R_e \ll R_{be}$ is satisfied for $R_s = 10\,\Omega$. The base-emitter noise source is effectively short-circuited by the source, and does not contribute to the collector noise current.

The noise measured must therefore originate from another source. From measurement of similar semiconductor layers, it was expected that the noise originates from the built-in emitter ballasting resistor. It is realized by a semiconductor layer that is part of the HBT layer stack. It exhibits Hooge noise that explains the excess noise measured under this condition. Indeed, the dash-dotted lines in Figure 10.21b were simulated assuming a pure $1/f$ noise source in the emitter resistance.

The dependence of the emitter noise voltage source on the source resistance is orthogonal to how the base-emitter source depends on it. In the case of a high-impedance source, $R_s = 10\,k$Ohm, the emitter branch is in fact practically open-circuited and the noise source does not contribute to the measured collector noise current.

This example highlights the importance of performing more than just a measurement of the low-frequency noise at a fixed source impedance. Omitting one of the two measurements presented here would not allow the parameters of a noise model to be determined that account for the impact of the base-emitter termination during circuit design. In oscillator design, however, proper base-band termination that minimizes the $1/f$ noise can be an appropriate measure to reduce phase noise [12].

In the low-frequency region, however, it is not feasible to perform the general source-pull measurement that covers the whole Smith chart, as it is commonly performed in the microwave region. But it is feasible to measure at a discrete set of resistances. Due to the limited number of measurements, it is, in effect, impossible to distinguish more than two orthogonal sources reliably.

10.6   Summary

Noise is present in every transistor, and for many circuit applications, it needs to be accurately modeled, since it determines, for example, the sensitivity of a receiver.

This chapter first introduces the concepts of noise in semiconductors, starting with the mathematical background, how to describe noise signals, and followed by the physical background, the origin of the noise signals. Then, how noise signals can be treated in the analysis of linear circuits is discussed. The correlation matrix method is introduced that allows for a very systematic way of analysis and calculation. Noise measurement techniques are also addressed.

Based on correlation matrices, it is relatively straightforward to de-embed the noise sources from transistor measurements, and subsequently to determine the parameters describing the sources. How to apply the algorithm is discussed in the example of the white noise model of a CMOS transistor, and in the example of the $1/f$ noise model of an HBT.

References

[1] R. Brown, "A brief account of microscopical observations mode in the month of June, July & August, 1927, on particles contained in the pollen of plants, and on the general existence of active molecules in organic & inorganic bodies," *Philosoph. Mag.*, vol. 4, no. 21, 1828, pp. 161–173. Available online at http://sciweb.nybg.org/science2/pdfs/dws/Brownian.pdf

[2] H. Nyquist, "Thermal agitation of electric charge in conductors," *Physical Review*, vol. 32, pp. 110–113, July 1928.

[3] A. van der Ziel, "Theory of shot noise in junction diodes and junction transistors," *Proc. IRE*, vol. 43, pp. 1639–1646, Nov. 1955.

[4] J. B. Johnson, "Thermal agitation of electricity in conductors," *Physical Review*, vol. 32, pp. 97–109, July 1928.

[5] J. A. L. Mc Worther, "1/f noise and germanium surface properties, semiconductor," *Surface Physics*, p. 207, 1957.

[6] F. N. Hooge, "1/f noise is no surface effect," *Physics Letters*, vol. 29A, no. 3, pp. 139–140, April 1969.

[7] H. Hillbrand and P. H. Russer, "An efficient method for computer aided noise analysis of linear amplifier networks," *IEEE Trans. Circuits Syst.*, vol. CAS-23, no. 4, pp. 235–238, April 1976.

[8] A. Pascht, M. Grözing, D. Wiegner, and M. Berroth, "Small-signal and temperature noise model for MOSFETs," *IEEE Trans. Microw. Theory Tech.*, vol. 50, no. 8, pp. 1927–1934, Aug. 2002.

[9] U. Basaran, N. Wieser, G. Feiler, and M. Berroth, "Small-signal and high-frequency noise modeling of SiGe HBTs," *IEEE Trans. Microw. Theory Tech.*, vol. 53, no. 3, pp. 919–928, March 2005.

[10] M. Rudolph, R. Doerner, P. Heymann, L. Klapproth, and G. Böck, "Direct extraction of FET noise models from noise figure measurements," *IEEE Trans. Microwave Theory Tech.*, vol. 50, pp. 461–464, Feb. 2002.

[11] P. Heymann, M. Rudolph, R. Doerner, and F. Lenk, "Modeling of low-frequency noise in GaInP/GaAs hetero-bipolar-transistors," *IEEE MTT-S Int. Microw. Symp. Dig.*, 2001, pp. 1967–1970.

[12] G. D. Vendelin, A. M. Pavio, and U. L. Rohde, *Microwave Circuit Design using Linear and Nonlinear Techniques*, 2nd ed. Hoboken, NJ: John Wiley, 2005, ch. 10.

Index